WHAT IS LIFE? ON EARTH AND BEYOND

Approaches from the sciences, philosophy and theology, including the emerging field of astrobiology, can provide fresh perspectives to the age-old question "What is life?". Has the secret of life been unveiled, and is it nothing more than physical chemistry? Modern philosophers will ask if we can even define life at all, as we still don't know much about its origins here on Earth. Others regard life as something that cannot simply be reduced to just physics and chemistry, while biologists emphasize the historical component intrinsic to life on Earth. How can theology constructively interpret scientific findings? Can it contribute constructively to scientific discussions? Written for a broad interdisciplinary audience, this probing volume discusses life, intelligence and more against the background of contemporary biology, and the wider contexts of astrobiology and cosmology. It also considers the challenging implications for science and theology if extraterrestrial life is discovered in the future.

ANDREAS LOSCH is an award-winning theologian, specialising in the dialog between the sciences, philosophy and theology, and he is currently coordinating the project "Life beyond our planet?" at the Center for Space and Habitability, University of Bern. Losch is a member of the Center of Theological Inquiry in Princeton and he serves in the councils of the European Society for the Study of Science and Theology and in the Karl Heim Society. He is also editor-in-chief of a German forum for dialog between the sciences and theology.

Editorial Advisory Board
Kathrin Altwegg, Claus Beisbart, Andreas Krebs, Antonio Lazcano

WHAT IS LIFE?
ON EARTH AND BEYOND

ANDREAS LOSCH
University of Bern, Switzerland

CAMBRIDGE
UNIVERSITY PRESS

University Printing House, Cambridge CB2 8BS, United Kingdom

One Liberty Plaza, 20th Floor, New York, NY 10006, USA

477 Williamstown Road, Port Melbourne, VIC 3207, Australia

4843/24, 2nd Floor, Ansari Road, Daryaganj, Delhi - 110002, India

79 Anson Road, #06-04/06, Singapore 079906

Cambridge University Press is part of the University of Cambridge.

It furthers the University's mission by disseminating knowledge in the pursuit of education, learning and research at the highest international levels of excellence.

www.cambridge.org
Information on this title: www.cambridge.org/9781107175891

© Cambridge University Press 2017

This publication is in copyright. Subject to statutory exception and to the provisions of relevant collective licensing agreements, no reproduction of any part may take place without the written permission of Cambridge University Press.

First published 2017

Printed in the United Kingdom by Clays, St Ives plc

A catalogue record for this publication is available from the British Library

Library of Congress Cataloging-in-Publication data
Names: Losch, Andreas, 1972– editor.
Title: What is life? on earth and beyond / [edited by] Andreas Losch, University of Bern, Switzerland.
Description: New York : Cambridge University Press, 2017. | Includes bibliographical references and index.
Identifiers: LCCN 2016059503 | ISBN 9781107175891 (alk. paper)
Subjects: LCSH: Life – Congresses. | Exobiology – Congresses.
Classification: LCC BD431 .W4525 2017 | DDC 113/.8 – dc23
LC record available at https://lccn.loc.gov/2016059503
ISBN 978-1-107-17589-1 Hardback

Cambridge University Press has no responsibility for the persistence or accuracy of URLs for external or third-party internet websites referred to in this publication, and does not guarantee that any content on such websites is, or will remain, accurate or appropriate.

To Christian Link
who taught me to see life with the eyes of a living being

Contents

List of Contributors		*page* ix
Preface		xi
KATHRIN ALTWEGG		
Introduction		1
ANDREAS LOSCH		

Science

1	Reflections on Origins, Life, and the Origins of Life	13
	MARIE-CHRISTINE MAUREL	
2	The Search for Another Earth-Like Planet and Life Elsewhere	30
	JOSHUA KRISSANSEN-TOTTON AND DAVID C. CATLING	
3	The Shape of Life: Morphological Signatures of Ancient Microbial Life in Rocks	57
	BEDA A. HOFMANN	
4	Precellular Evolution and the Origin of Life: Some Notes on Reductionism, Complexity and Historical Contingency	75
	ANTONIO LAZCANO	

Philosophy

5	Science and Philosophy Faced with the Question of Life in the Twentyfirst Century	97
	MICHEL MORANGE	
6	What is Life? And Why is the Question Still Open?	111
	CLAUS BEISBART	

7	Is the Origin of Life a Fluke? Why the Chance Hypothesis Should Not be Dismissed Too Quickly CHRISTIAN WEIDEMANN	132
8	Some Contemporary – and Persistent – Fallacies and Confusions about Astrobiology MILAN M. ĆIRKOVIĆ	156
9	Superintelligent AI and the Postbiological Cosmos Approach SUSAN SCHNEIDER	178

Theology

10	What Theology can Contribute to the Question "What is Life?" ANDREAS LOSCH	201
11	Autopoietic Systems and the Theology of Creation: On the Nature of Life ALEXANDER MAßMANN	213
12	Where There's Life There's Intelligence TED PETERS	236
13	Life in the Universe, Incarnation and Salvation: A Conversation between Christianity and the Scientific Possibilities of Extra Terrestrial Life JUAN PABLO MARRUFO DEL TORO, SJ	260
14	Talking Lions, Intelligent Aliens, and Knowing God – Some Epistemological Reflections on a Speculative Issue TAEDE A. SMEDES	287

Conclusion

15	What is Life? On Earth and Beyond: Conclusion ANDREAS LOSCH	303
16	A Skeptical Afterword ANTONIO LAZCANO	311
	Index	314

Color plates section can be found between pages 68 and 69.

Contributors

Professor Dr. Kathrin Altwegg
University of Bern, Physics Institute, Space Research and Planetary Sciences

Professor Dr. Claus Beisbart
University of Bern, Institute of Philosophy

Professor Dr. David C. Catling
Department of Earth and Space Sciences/Astrobiology Program, University of Washington

Dr. Milan M. Ćirković
Astronomical Observatory of Belgrade and Future of Humanity Institute, Faculty of Philosophy, University of Oxford

Professor Dr. Beda Hofmann
Naturhistorisches Museum der Burgergemeinde Bern and Institute of Geological Sciences, University of Bern

Joshua Krissansen-Totton
Department of Earth and Space Sciences/Astrobiology Program, University of Washington

Professor Dr. Antonio Lazcano
Miembro de El Colegio Nacional, Universidad Nacional Autónoma de México

Dr. Andreas Losch
Center for Space and Habitability, University of Bern, and Research Fellow, Center of Theological Inquiry, Princeton

Dr. Juan Pablo Marrufo del Toro, SJ
Society of Jesus, Oregon Province (Jesuits West)

Dr. Alexander Maßmann
Faraday Institute for Faith and Science, Cambridge

Professor Dr. Marie-Christine Maurel
University Pierre-and-Marie-Curie and Museum National of Natural History (Paris), and Fellow of the International Society for the Study of the Origin of Life (ISSOL)

Professor Dr. Michel Morange
Centre Cavaillès, République des Savoirs USR 3608, Ecole normale supérieure, Paris

Professor Emeritus Dr. Ted Peters
Center for Theology and the Natural Sciences, Graduate Theological Union

Professor Dr. Susan Schneider
Department of Philosophy and Cognitive Science Program, The University of Connecticut, and Technology and Ethics Group, Yale University

Dr. Taede A. Smedes
Nijmegen, the Netherlands

Dr. Christian Weidemann
Department of Christian Philosophy, Innsbruck

Preface

We all know very well what life is when we encounter other living beings. Why is it then so difficult to define life in the framework of the exact sciences? Why is it that there are no clear answers from philosophy or from theology? Maybe we should try to tackle this problem not separately, everybody alone in his field of science, but rather come to an answer looking at the question from very different angles. This was the idea behind the workshop on "What is life?" in January 2015 in Bern, organized by the Center for Space and Habitability of the University of Bern. Scientists from theology, philosophy, biology, biochemists and space researchers came together to share their view on "What is life?". This was certainly not an easy endeavor, as several hurdles had to be overcome: different "scientific" languages, very different methods, reservations and prejudices vis-à-vis the other sciences. About 50 scientists from the different fields rose to the challenge such a conference is posing and the question "What is life?" was discussed intensively during three days. This book is a selection of papers given during this conference and has the intent not to give a final answer (there probably is none), but to foster discussion on this and other truly interdisciplinary questions in the field of space and habitability. In order to get a more comprehensive view, several experts in the different fields not participating in the conference have also contributed to this work. One of the major outcomes was that we all learnt to listen to scientists from other fields, to appreciate their methods and to acknowledge the fact that such a topic needs a truly "universal" view.

This conference as well as the book would not have been possible without the initiative, enthusiasm and work of Dr. Andreas Losch. His contribution is herewith especially acknowledged. Our thanks go to all participants at the conference, to the authors of the papers of the book and to the referees. We would also like to thank the University of Bern, which made this possible through the Center of Space and

Habitability, as well as to the Swiss National Science Foundation, which financially supported the conference.

Professor Kathrin Altwegg
Former Director of the Center for Space and Habitability

Introduction

ANDREAS LOSCH

The aim was to search for microbial life on Mars. Each *Viking* lander robot carried three experiments to detect metabolism, which is perceived as an indicator of life, and a fourth experiment to find organic molecules to confirm potential findings. In one experiment, something strange happened. This experiment added certain nutrients to the soil and the nutrients were marked with radioactive carbon 14. Thus, if there are microorganisms on Mars, they should metabolize and emit gas containing radioactive carbon 14. This radioactive gas was *indeed emitted*, and another control experiment with sterilized soil released no radioactive gas – as if Martian microbes were now dead because of the sterilization. At face value, these results looked like a *positive detection of life*. The biology team was stunned. They couldn't explain these results in terms of any known chemical process, biological or non-biological.

Nevertheless, it was finally concluded that the cause of this result need not have been biological material, on the basis of the anomalous (for Earth microbes) results and, most importantly, the failure of the *Viking* landers to find any organic molecules in the soil. So, for now, no life has been considered found on Mars.[1]

Our imagination and definition of life highly influence what we look for, and hence what we can find. Any irregularity in experiment and observation can be a pointer to something for us to learn. Was this experiment an example of such a learning occasion on the nature of life? We would know better, if we would know what life is. Most life scientists assume they know or they avoid the issue altogether. For many, the secret of life has been unveiled and it is often regarded as nothing other than physical chemistry. But is this all that can be said about it? One could also ask if one could define life at all. NASA's working definition, "life is a self-sustained chemical system capable of undergoing Darwinian evolution" (Joyce *et al.*, 1994, xi) explains as much as it leaves open for discussion. At least, by mentioning evolution it incorporates the historical dimension intrinsic to life.

What is life, then? In this book, this question serves as the central topic where different attempts at answers will be given. The focus is of course scientific, yet includes perspectives from philosophical and theological angles as well. Some topics in the philosophical and theological part are more speculative, like the potential impact of postbiological superintelligence, the idea that all life is to some extent intelligent, or the implications of intelligent extra-terrestrial life for theology. We learned from Immanuel Kant that we need to balance speculation with critical thought. Yet we do not know what we will find if we reach for the stars. That's why the second part of the title of the book was chosen to be *On Earth and Beyond*.

The book is divided into three parts, beginning with a science section as part I, which provides the reader of the book with an up to date account of the scientific foundations of the topic. This part is followed by philosophical considerations in part II, and finally theological questions as part III. The body of the book is therefore compiled by an interdisciplinary group of authors from biology, astrophysics, philosophy and theology. The reader will realize that they have partly divergent opinions, which are encouraged by the editor; evidence of this is also the sceptic's afterword by Antonio Lazcano. Presupposition of a dialogue is to take the other one's discipline and position seriously, yet not to agree on everything. While a conflation of the disciplines – as in the case of creationism claiming to present scientific insights – has to be avoided, scholars involved in the dialogue of all three of these disciplines know about the differences, yet also tend to witness some fruitful interactions.

In this context, one thing needs to be mentioned: almost every academic theologian regards creationism or intelligent design as nonsense (not only from a scientific, but also from a theological point of view). However, scientists and philosophers don't always seem to know about this. Some even conflate *any* idea of creation with creation*ism*. Yet the problem lies in the early-twentieth-century phenomenon of creationism, not in the idea of creation, which on a fundamental level actually was even able to further the development of science, because it deemed "the book of nature" worthy of research while not taboo (in other religions it was regarded as Divine itself) (Hooykaas, 2000; McGrath, 2003, pp. 53–4). Only *after* Darwin's tremendous achievements it is of course unwise to stick to a literal reading of the beginning of the Bible, while at its time the Genesis account may have been a good approximation to "science" (Arber, 2012). The creation story of the Bible, focusing on one creator God, rendered almost any entity like Moon and Sun or the Leviathan as creation instead of regarding it like the Babylonians (in their *Emuna Elish*) as Gods themselves, so it had a *demythologizing* intent. Likewise, Earth was not regarded as a Goddess anymore, but considered to have received its obvious creative potential from the creator ("Let the earth bring forth grass", Gen 1:11 KJV). Similarly, Anglican clergyman and writer Charles Kingsley pointed out that God "made all things make themselves" and Charles Darwin liked this theological interpretation of evolution enough to refer to it in his *Origin of Species* from the second edition on (Browne, 2003, pp. 95–6). Also, generally, the idea of evolution entails a more or less linear progression through time, an imagination originating from Jewish and Christian thought.

As previously mentioned, the focus of part I of the present book, composed of the first four chapters, is scientific. The question of the origin of life, the search for another Earth in other solar systems, a way to find potential signs of extra-terrestrial life on Mars and the history of the research into the origins of life are presented and discussed in a way understandable for the non-scientist. This summary is further elaborated as follows.

We still do not know much about life's origins on Earth. Despite the immense interest in this topic, it remains an unanswered side-question in modern scientific research. In the first chapter of this book, Marie-Christine Maurel sets herself on the track to explore the traces of life's origins. By that, she understands the *source* from where matter got its primal actions. Maurel is trying to avoid the opposition of mechanism and vitalism. Is it possible to find another path close to biological reality, both material and living? She stresses that the origin and the evolution of life coincide; there was no creationist start from nothing. What about Darwin's famous "warm little pond"? Maurel presents several scenarios for how the beginning of life could have happened, derived from astrophysics, astrochemistry and geochemistry, furnished with further elements from speculation and from laboratory experiments.

Is there life beyond Earth? Positively answering this question, Joshua Krissansen-Totton and David Catling argue, would challenge our self-understanding dramatically. Earth would not be special any more but, in exchange, our universe would be much richer. Chapter 2 argues that the question of life elsewhere will likely be answered within the lifetime of readers of this book. This has to do with the relatively recent discovery of exoplanets, circling stars other than our Sun. For now, the vast majority of them have been detected by using indirect methods, yet the authors perceive direct imaging as a more promising technique for the potential discovery of life on other planets, as life may modify its environment on a planetary scale. Nevertheless, they suggest that the more traditional Search for Extra-Terrestrial Intelligence (SETI) using radio telescopes should not be dismissed too easily.

If we refocus our attention to our solar system, Mars is the closest candidate for potentially carrying or having carried life. Beda Hofmann argues in chapter 3 that there is a way Mars missions could hint at previous or present microbiological life on Mars, namely by searching for fossils in the extra-terrestrial matter. We are talking about microbial life here. On Earth, conditions for life exist in numerous places, not only on the surfaces of the continents and in the oceans, but also in sediments and hard rocks. Fossilized microbial fabrics, that is to say the remnants of slime, can indeed be used as an easily visible expression of fossil life. Hofmann discusses several common types of morphologies preserved in rocks and resulting from prokaryotic microbial life. Nevertheless, such fabrics typically need corroboration by other methods, a feature currently still unavailable in our attempts at Mars exploration.

In chapter 4, we turn back to the history of life on Earth and follow Antonio Lazcano's portrayal of the story of how biological research on the origins and nature of life made progress. An evolutionary framework derived from Darwin's ideas has led to the most fruitful approaches. Ernst Haeckel enlarged the scenery for this story by including cosmic evolution, and his idea of a *unity of nature* invoked the imagination of an evolutionary continuity between the inorganic world and living entities. It was Oparin who pointed out likewise that life is not characterized by any special properties but by a definite, specific combination of these properties that is the outcome of a historical process. The appearance of life on Earth should hence be seen as a non-progressive evolutionary continuum that seamlessly joins the prebiotic synthesis and accumulation of organic molecules in the primitive environment, with the emergence of self-sustaining replicative chemical systems capable of undergoing Darwinian evolution. History's contingency is of the essence to understand the nature of life.

Part II is about philosophical contributions to the theme of the book. It has already been asked: can one set up a definition of life? And is there a way philosophy can assist science today? We should recall that, initially, science was considered "natural philosophy". Also, the idea of life beyond our planet is philosophically an old one: the starry skies above us did not only induce Immanuel Kant's interest in extraterrestrial life (Losch, 2016b), philosophers from the early Atomists on have speculated on the issue (Crowe, 1986).

Yet how is the situation today? Chapter 5 is a contribution to the dialogue between biologists and philosophers at the beginning of the twentyfirst century. Michel Morange considers recent transformation of biological knowledge and shows how they raise major issues on the "nature of life". The emergence of molecular biology dramatically transformed the question "What is life?". The subsequent fading of the informational vision and the search for a minimal genome resulted in surprising outcomes: an ecological vision of life and the return of the question about whether viruses are alive. The border between living and non-living seems to be more porous than expected. Regarding the projects of synthetic biology, Morange predicts that separation between life and non-life will hence not be a simple border, but a territory the geography of which is still poorly defined. Life is a contingent process, so the characteristics of an organism are not the simple result of the action of natural selection, but also of the constraints linked to the progressive construction of this organism.

Chapter 6 returns to the most fundamental question, "What is life?". We can only detect life on other planets and gain insights about its origin, if we know to some extent what life really is. But this condition does not seem to be fulfilled; rather, various definitions of life have been proposed, none of which has achieved consensus. Claus Beisbart thus proposes to step back and to analyse the meaning of the question more closely. Starting from a few requirements that any answer to the question should obey, he arrives at the conclusion that attempts to define life face a dilemma. A definition of life is either unprincipled or to some extent stipulative. Beisbart proposes explication as suggested by Carnap as a sensible framework to discuss the question. His suggestion is to replace our notion of life for the purposes of *scientific* inquiry with a notion that strikes a fair compromise between Carnap's desiderata of similarity, exactness, fruitfulness and simplicity.

Is the origin of life a fluke? Differing from convictions of prominent biologists, our author Christian Weidemann argues, in chapter 7, why the chance hypothesis does not need to mean siding with creationists and why it should not be dismissed too quickly. Many believe that even if all hitherto proposed theories on the origin of life were shown to be false by new evidence, the justification for believing the emergence of life on Earth was not due to a mere fluke would remain largely unaffected. Weidemann is convinced that this belief is ill founded. He gives two reasons, the *vast cosmos argument* (given a sufficiently big and variegated ensemble of universes, any finite chance of life emerging will suffice to guarantee its formation) and the *indifferent nature argument*, meaning that there seems to be no conceivable reason to suppose that processes by which complex molecules arise are more likely to be biased towards life-producing molecules. Unless a plausible alternative explanation is available or other life that has emerged independently is discovered, the chance hypothesis remains a serious contender for him.

In chapter 8, Milan Ćirković defends the ongoing astrobiological revolution. He rejects several contemporary fallacies about astrobiology, such as those he labels as old and new varieties of Geocentrism, accompanying the astrobiological enterprise from its exobiological start ("There is no astrobiology because there is no instance of confirmed extraterrestrial life"), recently rearticulated in the form of the rare Earth hypothesis (Ward & Brownlee, 2000). Against critics who view the astrobiological efforts as pure science fiction, Ćirković asserts that literary discourse presents a treasure-trove of potentially useful scientific hypotheses. There also has been much confusion about the issue of whether the theory of biological evolution conflicts with the idea of extra-terrestrial life, which it, of course, does not. Ćirković regards the astrobiological revolution in a wider context of the Enlightenment struggle for a completely rational world view, so extra-scientific resistance to this extension of the Copernican revolution has to be expected.

There has rightly been a good deal of attention on the search for *microbial life*. This could, however, present an anthropocentric bias, for in doing so we are implicitly assuming humans to be at the top of life in the universe. Susan Schneider observes that the current work in astrobiology, as diversified as it may be, does not draw from the intriguing discussions of superintelligence in the literature about artificial intelligence. Chapter 9 pieces these two domains together. On the one hand, Schneider identifies new directions for the postbiological intelligence approach in astrobiology. On the other hand, while the discussion of superintelligent artificial intelligence has mostly focused on the control problem, Schneider takes a step back and considers under which circumstances we could understand its computations, if at all. Anticipating ways to understand such sorts of intelligence may assist our efforts to control it. Schneider ends with discussing the social impact of encountering superintelligence, be it on Earth or discovered elsewhere.

Part III is about theological approaches. Ian G. Barbour (1997) sketched a very helpful typology of potential relations between science and religion,[2] distinguishing *conflict* cases (like creationism or a scientism blindly fighting any sort of religious interpretation of nature), an *independence* model (like Stephen J. Gould's NOMA or a complementarity approach), a *dialogue* stance or even an *integration* of aspects of one discipline (usually science) into the other (usually theology). All stances except for conflict of course mean respecting the other discipline's territory from the start.

In the media, many times only the conflict cases are reported and could hence invoke the impression of being dominant within academic communities as well. This is not the case. Although some sociological studies seem to enforce the conflict view, with titles such as "Leading scientists still reject God" (Larson & Witham, 1998) or "Eminent scientists reject the supernatural" (Stirrat & Cornwell, 2013), one should, however, note that the authors of the first study, for instance, chose as the title of the same study in the previous year "Scientists are still keeping the faith" (Larson & Witham, 1997). This is also the result of a very profound study conducted by Elaine Howard Ecklund (2010): even half of US top academics stick to some sort of religious faith, despite outward appearances. Of course, freedom of belief or non-belief is an important societal achievement, but it does not help that many still believe in the nineteenth-century myth that science and religion are at constant war with each other. The truth is that professional science in the nineteenth century often emancipated itself from the profession that previously only had the time to perform natural study, which was actually the clergy: the "scientific parson" was a well-established social stereotype, hence there have been some necessities of scientific distinction from theology (McGrath, 2003, p. 46), just because the two have often been very close together.

Introduction 7

Andreas Losch asks, in chapter 10, what today's theology can contribute to the scientific question of "What is life?". He observes from the scientific beginnings a focus on the physical and chemical level of life and hence a potential bias. As the character of the dialogue between science and theology is, however, largely asymmetrical, theology can contribute more to the scientific setting than to science itself. First of all, scientists and theologians need to agree on some basic premises for dialogue, which actually means, as has been said, taking the other's contribution seriously. Elaborating on ideas from Michael Welker, Losch then argues that theology can prevent the sciences from developing false perspectives of theology and can correct mistakes and inconsistencies in scientific presentations of theological and religious issues. More constructively, it can also develop explorations of areas of knowledge common to both from multiple perspectives, and try to build small bridges at the boundaries of each side's areas of knowledge. It can remind science that completeness of knowledge has not yet been achieved and it can maybe serve in an auxiliary function for science, pointing out the existing gaps in the scientific account of the universe, and sometimes even propose a preliminary "filler". It may well be that theology one day has to retreat from the claim of a particular gap as it is closed by science. Yet until then, it leaves the floor open for new research.

One example of how such a contribution could look is presented in chapter 11. Most modern philosophy and especially molecular biology have typically looked at nature through mechanistic glasses. Alexander Maßmann, however, wants to set aside the inherited Cartesian dualism that isolates mind from matter and, with the help of philosophers Evan Thomson and Hans Jonas, attempts at regrinding our lenses to see more clearly physical self-organization as the external dimension of cognition, and cognition as the internal aspect of self-organization. He also enters into dialogue with Kant, stressing that only if the regulative idea of teleology is corroborated by a cogent connection to the phenomenal world can it appropriately define the referent of the discipline of biology. Maßmann combines the mechanist events of genetic mutation and natural selection with the teleological autopoiesis of the organism. Towards the end of his chapter, he shifts to more proper theological interpretation of what he has presented, developing a theology of creation which understands creation in systemic and in evolutionary terms, elaborating on the potential role of the Holy Spirit in autopoietic organization.

Chapter 12 by Ted Peters aligns with Maßmann's underlying idea and claims that *where there's life there's intelligence.* He provides a comprehensive scale of intelligence that includes simple single-celled organisms in continuity with the highest level of intelligence we know, viz. *Homo sapiens*. Elaborating on seven distinct traits of intelligence he demonstrates a spectrum within the single category of intelligent life, hoping to steer astrobiological eyes to see what might surprise us. Hence what happens at the human level is in continuity with all levels of life in our biological evolution. According to Peters, an organism is intelligent when it possesses interiority, intentionality, communication, adaption and finally mental activity, including problem-solving, self-reflection and theory of mind, and judgment. In his account of intelligence, Peters leaves consciousness, values and free will aside. He also discusses three potential holes in his argument's boat.

Juan Pablo Marrufo del Toro, SJ, zooms out of this particular focus and reinstates the great picture of life in the universe from a theological point of view. Chapter 13 takes science utterly seriously and hence starts with a recapitulation of the contemporary scientific findings. It then turns to theology proper, which discusses the implications of the potential existence of extra-terrestrial intelligent life for the Christian dogma of incarnation, which relates God deeply to creation. Would this mean for the theologian that he could conceive of multiple incarnations? Marrufo del Toro subscribes to Niels Henrik Gregersen's idea of "deep incarnation", which means that the presence of the Logos reaches not only humanity but every tissue of biological existence with its growth and decay, entailing "deep crucifixion" and "deep resurrection" as well. The search for extra-terrestrial intelligence has begun, so theology should be prepared to answer such questions.

What really drives the search for exoplanets and extra-terrestrial life is the central question of chapter 14. Taede Smedes is convinced it is actually the hope of finding *intelligent* (and self-conscious) extra-terrestrial life, so somebody more or less like us. Smedes, however, expresses some scepticism about this approach, as it reduces *otherness* to *sameness*. Following Wittgenstein's remark "If a lion could speak, we could not understand him", he emphasises the otherness of potential aliens inhabiting another ecological niche and reflects on the possible incommensurability between potential aliens and human beings, a position he calls "apophatic astrobiology". As with regard to the possible religious beliefs of aliens, he argues that if god concepts supervene on the cognitive functioning of human beings, then perhaps religion is limited to humans only. If intelligent extraterrestrial beings lack religion and god concepts altogether, this would obviously raise interesting theological questions for e.g. Christian theology.

The reader is welcome to regard chapter 14 in some sense as a critical counterpart to chapter 13. As previously mentioned, reasonable speculation and sound critique both need to be trained, and so this book will end after a summarizing conclusion with a sceptic's afterword. There are many more objections I can imagine: isn't the theological focus on Christian traditions witness of a bias? Why don't we learn more about the historical perspectives of interactions? Every book has its limitations, and in this book we set the focus mostly on contemporary interactions in the Western hemisphere. It can only be a beginning. "Astrotheology" has just been started to be re-developed (see the editorial of the collection of articles in the journal *Zygon*, Losch, 2016a). Who knows, maybe one day we will indeed find extra-terrestrial life? If not on Mars, maybe somewhere else in the solar system – or beyond. I hope this book will have helped us to be prepared.

I want to thank the editorial advisory board, consisting of Kathrin Altwegg (physics), Claus Beisbart (philosophy), Andreas Krebs (theology) and Antonio Lazcano (biology), who all had their share in making this book project real. The starting point was a conference co-sponsored by the Swiss National Science Foundation (SNSF), the Intermediate Staff Association of the University of Bern (MVUB) and the university's Center for Space and Habitability (CSH), of which Kathrin Altwegg is the visionary former Director. With Willy Benz she launched the project "Life beyond our planet?", which provided me with my current position, and where Andreas Krebs, Torsten Meireis, Silvia Schroer and Andreas Wagner served as theological supervisors. It was Claus Beisbart who motivated me to turn this book project into more than conference proceedings. Most of all, however, I have to thank Antonio Lazcano for his lasting support of this idea and his enduring friendship.

A project like this is not possible without the involvement of competent peer reviewers, to whom I want to express my gratitude. Not less, I wish to thank the wonderful authors who were willing to contribute to this book and to deal with the time it took to set up the peer review. Their numbers include established scholars as well as young researchers. Conferences and conversations within the European COST Action "Life-Origins", of whom I have the honor to have become a member, enriched my understanding of the subject matter, and also lead to a fellowship at the Center of Theological Inquiry (CTI) at Princeton in 2016/17. Finally, the openness of Cambridge University Press and of its commissioning editor Vince Higgs for the idea made this book become real. To Emanuel Graf I owe many thanks for helping to prepare the manuscript (and for setting up the index). I hope you enjoy reading it.

Notes

1 The paragraphs are partially taken from Catling, 2013, pp. 96–7, and from Carol Cleland's *The Quest for a Universal Theory of Life: Searching for Life as We Don't Know it*, see http://rintintin.colorado.edu/~vancecd/phil150/Cleland.pdf.
2 The drawback of Barbour's approach is to have left philosophy as a discipline of its own out of the picture, which is needed in almost any attempt of fruitful interaction bridging science and religion (Losch, 2011). This is one of the reasons why philosophy plays a crucial part in this book.

References

Arber, W. (2012). *Contemplation on the Relations Between Science and Faith*, Web: http://www.casinapioiv.va/content/accademia/en/academicians/ordinary/arber/contemplation.html Accessed: 9/3/2016.

Barbour, I. G. (1997). *Religion and Science. Historical and Contemporary Issues*, 1st HarperCollins rev. edn., San Francisco, CA: HarperSanFrancisco.
Browne, E. J. (2003). *Charles Darwin*, London: Pimlico.
Catling, D. C. (2013). *Astrobiology. A Very Short Introduction*, Oxford: Oxford University Press.
Crowe, M. J. (1986). *The Extraterrestrial Life Debate 1750–1900. The Idea of a Plurality of Worlds from Kant to Lowell*, Cambridge: Cambridge University Press.
Ecklund, E. H. (2010). *Science vs. Religion. What Scientists Really Think*, Oxford: Oxford University Press.
Hooykaas, R. (2000). *Religion and the Rise of Modern Science*, Vancouver: Regent College Pub.
Joyce, G. F., Deamer, D. W. & Fleischaker, G. R. (eds.) (1994). *Origins of Life. The Central Concepts*, Boston, MA: Jones and Bartlett Publishers.
Larson, E. J. & Witham, L. (1997). Scientists are still keeping the faith. *Nature* **386**(6624), 435–6.
Larson, E. J. & Witham, L. (1998). Leading scientists still reject God. *Nature* **394**(6691), 313.
Losch, A. (2011). *Jenseits der Konflikte. Eine konstruktiv-kritische Auseinandersetzung von Theologie und Naturwissenschaft*, Göttingen: Vandenhoeck & Ruprecht.
Losch, A. (2016a). Astrotheology. On exoplanets, Christian concerns, and human hopes. *Zygon* **51**(2), 405–13.
Losch, A. (2016b). Kant's wager. Kant's strong belief in extra-terrestrial life, the history of this question and its challenge for theology today. *International Journal of Astrobiology* **15**(04), 261–70.
McGrath, A. E. (2003). *Science & Religion. An Introduction*, reprinted, Oxford: Blackwell Publishing.
Stirrat, M. & Cornwell, R. (2013). Eminent scientists reject the supernatural: a survey of the Fellows of the Royal Society. *Evolution: Education and Outreach* **6**(1), 33.
Ward, P. D. & Brownlee, D. (2000). *Rare Earth. Why Complex Life is Uncommon in the Universe*, New York, NY: Copernicus. Available online at http://www.loc.gov/catdir/enhancements/fy0816/99020532-d.html.

Science

1

Reflections on Origins, Life, and the Origins of Life

MARIE-CHRISTINE MAUREL

> "Life is the memory of matter"
> *Roupnel, 1945*

Introduction

We owe it to Jean-Baptiste Lamarck to have developed the first biological hypotheses concerning the origins of living beings. In *Philosophie zoologique* (1809) Lamarck considers that the simplest living beings were formed in an appropriate milieu under the action of physicochemical laws. As a result of this spontaneous generation, living beings conditioned by the environment adapted themselves and became more complex. A few years later and long after the publication of the *Origin of Species* (1859), Charles Darwin wrote in a letter to Joseph Dalton Hooker (1871): "It is often said that all the conditions for the first production of a living organism are now present, which could ever have been present. But if (and oh! what a big if!) we could conceive in some warm little pond, with all sorts of ammonia and phosphoric salts, light, heat, electricity, &c. present, that a protein compound was chemically formed ready to undergo still more complex changes, at the present day such matter would be instantly absorbed, which would not have been the case before living creatures were found".

We can state that since the beginning of the nineteenth century the question of the origins of life has been at the forefront in science, but in spite of the immense interest it has aroused, it remains an unanswered side-question in most modern scientific research communities and institutions. Buried deeply in our thoughts, in the history of science and of humanity and in the personal history of each human being, the origins of life remain difficult to explore scientifically, especially as opposed to the more recent history of what is alive, in part because, due to the passage of time and geological events, there is no clear fossil trail.

The aim of this review is not to re-write the history of concepts and experiments on the origins of life, to which I have already contributed (Maurel, 1994; 1999; 2002; 2003; Maurel & Décout, 1999), but rather to develop a few subjects that have resulted from our reflections.

Several physicochemical parameters are provided by the astrophysical, geological, and chemical sciences that help propose that a primitive environment was present around 4–3.8 billion years ago.

But how was the living "tissue" that populates our planet formed and woven? Before discussing what recent scientific research has revealed to us about the first bodies, it is necessary to clarify certain terms, in particular what we mean by *origins* and by *life*.

Then we will ask ourselves several basic questions, namely the following.

What do we know about the past steps of life? Does a plan exist concerning the basis of the origins? Can one think of the occurrence of biological entities as the artisan thinks of his own project, his own plan?

Science demonstrates that a thing is not a Kantian thing-in-itself, and most current theories on the *origins of life* propose that organic molecules and some minerals present on the early Earth became the raw material from which biological systems developed. Then we will summarize our current understanding of the conditions that led from simple, abiotically synthesized monomers to some of the earliest chemical steps from which the basic metabolic steps first originated, and ultimately to macromolecules. A model is presented for the development of replicative systems from random polymers that evolved through mutation and selection. Hence one can say that the origins of life mark the start of biological evolution.

We present here a scheme based on simulation experiments and explanations in progress from which one can develop essential concepts of prebiotic synthesis, self-assembly of organic molecules, primeval catalysis and template-directed synthesis that can lead to general epistemological principles in the study of the origins of life.

1.1 The Concept of Origin

Over the centuries, various ways of thinking have attempted to answer the question of origins, be they origins of the universe, of the living world or of human beings.

The term origin has at least two meanings:

(i) that which is located at the origin, at the beginning, (ii) that which is the basis, the foundation on which something is based.

The universe and the solar system were formed billions of years ago in the greatest chaos. This is how Hesiode describes the origin of the world (Hesiode, 2001,

Théogonie, 114–22): "at first arises Beance [the void], and then the Earth (Gaia) with broad sides, forever a robust foundation for all of us".

In the 1970s, Jim Lovelock and Lynn Margulis summarized the Greek myth in an ecological sense, to try to redraw the "basis" and the living tissue written within it. In this "Gaia hypothesis" life is tightly connected to the geochemical environment; it is the main message that has become famous in the climatic evolution of today, on which the survival of numerous species depends. Accordingly, biodiversity draws its origins from the origins of life itself.

However, according to Cassirer, Greek philosophers created the concept of *archê*, that is the "principal" concept marking the boundary between the origin identified by the myth, and the principle or the foundation characterizing science. The advent of rational thought in ancient Greece is thus linked to disqualification of the myth. The abstract thought would have been formed by loosening itself, by tearing itself away from the origins as *archê*, from the archaic.

Now if we focus on the etymology of the word, origin brings us back to the beginning and by extension to the notions of birth, of "source". Three possible types of analysis arise: either a historical approach, in which one goes back step by step to the source, or an approach whose aim is to unveil the "essence" of the object studied, or even better, the basis, the "foundations" of life. We shall see this in more detail below when we examine the concept of life.

We focus on the *source*, that is to say from where, from what primeval driving force, matter got its primal actions. This is what life needs; some may call it the "efficient cause". As a biologist one speaks of the *source* of energy that became the *source* of information repeating the past over the ages, renewing the history of life in new ways. Hence the origin, the first, and the primitive refer to chronological time.

Finally, what is needed is to understand the concept, that is, how an entire organism is made. This last point is more or less related to life cycles, metabolism in a broad sense, which are exchanges with the environment as well as changes, growth, and development.

In all its aspects, the biological interest of the origins of life resides in the fact that it presents itself through its historical and fundamental aspects. The historical aspect is not retained by some physicists, who are fond of categories, rejecting time as a historical parameter. But all biologists know that life, which originated four billion years ago, is the result of a long historical process called evolution that began right at the outset.

If we went back four gigayears and life were to start again, no one can guarantee that it would take the same path. If life exists – or existed – elsewhere, benefitting from the same initial planetary conditions as Earth, it most likely would not have the same history, hence not the same shape, the same metabolism in the sense of exchange with the environment, storing and transmitting information, or reproducing itself, or what information is concerned within other circumstances, in other surroundings, under other influences.

1.2 The Concept of Life

The word "life" is a concept in the sense that it summarizes and renders empirical things abstract.

Defining what is alive undoubtedly constitutes a fundamental pillar of biological sciences. Still it is necessary to agree on the terms "to define" and "living", given that semantics is flexible regarding these two words. A wide consensus covers most biologists, who characterize current life as being represented by a compartment, the cell, in which biochemical transformations take place, and by the metabolism, meaning the changes allowing, among other things, reproduction and transfer with some hereditary variations, essential for evolution in a changing environment. However, this description of present-day life is partial in so far as it concerns only certain features, and furthermore does not take into account the expression of other entities such as viruses and other subviral particles or the immense biodiversity of species brought to the fore in recent years. This reveals the fact that although we have access to a great number of samples of living organisms coming from surroundings often considered exotic, we are incapable of "cultivating" them, thus observing and knowing their needs, behaviour, appearances, reactions, adaptations, and their life. The "living" are everywhere, in the smallest, most remote nook of the Earth, in the heart of rocks, not to mention the hundreds of species that inhabit our bodies, skin, mucosa, digestive tract, etc.

Indeed, we only know 1% of the species living today, which themselves represent just one thousandth of the living organisms that have existed over the past four gigayears. This underscores the weakness of the concepts and generalizations that surround the notions of the living, the biosignatures and the multiple and diverse definitions of life.

The reduction of life to certain common characteristics limits the understanding of the processes that distinguish the living. This highlights the abstract boundary between life and living that is often brushed aside. Must it be described? Or explained? The way the question is asked reveals the connections and directs the answers.

From another viewpoint, life is undoubtedly one of the fundamental concepts of philosophy. The history of philosophy is rich in texts questioning life, death, the soul and organized bodies, the "vital principle" and the vitalism studied by most major figures of philosophy. These terms are but a few of the numerous concepts related to thought about the living.

Basically, the question "What is life?" has often been discussed at length and, like all undefined questions, it does not really function. Furthermore, "we ask ourselves questions only where we always have the answer. This seems to limit the reach of the question, but it is the occasion to evaluate in each case the answer to the question. The answer differs for each". This is one of the lessons given by Jacques Lacan (1972) during one of his seminars. He did not think that he was saying the right thing so clearly, because indeed we know today of several hundred definitions of life as compiled by Gyla Pali, Claudia Zucchi and Luciano Caglioti (2002), Radu Popa (2004) and others.

In fact, there are many definitions dedicated to the characterization of the fact that life is limited to either its physicochemical characteristics or its spiritual characteristics. The physicochemical characteristics reduce life to mechanical things. The spiritual characteristics limit life to something insubstantial, ethereal, sometimes nebulous. In both senses it does not explain anything. The question sends back analogies to machines (apparatus, machineries, etc.), or to metaphysical or futile reflections.

The notions of enlivened and of non-enlivened are often used as demarcation criteria. To enliven comes from the Latin *animare*, from "breath of life", leading to the declinations of soul and animal.

Relationships with movement, mobility, immobility, communications (see the particular status of plants), activity, inactivity and, finally, the distinction between non-animated and non-enlivened, lead to two different states. Indeed, robots are enlivened objects, and rocks or metals do not answer to the same characteristics of enlivened as robots. Yet animists believe in a vital force that is present even in stones and adore Nature likened to God. Subjects are numerous and carry rich philosophical, anthropological, and mystical reflections.

To retreat from the reductionist-spiritual dead ends, it is necessary to distinguish what is alive from what is not. With this aim in mind, the word *alive* seems to be more practicable than the word *life*, probably because the former term includes, from my point of view, notions of dynamics and energy.

The notion of vitalism deserves our attention despite the fact that it is linked to "so many extravagances", according to the words of Georges Canguilhem (1992, 84), leading to creationist heresy. Vitalism was of particular interest to numerous philosophers and scientists who were concerned with the origins of life. The thesis of vitalism defended by proponents, and often poorly known, was most of the time treated as inconsistent, and this has contributed, and still contributes, to rendering it blurred. Two extreme situations are known: one insists of the existence of a vital principle, the soul; while the other, closer to the physical and chemical reality of the living, rests on the notion of dynamic that in etymological terms means strength and movement. A subtle analysis of all the historical, philosophical, political and scientific aspects of the history of the notion of vitalism is made by Canguilhem in his seminal text, "Aspects du vitalisme" (1992).

The oppositions between enlivened/inanimate and vitalism/mechanism would lead to the emergence theory. The alliance of physics and biology must not be made to the detriment of biology under the theoretical influence of physics. Mathematical formalizations are often necessary here. Sometimes a mathematical–chemical machine, an automaton (such as Ganti's *Chemoton*, 2003), is presented as a minimal living system whose aim is to explain the origins of life. Following the same trend, molecular evolution is modelled by way of mathematical applications leading to the theory of self-organization. On the other hand, for prebiotic chemistry, the origins of life would be a succession of functional organizations, but all these are models that idealize, simplify, and distort the reality. However, they are helpful in the sense that they provide elements of rigorous formulation.

Finally, it is interesting to note that today the term "molecular vitalism" is proposed to design research into integrative cell physiology. Although living units are obviously the products of genes, what promotes their function is often far removed from sequence information. The phenotype might depend as much on external conditions as on random events (Kirschner *et al.*, 2000). Out of such a non-determinist world, the organisms have fashioned a very stable physiology. According to Kirschner, this robustness suggests "vital forces", and it is this robustness that he and co-authors wish to explore in terms of chemistry. As an echo, from the scientific community dealing with the origins of life, some scientists involved in synthetic life are searching for the "essence of cell life" (Szostak *et al.*, 2001).

In conclusion, rather than ascribing "the origin of all things" to a spiritual authority that can be qualified as intrinsic, the materialistic trend has broken away from this interpretation of the world, in favour of a mechanistic and deterministic approach to the origins. Mechanism and vitalism are frequently opposed to one another and such opposition is still the basis of the history of scientific and epistemological thinking today. Is it possible to go beyond this opposition, to find another path close to biological reality, both material and living?

After August Weisman, the "father" of modern genetics, who stated that "The science of living nature does not have to wait for physics and chemistry to have reached the end of their development" (1892), Erwin Schrödinger claimed some years later (1947), that "Living matter works in an irreducible manner with respect to ordinary laws of physics. Consequently, we must not be discouraged by the difficulty we have in interpreting life by the ordinary laws of physics; we must be prepared to expect to discover new types of laws in physics...".

1.3 The Origins of Life

The biological interest in the origins of life resides in the fact that it presents itself through its historical and fundamental aspects. On the one hand, we are driven towards going back in time through each biological and physicochemical step to the origins, requiring the need to take into account the historical process of evolution that is characteristic of the living. However, one must simultaneously consider the supposedly smallest living entity known to date, the cell, which we attempt to define, describe, and understand. Reconstructing it would help us to understand the "being" of the living.

The origins of an object that has now disappeared force us either to find traces, fossils, or to "construct", in laboratory conditions, a *plausible* model. The same idea applies to all the biological objects studied in laboratories, which are, as stated by Canguilhem, "'superreal', nonnatural object[s]", like the crystal of DNA and the intermediate metabolite. "Life is now studied as far as possible as though it were nonlife, as devoid as possible of its traditional attributes" (1988, 117–18). This latter point constitutes an element of fragility that according to some scientists derives from a separate, non-scientific status. Indeed one must consider the status of a model only by simulating a reality that is still far from the model. Within this scope, the aim of biotechnology and synthetic biology is to make intermediate beings.

Yet, and this is never sufficiently stressed, such a constraint is not part of the study of the origins of life. All modern science has been constructed by human beings. The techniques and products engineered in the laboratory create their own object. Just as Marcelin Berthelot stated in 1887 that chemistry "*creates its object*",[1] later Canguilhem moved this idea forward and added that just as "*chemists had 'dematerialized' matter, biologists were able to explain life by 'devitalizing' it*" (1988, 122). A new Nature, a para-Nature, has been created in laboratories. Experiments designated as prebiotic are rather parabiotic experiments; they mimic living processes.

Some researchers argue that Nature has not explored all the possibilities permitted by the arrangements of atoms in molecules. Hence the proposal is to constantly do better in the construction of macromolecules, protocells and "organisms", performing what is not done today by the current biochemistry of the living. In fact, such assumptions ignore whether Nature did not explore those paths which would have failed or which would have broken off during evolution, submitted to the laws of natural selection or exalted towards other functions.

Furthermore, one must highlight the fact that these kinds of object are supposed to be the future of artificial living beings produced by engineering that is characteristic of a finalist approach. Moreover, this is related to automatisation, normalization of processes, and standardization that is far from the natural course of evolution. For instance, the ambition to build a "robosapiens" is more or less connected to "empowerment", meaning to a radical innovation in the history of life that would derive not from energy, but from information. What is known as transhumanism is a very fashionable approach today because of the development of artificial life and of the sciences of information and their technologies that have invaded the life sciences.

1.4 When Does Evolution in the Origins of Life Come into Play?

No living form is known that is incapable of assuming elements of its own production, reproduction, movements, or exchanges, transforming itself, to become capable of variability, and hence of evolution. At this point, and despite numerous debates and provoking effects, we must ask ourselves if viruses can be considered to be "living creatures".

Indeed we know more than 10^{30} viruses, hence they possess a huge genetic variability and a diversity greater than that of the three domains of life combined. The discovery of parasitism of some bacteria shows that this characteristic is not the prerogative of viruses alone; this was long considered a criterion of demarcation of viruses as not alive. Like seeds, some viruses need favourable conditions to multiply and some of them change their morphology/shape outside the cellular context. For instance, the archaeal virus ATV (Acidianus two-tailed virus) produces, outside any host, two protein appendices of 800 amino acids each (Häring *et al.*, 2005).

We know (David Prangishvili, personal communication) that the development of the extracellular tail of ATV is well described and perfectly reproducible; however, we do not know any mechanism behind it. It can hardly result from metabolic activity. It seems more likely that what happens is subunit polymerisation. Concerning ATV virus growth in acellular media, there are examples of virion morphogenesis – actual formation of mature infectious virus particles – in cell lysates, which can be considered an acellular medium. In a similar manner, some colleagues see the possibility of virus morphogenesis in prebiotic, acellular environments, which they consider to be an ancestor of protocells.

Hence it is obvious today that the boundary between cell and virus has become blurred.

As an experimentalist in biology, I retain an absolutely heuristic approach. We must keep in mind the lesson of the discovery of the catalytic activity of RNAs suspected since the 1960s by Woese, Crick and Orgel (who in 1968 explained it very precisely) and still considered suspicious. For 20 years, until the discovery of ribozymes by Zaug, Grabowski and Cech in 1983, it was considered bad taste to refer to this possibility. And even today, 30 years later, in many universities the existence of ribozymes is not discussed. It is common to encounter students at the level of a Master's or a Ph.D. degree in biology and chemistry, who have never heard of them!

We must prevent dogmas. Each discovery must keep us awake and the necessary caution must not be used to slow down discovery and thought-provoking ideas.

If, as is common, one now considers the cell to be the smallest living unit, it is characterized by growth, exchanges, and interactions between an interior and an exterior environment, between a "self" and a "non-self", the transformation and the creation of new forms, by sexual or asexual reproduction.

It is easy to reproduce in laboratory conditions the morphological and mechanical processes of encapsulation, assimilation, association, and fission of a simple vesicle, compartment or liposome submitted to mechanical action. Interestingly, the proliferation of bacteria cells devoid of a wall, known as L-form bacteria, such as L-*Bacillus subtilis*, is similar to *in vitro* artificial vesicle replication. The L-form *B. subtilis* is surrounded by recently generated progeny blebs that look very similar to *in vitro* proliferating lipid vesicles (Leaver *et al.*, 2009; Errington, 2013*)*.

Such aptitudes might have led to the birth of diverse forms of life, with various phenotypes. From numerous interactions, there arises the formation or extinction of species from adaptative responses by means of sorting or of selections. In other words, it is well-known that selection supposes the production of diversity prior to the filter of experience. Natural selection only operates between what is viable, that is everything that for a time at least interacts with the environment. Thus, as soon as it appears, life is confronted with the problem of survival.

Certain structures can chemically or physically reproduce themselves (crystals, for instance); can such structures qualify as living? Crystal or mineral structures are not alive because they do not possess the variability required to adapt to a changing environment. Therefore, crystals do not have to solve any "survival" problem. And finally, everything that is living is unique, which is not the case with a mineral object, even if it is natural.

1.5 How Does this Translate Biochemically?

Every new biochemical molecule creates new possibilities that in turn will allow the synthesis of new possibilities that previously did not exist. This ensemble of alternatives is infinite, and most steps of evolution have consisted of making exclusive choices and eliminating numerous other possibilities. Biochemical evolution resembles a tree shaped like a bush, many of its branches being mysteriously lost – there are origins, cracks, ramifications, and transitions between different pathways. Many adaptations have appeared several times in the course of evolution, as is particularly well illustrated by physiology in the case of the formation of the eyes and of certain communication capacities (e.g. the echolocation of bats).

At the molecular level, with the four bases of our nucleic acids, AUGC in RNA, positioned randomly, a million different arrangements exist when they constitute a short segment ten monomers in length ($3 \cdot 10^{-6}$ mm),[2] and for a fragment of 20 monomers, 10^{12} possibilities exist! In these conditions, it is difficult to envision how to rationally determine *the* possibility of *the* functional sequence(s) that favoured the advent and the demonstration of the "good" sequence required for life. It is obvious that the installation of life on Earth cannot result from the absolute and inescapable determinism of ordered chemical bonds. Life can only have appeared following an important process of trial and error, in a combination involving a large number of molecular species capable of interacting with one another in a given environment, in given conditions and situations. Finally, strong interactions as encountered in a tissue occur between the atmosphere, the hydrosphere, the lithosphere and the living, and this allows us more easily to identify the parameters and the environments in which an organized biosphere was able to develop. In this context, evolution is the result of optimized encounters and successive adaptations selected because they were viable in a particular environment.

One can therefore say that the origin and the evolution of life coincide, or as stated by Popper (1974) "...the origin of life and the origin of problems coincide...in other words, the essence of living matter consists (from my viewpoint) in the solving of problems"; by problems one can imply solutions (in response to needs). There was no unique starting-point, no spark, no D-day, nor the formation of a finished product or of a finalized creature, *ex nihilo*, that would be life, no generation starting from nothing, thus no creationism.

Life results from several attempts, developments, and associations that occurred and were produced in different contexts. The origins of life on Earth (and not the origin) thus appear as the result of adaptation leading to the invention of what is new. Each original invention, even partial and always piecemeal, was a major discovery; most of them were preserved, allowing the installation of a metabolic pathway, of a given specific sequence endowed with a function, of a species, while another part continued to evolve. Every fragment may be considered to be alive. One only needs to see the considerable genomic contribution brought by the viruses and by small RNA pieces in the construction of current genomes, in metabolic pathways and in their modern regulations. The evolution of life is an evolution in the form of compatible biochemical networks. Grafts can occur in living systems, opening the way to new avenues. Throughout his work, the great biochemist Florkin stated that in metabolic networks it is not rare to find ancient metabolic pathways, several million years old, that have made ulterior grafts possible.

If one observes the present limits of microbial life, one is struck by the extent of the possibilities. This allows us to propose that in different environments, other attempts and other living objects have developed and continue to develop. Whatever their shape, whether we consider them to be cellular or only partially answering our canonical criteria for cell life, they can be considered living beings. With respect to time, another appearance, another posture of life, must also be considered.

1.6 An Extrapolation: the Recapitulation Theory?

During the course of the embryogenesis of vertebrates, we can find traces of archaic ancestors, several million years old. For instance, the *lancelet*, a small 5-cm-long animal, a chordate that looks like a worm (with a cylindrical tube-like body) that lives in sandy coasts, feeding on micro-organisms and organic particles, possesses an organisation plan similar to that of vertebrates. This small marine animal is the ancestor of vertebrates, having established, about 700 million years ago, what was to become our brain and our skeleton. It is thus a kind of passage between invertebrates and vertebrates. Also, the observation by microscopic imagery of the embryonic development of small zebrafish allows us to follow the birth and beginning of the life of macrophages, those defence cells of the organism that learned to phagocytose apoptotic bodies. This function is not innate. One can observe and follow an important apprenticeship performed on the approach, exploration, and contact before these cells acquire the maturity for their full role.

It is striking to discover today that stem cells can differentiate where they are grafted; they adopt the morphology and function of the cells in whose environment they are placed. Functional adaptation is one of the characteristics of the living, and the study of the origins of the living has as much to learn from physicochemical laws as from the reasons that render a stem cell totipotent. Similarly, the evolutionary pathways of the metabolism can reveal a part of their history.

Bases and modified nucleosides would have served as building blocks for the construction of genetic material and of the present catalysts. More than 100 modified nucleosides obtained in a living cell from the purine A, G and pyrimidine T (or U) and C bases have been identified to date, and this greatly expands the diversity of the four canonical nucleosides (A, U, (T), G, C). Certain nucleosides are coenzyme moieties indispensable for the activity of more than 50% of our proteins; others are modified by the addition of amino acids, or of functional groups adapted to amino acids. Short primitive nucleic acids on which a modified nucleoside would have been grafted would have allowed a correspondence between nucleic acids and proteins to be established; in other words, the first steps or the learning of the elementary genetic code. Finally, a phylogenetic examination of the modified nucleosides reveals their extraordinary conservation in the three domains of life and, according to their chemical structure, their possible presence in a prebiotic environment (Friedmann et al., 1971).

1.7 Opportunism of the Living

Today, most theories on the origins of life postulate that, on the primitive Earth, mineral and organic molecules were present from which organized biological systems could have evolved. Conditions have been described for the synthesis of macromolecules starting from simpler monomers, within the framework of the hypothesis of the *"primitive ocean"* or *"prebiotic soup"* of the first metabolic steps. Finally a model is presented for development, starting from random polymers of replicative systems that would have evolved by way of successive mutations and selections.

The best-known scenarios rest on observations derived from astrophysics, astrochemistry, and geochemistry. Other elements come from speculations and laboratory experiments. We are familiar with the hypotheses of the primitive ocean, or "prebiotic soup", of the "world of surfaces", of the "sulphur world" that in some ways are related to the world of surfaces, the "RNA world" and the "lipid world".

The thesis of the *"prebiotic ocean"* reminds us of an essential point, which is that 90% of life on our planet has occurred in oceans. No form of life is known that does not require liquid water. Organic molecules produced by chemical combinations of carbonated molecules present in the primitive atmosphere would have been deposited and assembled in the primitive ocean to gradually produce living organisms appearing in the prebiotic soup. This model (described by Alexander Ivanovitch Oparin and John Burdon Sanderson Haldane) has led in recent decades to a large number of experiments, frequently with spectacular results, such as the production of amino acids by Stanley Miller in 1953 starting from gaseous molecules present in the primeval atmosphere, then the synthesis of purine bases of nucleic acids by Juan Oro thanks to the pentamerization of cyanhydric acid under UV irradiation, and then the template-directed synthesis by Leslie Orgel describing the first steps of the replication of nucleic acid fragments without the intervention of enzymes, as well as the synthesis of diverse macromolecules in hydrothermal conditions, etc. One cannot summarize here the huge number of successful experiments performed since the pioneering ideas of Oparin.

The theory of the *"surface-world"* initially proposed by Desmond Bernal (1951), then by Graham Cairns-Smith (1966) involves mineral surfaces, such as clays. According to Cairns-Smith, the very first replicative system must have been mineral or geochemical (1982); this could have meant crystals on which, and thanks to which, the present organic genetic system could have developed and would have "learned" to replicate. This genetic "take-over" occupies a central position in the process of evolution. It is what we observe with molecules such as PNA (peptide nucleic acids), HNA (hexitol nucleic acids), TNA (threose nucleic acids), and XNA (xeno nucleic acids), which are artificial constructs and are good templates for template-directed synthesis (Pinheiro & Holliger, 2012). It does not mean that these molecules existed in the past. Apart from the discovery by Banack *et al.* (2012) of the existence of PNA in cyanobacteria, most of them are pure chemical laboratory constructs. Yet their efficiency in template-directed synthesis for the formation from nucleotides of a complementary strand of nucleic acids allows us to think that a transfer of information is possible whatever the initial system. It is now possible to perform in a laboratory the simple replication of a template to which monomers belonging to another molecular system belong, provided that weak pairings of Watson–Crick type hydrogen bonds can form between two categories of molecules. In this case, based on this hypothesis of genetic take-over, one observes the transfer of information from one molecular system to another totally different system, and this without loss of genetic information.

Clays that are frequently submitted to hydration–dehydration cycles have demonstrated their efficiency as catalysts during the course of peptide synthesis and elongation of nucleic acids (Paecht-Horowitz *et al.*, 1970; Ferris *et al.*, 1996). Since they can store, position, and concentrate substrates, clays that facilitate polymerization can be considered to be primitive enzymes.

Finally, over several years, Günter Wächtershäuser (1988) developed the idea that pyrite-like mineral surfaces can adsorb mineral substrates such as carbon dioxide, thereby allowing a surface metabolism of the autotroph type to develop. A mineral environment rich in carbon monoxide, sulphur, and iron, simulating the composition of the surrounding warm submarine environments, favours the synthesis of compounds of biological interest obtained in abiotic conditions.

More recently, the hypothesis of an "RNA world" (Woese, 1965; Crick, 1968; Orgel, 1968; Gilbert, 1986) that would have preceded the present-day world based on DNA, RNA, and proteins allows us to propose that primitive nucleic acids were able to accomplish present-day performances of the living, to store and transmit genetic information and catalyze metabolic reactions. For most biochemists and biologists, this hypothesis possesses a strong explanatory power (Atkins *et al.*, 2011). Thanks to this evolutionary scenario that postulates the existence of an ancient RNA ancestor common to all forms of present-day life, many experimental and theoretical results are being obtained in laboratories (Leclerc *et al.*, 2016). With the help of quasi-systematic methods, it is possible to produce selected chemical automates that perform this or that auto-catalytic, auto-replicative function

(Attwater & Holliger, 2014). Analogues are then constructed, self-assembled and simulated.

It is becoming easy to mimic, optimize (and automate) most functions characterising the living; consequently, co-polymers are being produced with therapeutic aims. For instance, plastic substances can now repair themselves, provided a polymerisable chemical compound was previously included in the body. The production of ever more sophisticated chemical robots is under way, and one frequently forgets that what characterizes a robot is its specialization. A robot that brings together objects for their assembly is not the same as a robot that uses a carpet sweeper. A robot would be deemed a generalist that was truly close to achieving the capacities of the living only if it could embrace problems raised by its environment. However, we do not know how this can be programmed.[3]

The problem of the origins of the living comes to us in similar terms. How did we pass from specialized chemical reactions that are efficient and autonomous in a given environment, to an ensemble of new biochemical reactions dependent on one another and capable of interpreting new problems raised by a changing environment? How did the first biochemical networks appear, the first compartments, the first viable cells? This type of transformation, immediately endowed with adaptability, is hence historically part of the origin, that is, in biological evolution. Apprenticeship, irreducible to logics, physics, and even to chemistry, becomes the property of the living.

1.8 Can One Experimentally Recreate the Passage between Non-Life and Life? Where is the Discontinuity Located? What Does it Consist of?

The living cannot be reduced to automation of a chemical automate, even to the self-organization of structures or functions. Self-organization, as well as "dissipative" structures resulting from chance encounters between molecules, which irreversibly orient the effects of physicochemical laws supplying an explanation for the evolution of structures and making it possible to understand why the creativity of life is possible, do not answer the rules of physics, of logic, or of anything else.

According to the definition given by its conceptors (Atlan, 1972), self-organisation is the finite internal term. *Self* is whatever is *self-sufficient*. Therefore, by definition, the living is closely linked to what surrounds it, to its environment, and as soon as it is constituted, it must resolve in a split second the problems linked to its survival. The notion of self-organization remains linked to that of the mechanical automat and refers to the Monadology of Leibniz, for whom "... each organic body of the living is a type of divine machine or of natural automat that infinitely surpasses all artificial automates". After the prefix "auto", one remains in the order of a physical concept of inert matter (from the Latin *iners*: without capacity, inactive, without energy). Indeed, the various pieces that constitute the organization of the living interact cleverly with one another.

It is the principle of interaction between A and B. It is an adaptive relationship from A to B or vice versa, and it is not a relationship of transformation and of creation that characterizes the living and that is, for instance, produced by a symbiotic relationship. Symbiosis transforms the organism; heterogeneity is introduced into the system, whereas the interaction between detached pieces is a construction process that can be entirely automatic.

To me, tinkering with evolution, as discussed by François Jacob, seems closer to the "artificial cleverness" of the living than to determinism, even if a little probabilism is included, that is functional in logical automates, whether built by the construction of order, "by noise" or by the export of entropy into the environment.

1.9 Provisional Conclusions

We possess today a series of explanations and partial experiences concerning syntheses said to be prebiotic, phenomena of auto-assembly of directed replication of primitive catalysts, etc., from which one can extract a few general epistemological principles. One can apply two ways of reasoning of experimental problems regarding the origins of life as presented here. One of them proposes an evolution from the prebiotic to the biotic, thanks to the refinement of details, each step developing gradually from an anterior step. It is logic that presides over the theory of the appearance of life in the primitive ocean.

To this gradual principle, one can oppose that of evolutionary usurpation, the take-over of Cairns-Smith, thanks to which one can understand the present only from a very different past. The reconstitution of past periods can in this case only be indirect.

Another example is supplied to us by symbiosis, thanks to which eukaryotic cells appeared. Symbiotic organisms do not result from the progressive evolution of anterior organisms (which does not fit into the strictly gradualist Darwinian scheme, but rather corresponds clearly to a take-over type of scheme).

In another field, the palaeontologist Stephen Jay Gould also contests the gradualist vision of evolution. Backed by documents regarding fossils, he shows how evolution occurs stepwise. The rhythm of evolution is not regular; one observes very long periods followed by brutal variations, and for those who are in favour of this theory, known as punctuated equilibrium, living beings evolve by shifting from one equilibrium to another, as a result of constraints that limit the field of possible variations.

Opportunism, active in symbiosis as well as in genetic take-over, appears to be a fundamental property of the living, allowing new properties to arise that are inherent to a situation, similar to the propensities proposed by Karl Popper.

Acknowledgements

I thank an anonymous referee for the sound remarks, as well as the editor Andreas Losch for his patience and Helen Tsolaki for English improvements. This chapter is based in part on previously published reflections.

Notes

1 « La chimie crée son objet. Cette faculté créatrice, semblable à celle de l'art lui-même, la distingue essentiellement des sciences naturelles et historiques. »
2 One nucleotide measures 3 angströms (3 Å), that is $3 \cdot 10^{-4}$ microns. The totally unfolded DNA of a small, simple prokaryotic cell measures 1 mm in length!
3 The production of hybrids, the combination of biological tissues (for example grafting of animal retinas) and of microprocessors to produce bio-robots, demonstrates that it will soon be possible to endow robots with a capacity that we do not know how to program. Nevertheless, the construction of a mechanism always supposes the pre-existence or the help of a living original organism.

References

Atkins, J.F., Gesteland, R.F. & Cech, T.R. (2011).*The RNA World*, 3rd edn, Cold Spring Harbor, NY: Cold Spring Harbor Laboratory Press.
Atlan, H. (1972). *L'organisation biologique et la théorie de l'information*, Paris: Hermann.
Attwater, J. & Holliger, P. (2014). A synthetic approach to abiogenesis. *Nature Meth.*, **11**, 495–8.
Banack, S.A., Metcalf, J.S., Jiang, L., Craighead, D., Ilag, L.L. & Cox, P.A. (2012). Cyanobacteria produce N-(2-aminoethyl) glycine, a backbone for peptide nucleic acids which may have been the first genetic molecules for life on Earth. *PLoS ONE*, **7**(11).
Bernal, J.D. (1951). *The Physical Basis of Life*, London: Routledge and Kegan Paul.
Berthelot, M. (1887). *La synthèse chimique*, 6th edn, Paris: Félix Alacan.
Cairns-Smith, A.G. (1966). The origin of life and the nature of the primitive gene. *J. Theor. Biol.*, **10**, 53–88.
Cairns-Smith, A.G. (1982). *Genetic Takeover and the Mineral Origin of Life*, Cambridge: Cambridge University Press.
Canguilhem, G. (1952 (1992)). Aspects du vitalisme. In: *La connaissance de la vie*, 2nd edn, Paris: J. Vrin.
Canguilhem, G. (1988). *Ideology and Rationality in the History of the Life Sciences*, Cambridge, Mass.: MIT Press.
Crick, F. (1968). The origin of the genetic code. *J. Mol. Biol.*, **38**, 367–79.
Darwin, C. (1871). Letter to J.D. Hooker 1 February, Darwin Correspondence Project, "Letter no. 7471," accessed on 28 September 2016, http://www.darwinproject.ac.uk/DCP-LETT-7471
Darwin, C. (1859 (1992)). *On The Origin of Species. L'origine des espèces*, trans. Edmond Barbier, revue par Daniel Becquemont, Paris: GF- Flammarion.
Errington, J. (2013). L-form bacteria, cell walls and the origins of life. *Open Biol.*, **3**(1).

Ferris, J.P., Hill, A.R., Liu, R. & Orgel, L.E. (1996). Synthesis of long prebiotic oligomers on mineral surfaces. *Nature*, **381**, 59–61.
Friedmann N., Miller S.L. & Sanchez R.A. (1971). Primitive Earth synthesis of nicotinic acid derivatives. *Science*, **171**, 1026–7.
Ganti, T. (2003). *Chemoton Theory*, 2 vols, New York: Kluwer Academic/Plenum Publishers.
Gilbert, W. (1986). Origin of life: the RNA world. *Nature*, **319**, 618.
Häring, M., Vestergaard, G., Rachel, R., Chen, L., Garrett, R.A. & Prangishvili, D. (2005). Virology: Independent virus development outside a host. *Nature*, **436**, 1101–2.
Hesiode (2001). *Théogonie*, trans. Jean-Louis Backès. Paris: Gallimard.
Kirschner, M., Gerhart, J. & Mitchison, T. (2000). Molecular vitalism. *Cell*, **100**(1), 79–88.
Lacan, J. (1972). In *La Conférence de Louvain 1972* (extraits), ARTE France, INA. 2001.
Lamarck, J.B. (1809 (1994)). *Philosophie zoologique*, Paris: Flammarion.
Leaver, M., Domınguez-Cuevas, P., Coxhead, J.M., Daniel, R.A. & Errington, J. (2009). Life without a wall or division machine in *Bacillus subtilis*. *Nature*, **457**, 849–53.
Leclerc, F., Zaccai, G., Vergne, J., Řihovà, M., Martel, A. & Maurel, M.C. (2016). Self-assembly controls self-cleavage of HHR from ASBVd (−): a combined SANS and modeling study. *Sci. Rep.*, **6**(30287).
Maurel, M.C. (1994). *Les origines de la vie*, Paris: Syros.
Maurel, M.C. (1999). *August Weismann et la génération spontanée de la vie*, Paris: Kimé.
Maurel, M-C. (2002). Notion d'origines. Actes du colloque, 15 Mai 2001, MNHN, Paris: "Exobiologie, aspects historiques et épistémologiques", Cahiers François Viète, n°4, 2002.
Maurel, M.C. (2003). Origines de la vie, originalité du vivant. In M.C. Maurel & P.A. Miquel (eds.), *Nouveaux débats sur le vivant*, Paris: Kimé, pp. 9–21.
Maurel, M.C. & Décout, J.L. (1999). Origins of life: molecular foundations and new approaches. *Tetrahedron*, **55**(11), 3141–82.
Orgel, L.E. (1968). Evolution of the genetic apparatus. *J. Mol. Biol.*, **38**(3), 381–93.
Paecht-Horowitz, M., Berger, J. & Katchalsky, A. (1970). Prebiotic synthesis of polypeptides by heterogeneous polycondensation of amino acid adenylates. *Nature*, **228**, 636.
Pali, G., Zucchi, C. & Caglioti, L. (2002). *Fundamentals of Life*, Paris: Elsevier.
Pinheiro, V.B. & Holliger, P. (2012). The XNA world: progress towards replication and evolution of synthetic genetic polymers. *Current Opinion in Chemical Biology*, **16**(3–4), 245–52.
Popa, R. (2004). *Between Necessity and Probability. Searching for the Definition and Origin of Life*, Berlin: Springer.
Popper K. (1974 (1981)). *La quête inachevée. Autobiographie intellectuelle*, trans. Renée Bouveresse, Paris: Calmann-Levy.
Roupnel, G. (1945). *La nouvelle Siloë*, Paris: Grasset 1945.
Szostak, J.W., Bartel, D.P. & Luisi, P.L. (2001). Synthesizing life. *Nature*, **409**, 387–90.
Wächtershäuser, G. (1988). Before enzymes and templates: theory of surface metabolism. *Microbiological Review*, **52**(4), 452–84.
Woese, C.R. (1965). On the evolution of the genetic code. *PNAS*, **54**(6), 1546–52.
Zaug, A.J., Grabowski, P.J. & Cech T.R. (1983). Autocatalytic cyclisation of an excised intervening sequence RNA is a cleavage-ligation reaction. *Nature*, **301**, 578–83.

2

The Search for Another Earth-Like Planet and Life Elsewhere

JOSHUA KRISSANSEN-TOTTON AND DAVID C. CATLING

Introduction

Is there life beyond Earth? Unlike most of the great cosmic questions pondered by anyone who has spent an evening of wonder beneath starry skies, this one seems accessible, perhaps even answerable. Other equally profound questions such as "Why does the universe exist?" and "How did life begin?" are perhaps more difficult to address and must have complex explanations. But when one asks, "Is there life beyond Earth?" the answer is "Yes" or "No". Yet despite the apparent simplicity, either conclusion would have profound implications.

Few scientific discoveries have the power to reshape our sense of place in the cosmos. The Copernican Revolution, the first such discovery, marked the birth of modern science. Suddenly, the Earth was no longer the center of the universe. This revelation heralded a series of findings that further diminished our perceived self-importance: the cosmic distance scale (Bessel, 1838), the true size of our galaxy (Shapley, 1918), the existence of other galaxies (Hubble, 1925), and finally, the large-scale structure and evolution of the cosmos. As Carl Sagan put it, "The Earth is a very small stage in a vast cosmic arena" (Sagan, 1994, p. 6).

Darwin's theory of evolution by natural selection was the next perspective-shifting discovery. By providing a scientific explanation for the complexity and diversity of life, the theory of evolution replaced the almost universal belief that each organism was designed by a creator. Every species, including our own, was a small twig in an immense and slowly changing tree of life, driven by variation and natural selection.

Answering the question "Is there life elsewhere?" would be a third shift. An affirmative answer would fuse the Copernican Revolution with Darwinian evolution. Earth would not be special. We would live on one of billions of planets teeming with life. Yet, with this new outlook, the universe would abruptly become immensely richer. The night sky would no longer be a theatre of sterile physics and chemistry, but would instead be full of living worlds, evolving creatures, and perhaps even conscious beings.

If no life exists elsewhere then we would likewise have reason to reconsider our place in the cosmos. Trillions of other planets in existence would be wholly barren, and if life were extinguished on Earth then it would be extinguished everywhere and perhaps forever. What greater value would an individual human life or a species have? Ongoing anthropogenic extinctions would become cosmic losses.

The profundity of such possible implications has fueled speculation about life beyond Earth for millennia. The first recorded musings are those of the Greek philosopher Anaximander whose arguments for a "plurality of worlds" predates our understanding of planets, and indeed precede scientific thought completely (Preus, 2001, p. 58). If we fast-forward to today, popular culture is full of caricatures of life elsewhere, from gray humanoids with big almond eyes and enormous fingers, to Hans Giger's nightmarish, insectoid xenomorphs.

The purpose of this chapter is to convince you that we may not have to speculate about life elsewhere for much longer. We are on the verge of the aforementioned third shift in cosmic understanding. The science of exoplanets – planets around other stars – has exploded in recent decades. We now know that there are at least as many planets in our galaxy as there are stars (Cassan *et al.*, 2012). Many of these planets are rocky, Earth-sized planets that orbit at just the right distance from their host star to permit liquid water on their surfaces. It will soon be possible to detect life on these habitable exoplanets. The necessary technology is well understood and could be employed on a space telescope within the next 10–20 years. Alternatively, clever techniques using ground-based extremely large telescopes might also detect signs of exoplanet life (Rodler & López-Morales, 2014; Snellen *et al.*, 2015). Not only is the question "Is there life elsewhere?" answerable, it will likely be answered within the lifetime of readers of this book.

Detecting Planets Around Other Stars

Exoplanets are difficult to study. By 1995, astronomers had elucidated the structure and history of the universe: they knew that the universe began about 14 billion years ago in the fiery expansion of the Big Bang, they recognized that the universe has billions of galaxies each with tens to hundreds of billions of stars, and from the abnormal rotation of distant galaxies they deduced that most of universe was made of mysterious "dark matter" (Rubin *et al.*, 1980). Astronomers had discovered how the elements were forged in the thermonuclear furnaces of stars, and the entire lifecycle of stars, from formation to supernova to black hole, was well understood. This was all common scientific knowledge by 1995, and yet nobody knew for sure if other stars had planets. In fact, some astronomers even speculated that the Sun might be the only star with planets (Dick, 1993; Jeans, 1942).

Why is it so difficult to learn anything about exoplanets? The problem is that exoplanets are extremely faint when compared with their host stars. Suppose a distant civilization pointed their telescopes at us. For every meager photon received from the Earth, their telescope would be swamped by ten billion photons from the Sun. It is difficult to separate the planetary light from this blinding glare because planets and their host stars are close together compared to the vast distances between the stars. One of us (DC) remembers being confidently told as an undergraduate by a professor that astronomers would never know anything about exoplanets!

Fortunately, several techniques avoid or overcome the problem of the relative faintness of exoplanets. Broadly speaking, exoplanets can be studied by using either indirect methods or direct imaging. The indirect methods exploit a planet's subtle influences on its star to infer the properties of the planet. Almost everything that has been learned about exoplanets to date has made use of these indirect methods. The alternative approach is to design a telescope that can isolate the planet's light from the host star to obtain an image of the planet. Direct imaging is technically challenging, but it is the best way to systematically search for signs of life around nearby stars.

Indirect Methods for Detecting Exoplanets

A simple view of planets orbiting an immobile central star is incorrect. Planets have mass, and so they exert a gravitational force on their star causing both planet and star to orbit their common center of mass. Because stars are much more massive than planets, the star's orbit is a slow wobble around a point interior to, or slightly exterior to, the star.

Two indirect methods exploit the fact that the motion of a star due to the presence of a planet is much easier to detect than the planet itself. The *radial velocity* technique measures the star's back-and-forth motion relative to the observer. This is done using Doppler spectroscopy: starlight is split into its component wavelengths, which shift with very slight changes in the star's radial velocity. From the small swing in wavelengths, the speed of the star's orbit and the presence of the planet can be inferred.

The first exoplanet around a normal star, 51 Pegasi b, was discovered in 1995 by using the radial velocity method (Mayor & Queloz, 1995). This method is most sensitive to large planets that are close to their host stars because these planets induce the largest stellar wobble. Consequently, many of the first exoplanets discovered were "Hot Jupiters": Jupiter-sized planets that orbit their host stars in several days or less and have atmospheric temperatures in excess of 1000 °C. Despite being easy to detect, more recent studies have revealed that "Hot Jupiters" are less than 1% of planets (Howard, 2013).

The most precise radial velocity instruments can currently measure variations in stellar velocities of about 1 m/s (Fig. 2.1). To detect an Earth-like planet around a Sun-like star requires precision of 0.1 m/s (e.g. Catling & Kasting, 2017, p. 433), and so no such planets have been found using radial velocity. However, the Eschelle SPectrograph for Rocky Exoplanet and Stable Spectroscopic Observations (ESPRESSO) at the European Southern Observatory's Very Large Telescope (VLT) will soon begin operation (Pepe *et al.*, 2014). This new instrument will have <0.1 m/s measurement precision and so can find nearby Earth-like planets around Sun-like stars.

An alternative approach that also makes use of a star's wobble is *astrometry*. This technique makes precise measurements of a star's position in the sky over time. This allows the complete orbital motion of planets to be calculated, rather than just the radial motion. Unfortunately, the full potential of astrometry in detecting Earth-sized planets has yet to be realized because NASA's astrometry mission concept called the *Space Interferometry Mission* (*SIM*) was canceled due to budget constraints.

Not all indirect detection methods make use of a star's motion. *Gravitational microlensing* takes advantage of the gravitational effect of a planet on the path of light, as predicted by Einstein's theory of general relativity. When two stars align relative to Earth, the gravitational influence of the foreground star focuses the light from the background star like a lens (Fig. 2.1). Such an alignment temporarily magnifies the background star as seen from Earth. If the foreground star has a planet orbiting in the right position, then its gravitational influence may also distort the lensed light. The planet causes a pulse in background star magnification (Fig. 2.2). Einstein (1936) himself predicted that gravitational microlensing events involving stars would never be observed because stellar alignments are extremely improbable. However, modern technology allows large numbers of distant stars to be monitored, and numerous exoplanets have been detected using this method. In fact, microlensing reveals that the majority of stars in our galaxy have one or more planets (Cassan *et al.*, 2012).

Finally, exoplanets can be detected when they *transit* their host stars (Fig. 2.1). If the orbital plane of an exoplanet aligns with the Earth, then astronomers can observe the exoplanet crossing the disk of its host star once per orbit. This primary transit obscures some starlight, and with sufficiently sensitive telescopes a periodic dimming of the host star can be detected. The chance of seeing any specific planet transiting its star is small (about 0.5% for Earth-like planets), but by continuously monitoring a large number of stars many planetary transits can be detected. NASA's *Kepler* telescope and the European *CoRoT* spacecraft have been spectacularly

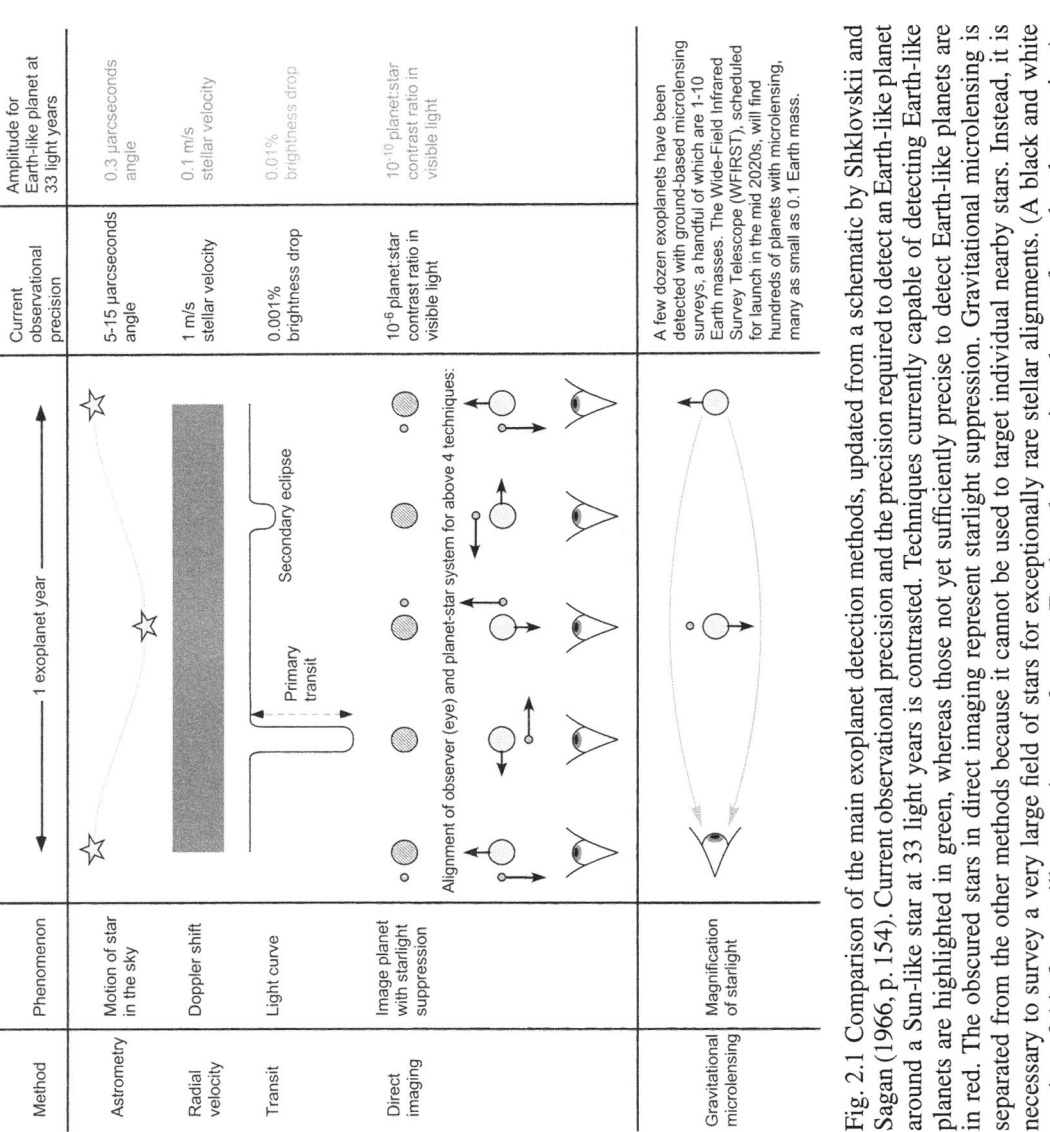

Fig. 2.1 Comparison of the main exoplanet detection methods, updated from a schematic by Shklovskii and Sagan (1966, p. 154). Current observational precision and the precision required to detect an Earth-like planet around a Sun-like star at 33 light years is contrasted. Techniques currently capable of detecting Earth-like planets are highlighted in green, whereas those not yet sufficiently precise to detect Earth-like planets are in red. The obscured stars in direct imaging represent starlight suppression. Gravitational microlensing is separated from the other methods because it cannot be used to target individual nearby stars. Instead, it is necessary to survey a very large field of stars for exceptionally rare stellar alignments. (A black and white version of this figure will appear in some formats. For the colour version, please refer to the plate section.)

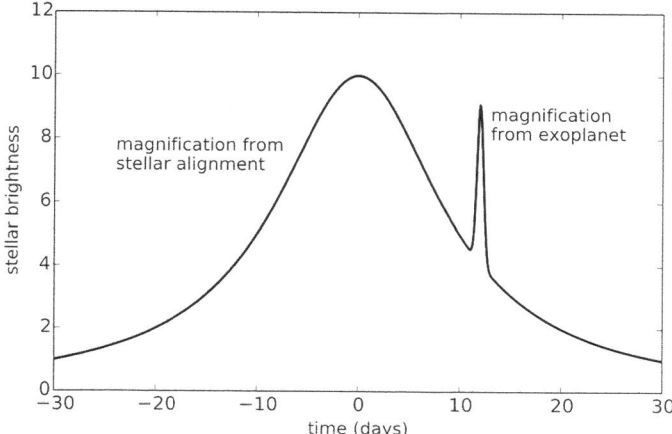

Fig. 2.2 The detection of an exoplanet using gravitational microlensing. The graph shows the magnification of a background star as it aligns with a foreground star and is lensed by its gravitational field. If the foreground star has a planet, then this may cause a brief but sizeable magnification in the light curve.

successful in detecting planets using transits. *Kepler* has discovered over two thousand exoplanets and a few thousand more probable planet candidates. Moreover, transits reveal that Earth-sized planets are common: Sun-like stars have on average at least 0.78 planets with a radius between 0.75 and 2.5 times the Earth (Burke *et al.*, 2015). Figure 2.1 summarizes all the different exoplanet detection methods.

It is worth pausing here to consider what can be learnt from indirect exoplanet detection. Speculative news articles frequently contain sweeping statements about the possible presence of life on exoplanets, and are accompanied by vivid artistic impressions of habitable worlds. These articles capture the imagination and convey the idea that exoplanets are not merely abstract; they are instead real places – worlds that could be mapped, explored, and maybe even settled. But such over-enthusiasm can be misleading.

In practice, indirect methods provide very limited information about exoplanets: radial velocities reveal planet mass, and transits reveal planet radius. For planets with both transit and radial velocity observations, mass and radius measurements can be combined to infer average density. Density provides crude information about whether the planet is gaseous, icy, or rocky. For rocky planets, the planet-star separation and the stellar radiation may tell you if it's possible for liquid water to exist on the surface. But with the lone exception of transit transmission spectroscopy (see below), that is the extent of what can be deduced from indirect detection methods – anything more is speculation.

The indirect methods provide no way of knowing whether an exoplanet is truly Earth-like. We cannot say for sure if an exoplanet has an atmosphere, ocean, or continents, and there is no way of knowing whether a planet has a biosphere. Other observations are required to address these questions, which brings us to the subject of direct imaging.

Direct Imaging of Exoplanets

From the 1990s to mid-2000s radial velocity techniques produced stunning new exoplanet detections. Then there was about a decade of new discoveries with transits. Both techniques will continue to expand our knowledge. But we anticipate that startling new advances will soon be made with direct imaging. As mentioned previously, faint exoplanets are difficult to see amid the glare from their luminous host stars. But with impressive optical know-how, starlight can be suppressed to isolate the light from orbiting planets.

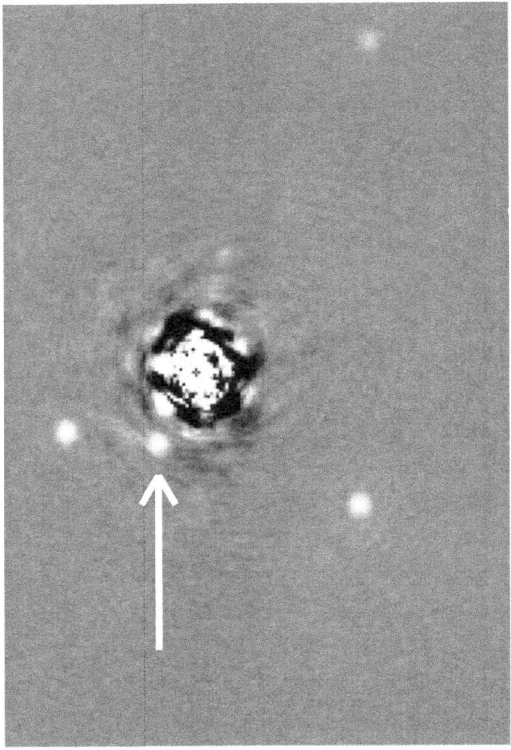

Fig. 2.3 An image of four giant exoplanets around the star HR 8799 – 130 light years away – taken using the Keck telescope in Hawaii (Marois *et al.*, 2008). The central star looks unusual because an internal coronagraph was used to reduce the brightness of the star and reveal the relatively faint planets. Reproduced with permission from Nature Publishing Group.

Several space telescope designs can overcome the glare of starlight. The first is a starshade. A thin, opaque disc with a pattern of notched edges is positioned in front of a space telescope. The starshade blocks the starlight and its patterned edge compensates for the spread of starlight (diffraction), so that only light from the planet enters the telescope. However, the starshade must be placed thousands of kilometers away from the telescope to be effective (Stark *et al.*, 2016). An alternative to the starshade is an internal coronagraph whereby the starlight and planet light are separated within the optics of the telescope itself. In principle, both designs could suppress visible starlight to take pictures of Earth-like planets around other stars. In fact, this has already been done for Jupiter-sized planets using ground-based telescopes. Figure 2.3 is a real image of a distant star system with four planets orbiting around a central star. A coronagraph reduced the brightness of the star so that the planets were revealed.

Direct imaging can do much more than take blurry pictures of exoplanets. As we'll see, planetary light can be split into a spectrum of its component wavelengths, and the atmospheric composition and perhaps even surface properties of the planet can be deduced. This brings us to the subject of exoplanet characterization and habitability.

Characterizing Exoplanets and Habitability

What Makes a Planet Habitable?

Broadly speaking, a planet is habitable if it can support life. However, exoplanet habitability means something more specific: a rocky planet that can sustain liquid water on its surface. The reasoning behind this definition is that liquid water appears necessary for life as we know it. The habitable zone of a star is the range of distances between planet and star that permit liquid water to persist on a rocky planet's surface. Planets closer than the habitable zone are unsuitable for life because they will undergo a hellish *runaway greenhouse* whereby oceans boil away and are lost to space. This was the fate of Venus, which today has only trace amounts of water in its atmosphere and a scorching surface at 460 °C. Planets beyond the habitable zone are also barren because surface water is permanently frozen, such as on Mars (although we will see the fate of Mars is more complex than merely being too far from the Sun).

In the late 1970s, some scientists believed Earth's habitability was a fluke. Had Earth formed merely 5% closer to the Sun, they argued, it would have been rendered uninhabitable by a runaway greenhouse, and if it had formed just 1% further away from the Sun it would be permanently locked in a global ice age (Hart, 1978). Such a narrow habitable zone would imply that life is rare in the cosmos.

Fortunately, this picture of the habitable zone is incorrect. It turns out that the cycling of carbon between the atmosphere and the solid Earth acts a planetary thermostat (Walker *et al.*, 1981). Carbon dioxide (CO_2) is a greenhouse gas that warms the planet's surface. CO_2 is continuously added to the atmosphere by volcanism and removed from the atmosphere through weathering of rocks and subsequent burial of carbon-bearing rocks. Specifically, atmospheric CO_2 and water react to form acid that gradually dissolves surface rocks into products that are carried to the ocean where they eventually form carbonate rocks. The rate of weathering and carbon burial depends on temperature – the warmer the atmosphere, the greater the rate of carbon removal. This feedback helps maintain the planet's temperature in a habitable range. If the Earth plummets into a global ice age, as has occurred several times over Earth's history, then the removal of CO_2 from the atmosphere will cease because conditions are too cold and dry for weathering. However, volcanoes will continue to erupt, causing CO_2 to build up in the atmosphere, thereby warming the planet until the ice melts. Conversely, if the Earth becomes warmer, then rates of weathering will be enhanced due to greater atmospheric water content, thereby reducing atmospheric CO_2 and cooling the climate. In fact, the Sun has increased in brightness by roughly 10% every billion years, but the Earth's temperature has remained comparatively stable because of this planetary thermostat (unfortunately this feedback takes hundreds of thousands of years to take effect, and so it will not help mitigate anthropogenic global warming). Recent habitable zone estimates suggest the Earth would be safe anywhere between its current position and 70% further away from the Sun (Kopparapu *et al.*, 2013).

By continuously observing about 150 000 stars for transits, the *Kepler* telescope was designed to find the fraction of stars that have an Earth-sized planet in their habitable zones. *Kepler* discovered that habitable Earth-sized planets are fairly common – by extrapolation, about 5%–10% of Sun-like stars have a rocky planet orbiting in their habitable zone (Silburt *et al.*, 2015). Unfortunately *Kepler* broke down before enough data were gathered to pin down an exact number.

At this point you might be wondering why our notion of the habitable zone is so restrictive. Within our own Solar System, both Europa (a moon of Jupiter) and Enceladus (a moon of Saturn) have potentially habitable oceans of liquid water beneath their icy surfaces, but are far outside the Sun's habitable zone. Their internal oceans are maintained by tidal forces: the immense gravitational forces from the gas giants they orbit, in combination with continuous tugging from other moons, create tides that heat the interiors of Europa and Enceladus. Indeed, it could be argued that there is no true outer edge of the habitable zone because tidally heated oceans can be maintained at any distance from a host star.

However, practical reasons restrict the habitable zone to the region where *surface* oceans can persist. Icy exoplanets with subsurface oceans may host life, but it would be extremely difficult to detect this life remotely because there are no clear atmospheric or surface biosignatures. Several spacecraft have visited Europa and Enceladus and we still do not know if these moons host life! In short, the "habitable zone" is not intended to rigorously define where life is possible, but is instead a pragmatic guide to detectable life.

Residing within the habitable zone is an important criterion for habitability, but it is certainly not the only one. Whether or not a planet's surface can sustain liquid water depends on many other factors, such as size. By some estimates, Mars orbits within the Sun's habitable zone, and yet it is a frigid, polar desert with no surface water. How is this possible? Mars' small size and low gravity means that it quickly lost most of its atmosphere to space. Furthermore, Mars' interior cooled rapidly due to being small. This cold interior does not generate enough volcanism to replenish the lost atmosphere. Consequently, Mars today has a very thin atmosphere with insufficient greenhouse warming for a habitable surface. Had Mars been larger it might have retained a thick atmosphere and still be habitable.

To appreciate the surprising variety of factors that influence habitability, consider Proxima Centauri b, an approximately Earth-mass planet that orbits within the habitable zone of its host star (Anglada-Escudé *et al.*, 2016). The discovery of this planet generated lots of excitement because it's only 4.24 light years from Earth, and will thus be comparatively easy to image with future telescopes. However, it is uncertain whether Proxima Centauri b is actually habitable because it orbits an M-dwarf star. After their formation, M-dwarfs undergo a long contraction phase whereby they are $\sim$10–100 times more luminous than later. Consequently, planets that reside within the habitable zone of mature M-dwarfs were effectively within the inner edge of the habitable zone for $\sim$100 million years after formation, and may have lost all their water during this time (Luger & Barnes, 2015). Whether or not Proxima Centauri b suffered this fate will depend on the details of planetary formation such as whether it formed in its current location and how much water it started with (Barnes *et al.*, 2016; Ribas *et al.*, 2016). Similar issues apply to a system of seven planets around the small TRAPPIST-1 ultracool M-dwarf star (Gillon *et al.*, 2017).

In fact, many potential influences on planetary habitability exist, including stellar properties, orbital dynamics, galactic position, atmospheric properties, and interior properties. Finding a truly habitable planet is more complicated than merely finding a rocky planet in the habitable zone.

Identifying Habitable Planets

Telescope observations can reveal a lot about an exoplanet's habitability. At this point we must make a brief digression into the physics of light, atoms, and molecules. Surprisingly, the physics of very small things is important on the scale of worlds. Every molecule absorbs electromagnetic radiation at a specific set of frequencies. These frequencies depend on the configuration of electrons and the ways in which a particular molecule rotates and vibrates, which in turn depend on

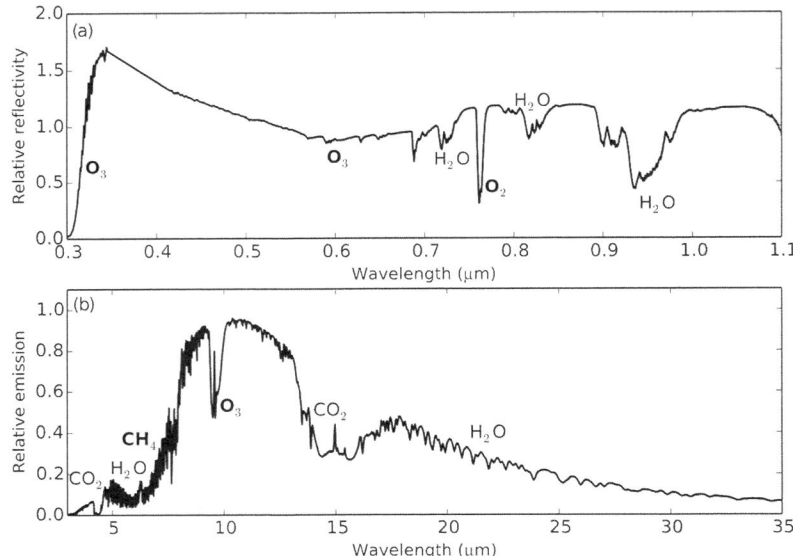

Fig. 2.4 Earth's spectrum in both (a) reflected visible light and (b) emitted infrared light. Atmospheric gases absorb light at characteristic wavelengths, and these spectral fingerprints are labeled in the figures above. Gases that are potential biosignatures are highlighted in bold font. For example, the oxygen (O_2) absorption feature at 0.76 μm is a biosignature because on Earth almost all atmospheric oxygen is produced by photosynthesis. Similarly, ozone (O_3) is a byproduct of biogenic O_2. The signatures of water vapor and carbon dioxide are also clearly visible. Courtesy of Ty Robinson and the Virtual Planetary Laboratory.

the molecule's shape, symmetry, and its chemical elements. For example, a carbon dioxide molecule absorbs infrared radiation because its O=C=O structure vibrates in several ways around the central carbon atom. Consequently, CO_2 is a good greenhouse gas – it absorbs outgoing infrared radiation from the Earth and radiates some of this energy back to the surface, thereby warming the surface. In contrast, there are not many ways a simple molecule like N_2 (molecular nitrogen) can vibrate, and consequently the absorption features from N_2 are few.

Light is thus preferentially absorbed at particular wavelengths when it passes through a planet's atmosphere, depending on the type and amount of gases present. The light that remains has diagnostic absorption features, which are the fingerprints of the constituent gases. From the size and position of these spectral features, it is possible to work backwards to derive the composition of gases present from direct imaging observations (Crossfield, 2015). Figure 2.4 shows Earth's spectrum with the fingerprints of different atmospheric gases labeled.

It is even possible to deduce atmospheric composition without the need for starlight suppression by coronagraphs or starshades for exoplanets that transit their host star. When a transit occurs, some starlight passes through the planet's atmosphere. Gases in the atmosphere may absorb some of this planet-grazing light, so that the amount of blocked light depends on wavelength (Seager & Sasselov, 2000). Thus, transits observed at different wavelengths reveal the spectral fingerprints of gases. This "transmission spectroscopy" technique has already been used to identify gases in the atmospheres of transiting "Hot Jupiter" exoplanets. This method might be suitable for investigating transiting rocky planets with NASA's upcoming *James Webb Space Telescope* (*JWST*).

A wealth of information can be gleaned about an exoplanet through direct imaging and spectroscopy. One can apply climate models and calculate a planet's surface temperature once atmospheric composition is known. Subtle spectral fingerprints from the atmosphere can be used to infer total atmospheric pressure. Then from temperature, pressure, and the amount of atmospheric water vapor, we can paint a more complete picture of a planet's habitability.

Although exoplanets appear as blurry blobs in telescope images, a surprising amount of information is revealed by variations in brightness over time. The length of a planet's day and the presence of variable cloud cover could be evident. It may also be possible to detect the glint of a surface ocean. This effect is most clearly visible in an exoplanet's crescent phase because the bright glint spot is large relative to the rest of the illuminated crescent surface (Robinson *et al.*, 2010, 2014). By measuring changes in exoplanet brightness over several months, it may even be possible to crudely map the planet's surface and differentiate continents and oceans. Figure 2.5 shows surface maps of Earth that could be obtained from nothing more than brightness and color observations over time.

Clearly, direct imaging could identify habitable exoplanets. However, simply finding a habitable planet will not necessarily change our cosmic perspective. If the origin of life was exceedingly improbable, billions of Earth-like exoplanets with oceans, clouds, and continents might be completely sterile, like global ghost towns for all life. Merely *detecting* a habitable world is not enough; we need to find life.

Detecting Life on Exoplanets

A back-of-the-envelope calculation reveals that any telescope capable of photographing whale-sized organisms on an exoplanet's surface would need to be the size of the Solar System! Indeed, distant planets will appear as faint, unresolved blobs to the first telescopes capable of taking pictures of them (similar to Fig. 2.3). How then, will it be possible to detect life?

To answer this question we must go back to the dawn of the Space Age, and the search for life on Solar System planets. Before the 1970s, life on Earth's surface –

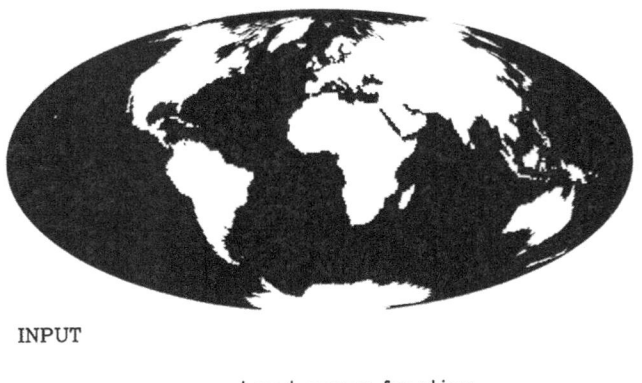

INPUT

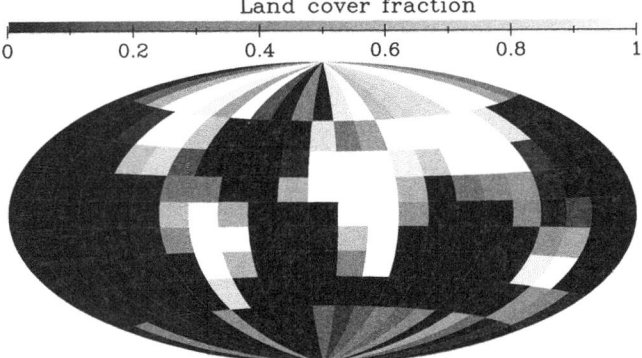

RECOVERED MAP : $\zeta=90.00°$, $\sigma=0.0\%$

Fig. 2.5 Surface maps like this one could be obtained for Earth-like exoplanets using only observations of planet brightness. The position of continents and oceans can be extracted from precise observations of brightness variations over a planet's rotation and orbit (Kawahara & Fujii, 2010). Reproduced with permission from IOP publishing. (A black and white version of this figure will appear in some formats. For the colour version, please refer to the plate section.)

or on any planet for that matter – was believed to be an inconsequential veneer of organic scum. The idea that biological processes could substantially modify a planet was rarely discussed, with some exceptions such as the work of Vladimir Vernadsky (1926). James Lovelock and Lynn Margulis broke with orthodoxy by popularizing their Gaia hypothesis – the idea that life modifies its environment to maintain conditions suitable for life (Lovelock & Margulis, 1974). Gaian ideas have had a mixed reception in the scientific literature (e.g. Kirchner, 2003). Some posit that the stability of Earth's temperature over geological time can be explained by Gaian climatic regulation by organisms, but as we saw earlier, a key planetary thermostat is the inorganic carbon cycle, which would still operate (albeit differently) in the absence of life. However, the idea that life modifies its environment on a planetary scale turned out to be prescient.

Nowhere is this concept more evident than in Earth's atmosphere. Most life on Earth produces waste gases as byproducts of metabolism and these gases accumulate and change atmospheric composition. In fact, every gas in Earth's atmosphere except the noble gases is cycled by life. In the 1960s and 1970s, it was argued that the presence of life on other Solar System planets could perhaps be deduced from biogenic gases in atmospheres (Lovelock, 1975). Spectroscopic detection of these so-called biosignature gases could also be a sign of life on exoplanets if non-biological sources like geological activity or photochemistry can be ruled out.

The most promising biosignature gas is molecular oxygen (O_2). Almost all the O_2 in Earth's atmosphere has been produced by photosynthesis. Indeed, it is difficult to make large quantities of oxygen without life because lots of energy is required to break apart water molecules and liberate oxygen atoms. Life overcomes this energy barrier by combining the energy of many photons from the Sun in complex biomolecular machinery. It is highly unlikely that any naturally occurring mineral could produce much molecular oxygen from sunlight (Léger *et al.*, 2011). In the upper atmosphere, high-energy UV photons can break apart molecules and produce small amounts of oxygen when the accompanying hydrogen escapes to space. But oxygen typically can't accumulate this way because reactions with volcanic gases mop-up the oxygen, leaving only trace amounts.

With that said, there are a few unusual scenarios whereby large amounts of oxygen could accumulate in the absence of life. Planets in the process of losing their oceans, planets too small to sustain volcanism, or even planets around unusually UV-luminous stars might all build up oxygen in their atmospheres (Harman *et al.*, 2015). These ambiguities have led some to suggest that a better indicator of life would be the coexistence of atmospheric oxygen and some other biosignature gas that we wouldn't expect to find in the absence of life (Krissansen-Totton *et al.*, 2016; Lovelock, 1965, 1975). For example the coexistence of oxygen and methane is a strong signature of life in Earth's atmosphere. Without life, all the methane in Earth's atmosphere would be destroyed by chemical reactions with oxygen in about 10 years. Methane persists in the atmosphere because it is continuously replenished by biology. In fact, the coexistence of oxygen and methane implies both gases are being generated in large quantities, which is almost impossible to explain without life.

There is another way that life may change its environment on a planetary scale. If organisms cover a significant fraction of a planet's surface then their color might be detectable in the reflected light from that planet. The chlorophyll pigment that plants use to carry out photosynthesis has a distinctive reflectance spectrum; the leaves of plants reflect a lot more infrared light than visible light, and so there is a large increase in reflectivity between red and infrared wavelengths. This so-called "red edge" is visible in spectra of Earth from space (Arnold *et al.*, 2002; Sagan *et al.*, 1993). Other biological pigments, such as those found in many prokaryotes, might also be visible in an exoplanet's reflected light (Schwieterman *et al.*, 2015).

Geocentrism in Astrobiology?

At this point the skeptical reader might argue that our methods for life detection are myopically focused on Earth-like life. What if life on exoplanets is very different? Naturally, it is impossible to rule out weird forms of life that we cannot imagine. But there are good reasons to focus on the habitability criteria and biosignatures described above. For instance, liquid water is arguably necessary for all life in the universe (Pohorille & Pratt, 2012). Any life based on chemistry will require a liquid solvent to mediate its chemical reactions. Solid phase life is unlikely because atoms are fixed in lattices, making chemical reactions extremely slow. Life-like reactions could occur in the gas phase, but high temperatures are required to vaporize large molecules, and heat tends to cause such molecules to react and decompose. This is problematic since large molecules are necessary for chemical life.

The next question to consider is whether there are any viable alternatives to water as the liquid solvent. Even if we knew nothing about Earth life, we might suspect water as a likely candidate for life's solvent based purely on cosmic abundances. Hydrogen is the most abundant element in the universe whereas oxygen is third (helium is second). Water is also a liquid over a much wider temperature range than most other substances.

Crucially, the properties of water are ideal for sustaining information processing in life. Even if extraterrestrial life has very strange biochemistry, it must have complex molecular machinery capable of inheritance and Darwinian evolution. Non-polar solvents like hydrocarbons are poorly suited to this because the chemical bonds formed in solution are incredibly hard to break, and rapid making-and-breaking of bonds is necessary for biological information processing such as replication, transcription, and translation (Pohorille & Pratt, 2012). These arguments do not definitively rule out unconventional solvents – indeed some astrobiologists have proposed exotic alternatives (Baross *et al.*, 2007) – but they suggest that liquid water is likely a commonly used solvent for extraterrestrial life.

If we accept water as a likely solvent for life, then searching for biosignature gases such as oxygen, which comes from the biological decomposition of water, is credible. Organisms that carry out oxygen-producing photosynthesis (and organisms that eat photosynthetic life or its dead remnants) dominate the Earth's biosphere because the necessary materials are virtually unlimited. Earth's surface is covered in life because oxygen-producing photosynthesis requires only water, carbon dioxide, and visible light.

Available materials limit other known metabolisms. Anoxygenic photosynthesizers – organisms that use sunlight to get energy but don't release oxygen – are limited by the amount of iron and sulfur in their local environment, whereas chemoautotrophs – organisms that get their energy from materials in their environment rather than from sunlight – are limited by the rate at which volcanic gases are released. In contrast, the materials required for oxygenic photosynthesis will all be readily available on habitable exoplanets. Therefore if organisms evolve oxygenic photosynthesis on such planets, then we would expect these organisms to similarly dominate the biosphere.

With that said, oxygenic photosynthesis probably only evolved once on Earth (Knoll, 2008; Lane, 2002, p. 145). If this innovation is highly improbable, then perhaps oxygenic photosynthesis is rare elsewhere. Additionally, Earth's atmosphere has only contained appreciable levels of oxygen for about half its history, and oxygen levels have only been comparatively high (between about 10% and 30% of atmospheric composition) for the most recent eighth of Earth's history. Scientists are still debating why it took such a long time for oxygen levels to rise because oxygenic photosynthesis probably evolved long before atmospheric oxygen increased (Catling, 2014; Kasting, 2013; Lyons *et al.*, 2014).

In light of the early Earth having negligible oxygen, it might be wise to consider alternative biosignatures. It may be possible to detect sulfur and organic compounds produced by life in atmospheres like that of the early Earth (Domagal-Goldman *et al.*, 2011). Another exotic possibility is the detection of ammonia (NH_3) on planets with N_2- and H_2-rich atmospheres. The formation of NH_3 from N_2 and H_2 occurs in the Haber process: the industrial reaction used to make fertilizer. If organisms could evolve catalysts to carry out the Haber reaction, then we might be able to detect them remotely (Seager *et al.*, 2013).

The Future

Suppose a future direct imaging mission finds a habitable Earth-like planet. Upon closer investigation it is revealed that this planet has both oxygen and methane in its atmosphere, a clear indicator of biological activity. Perhaps there are also hints of a "red edge" in the reflected light, subtle features suggestive of a surface covered in photosynthetic pigments. Once the effect of champagne and excitement from the discovery has subsided, the most obvious next question is "How do we learn more?"

The next generation of direct imaging space telescopes will not reveal the surface features of exoplanets – habitable exoplanets will initially be imaged as unresolved pale blue dots. However, improved telescope designs have been proposed. The hyper-telescope (Labeyrie, 1999) would involve a flotilla of space telescopes

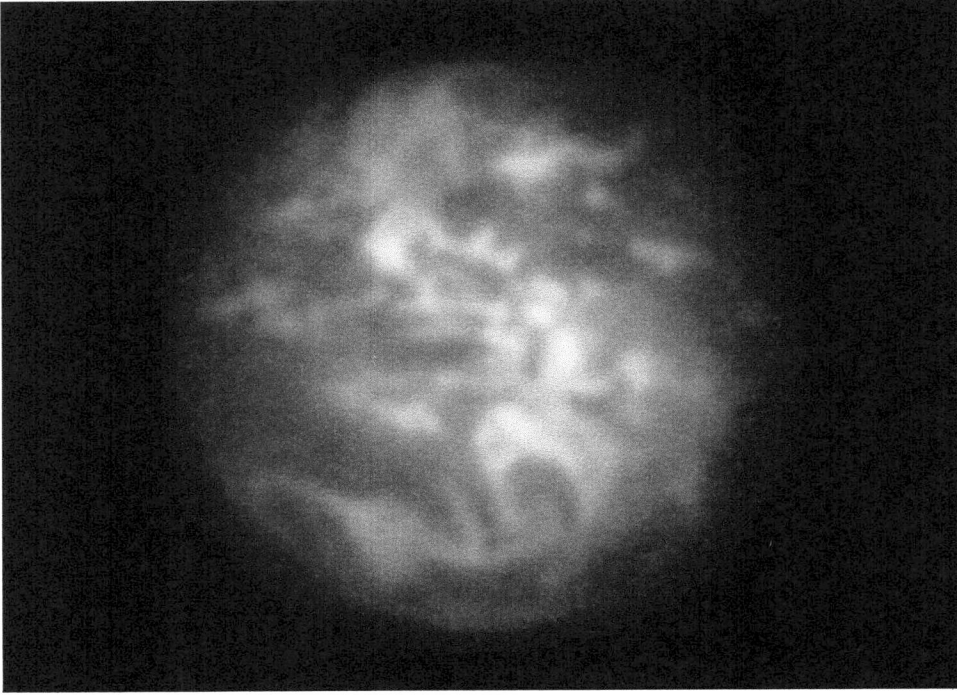

Fig. 2.6 In the more distant future, a flotilla of space telescopes could obtain images of Earth-like exoplanets, revealing continents, oceans, clouds, and vegetation. This image shows how the Earth would appear using such a telescope from a distance of 10 light years (Labeyrie, 1999). Reproduced with permission from AAAS. (A black and white version of this figure will appear in some formats. For the colour version, please refer to the plate section.)

dispersed over 100 km. Such an array of telescopes could image an exoplanet's surface. Figure 2.6 shows what Earth would look like from a distance of 10 light years using this telescope. Continents, oceans, vegetation, desert, weather patterns, and mountain ranges could all be revealed.

With large space telescopes, more subtle signs of life would also be detectable. For example, the CO_2 concentration in Earth's atmosphere oscillates annually due to the asymmetric distribution of continents. Most of the Earth's landmass is in the Northern Hemisphere, and so the annual uptake of CO_2 by plants in the northern spring and release of CO_2 by plants in the northern autumn dominates, producing an annual oscillation superimposed upon the steady increase due to anthropogenic carbon emissions (Fig. 2.7). Annually fluctuating carbon dioxide on an exoplanet would be a compelling biosignature. Large telescopes could also detect trace biosignature gases such as nitrous oxide (N_2O). In short, with a large space

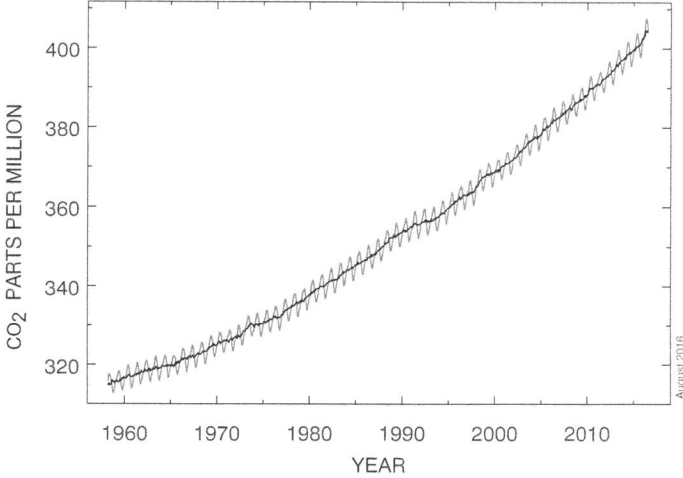

Fig. 2.7 Atmospheric CO_2 since 1960 measured at Mauna Loa observatory in Hawaii. The atmospheric CO_2 concentration oscillates every year due to the annual growth and die-off of vegetation in the Northern hemisphere, which has most of Earth's landmass. If habitable exoplanets also have asymmetric landmass then the seasonal variation of CO_2 could be a potential biosignature. Courtesy of NOAA.

telescope it would be possible to confirm the presence of life beyond reasonable doubt.

Ultimately, the best way to learn more about a planet is to visit it with a spacecraft. Although plausible concepts have been proposed for small interstellar probes (Long et al., 2010; Martin, 1978), with current technology the cost would be prohibitively large. But perhaps the discovery of a nearby exoplanet teeming with life would be sufficient impetus to seriously consider such a mission.

The detection of life on an exoplanet with telescopes is a feasible, albeit technologically challenging, approach to answering the question "Is there life elsewhere?". However, there is a complementary approach that could bypass the difficulties described above.

The Search for Extraterrestrial Intelligence (SETI)

Rather than search exoplanets for biosignatures, we could instead look for signs of extraterrestrial intelligence. More precisely, we could search the stars for evidence of alien technology. The Drake equation is a useful conceptual framework for "organizing our ignorance" on SETI. The astronomer Frank Drake originally formulated the equation in 1961 as a way of stimulating discussion on SETI. The equation, written below, quantifies the number of communicating technological civilizations if we knew the values of all the terms:

$$N = R_* \times f_p \times n_e \times f_l \times f_i \times f_c \times L. \tag{2.1}$$

Here, N is the number of technological civilizations in the galaxy that we could communicate with, which is the quantity we want to calculate. R_* is the rate of star formation in the galaxy, f_p is the fraction of stars with planets, and n_e is average number of habitable planets per star with planets. Astronomers have measured all three of these terms, although there is still some uncertainty in n_e. The remaining terms are f_l, the fraction of habitable planets upon which life emerges, f_i, the fraction of those planets upon which intelligence evolves, f_c, the fraction of intelligence civilizations that develop the technology to communicate with us, and L, the average lifetime of a technological civilization. There are no hard constraints on any of these last four variables, and depending on the values you choose, the number of communicating technological civilizations in the galaxy could be anywhere from close to zero, suggesting we're the only civilization in the observable universe, to millions of civilizations. If we were to detect biosignatures on nearby exoplanets this would tell us that f_l is close to 1, but there is probably no way of definitively constraining the remaining three terms without a SETI programme.

Motivated by this need for observations, the conventional approach to SETI – popularized by the film *Contact* – is to search for narrowband radio waves that could only be produced by a radio transmitter. The observer can only guess what frequency extraterrestrial intelligences would use, and early SETI searches focused on a region of the radio spectrum dubbed the "waterhole" (Fig. 2.8). Within the waterhole are frequencies corresponding to the spectral fingerprints of hydrogen (H) and the hydroxyl molecule (OH). Taken together these form water, and because water is cosmically abundant and perhaps even necessary for life, extraterrestrial civilizations might choose to broadcast at this universally recognized frequency (Oliver, 1979). The lack of galactic noise or atmospheric absorption around this range of frequencies thus makes the appropriately named "waterhole" region an ideal place for interstellar correspondence.

Modern computing can simultaneously search millions of radio frequencies, which eliminates the need to guess the precise broadcast frequency of other civilizations. Even so, the task is daunting. Searching for extraterrestrial broadcasts is sometimes described as a multi-dimensional cosmic haystack. The observer must find the right frequency, spatial location, moment in time (signals may be pulsed rather than continuous), bandwidth, modulation, and perhaps even polarization; the cosmic haystack has at least eight dimensions! Most of the dedicated SETI radio searches to date have only thoroughly surveyed the nearest few hundred stars for a specific type of narrowband radio signal (Tarter, 2001). Other SETI surveys

2 The Search for Another Earth-Like Planet and Life Elsewhere

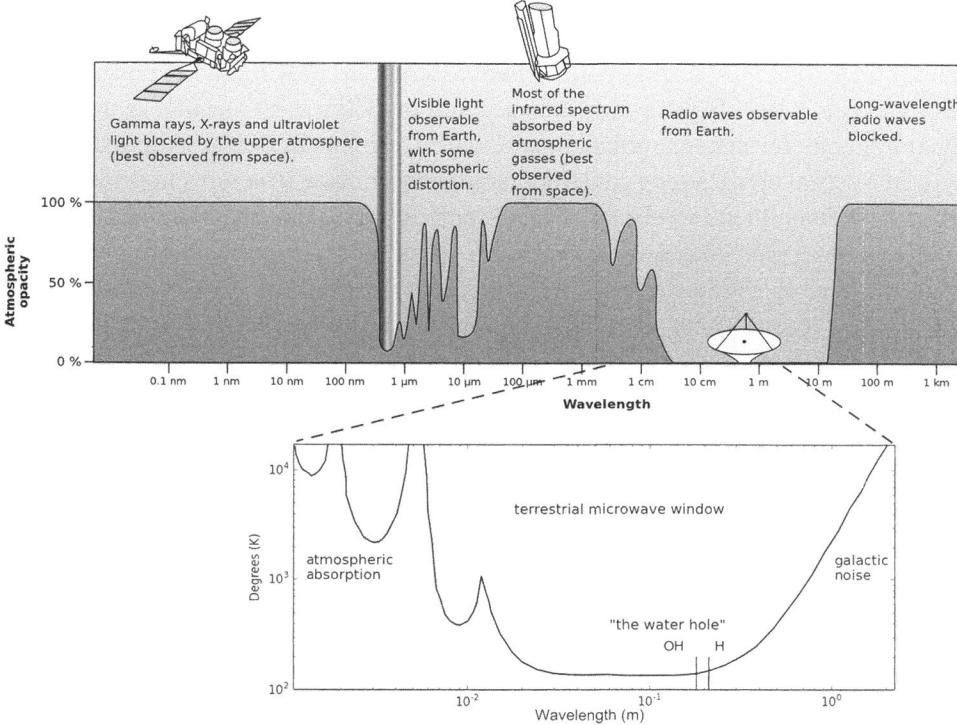

Fig. 2.8 Most of the electromagnetic spectrum does not make it through the Earth's atmosphere and so is unsuitable for ground-based SETI. Radio wavelengths and visible light are ideal for SETI since both penetrate the atmosphere. Within the radio spectrum there is a region around 10 cm where natural background radio noise is minimal. The "water hole" lies within this window. Figures adapted from NASA and Wikimedia commons. (A black and white version of this figure will appear in some formats. For the colour version, please refer to the plate section.)

have surveyed broader swaths of sky, but these searches are limited to detecting extremely powerful signals from deep space.

Of course there is no guarantee that extraterrestrial civilizations would use radio waves to signal their presence. Nonetheless, there are good reasons to focus on radio waves because Earth's atmosphere and the atmospheres of other habitable planets block X-rays, gamma radiation, and most of the infrared spectrum (Fig. 2.8). Earth's oxygenated atmosphere is transparent to visible light, and so extraterrestrial civilizations might choose to signal their presence with pulsed optical lasers, assuming they live under similarly oxic, haze-free atmosphere.

Even if extraterrestrial civilizations choose not to deliberately broadcast, it might be possible to detect their presence through other technosignatures because all technology – no matter how advanced – produces waste heat. If a civilization decided to use the majority of their planet's stellar energy, then the waste heat might be visible to infrared telescopes (Kuhn & Berdyugina, 2015). In particular, if they built a so-called Dyson sphere – a spherical structure to harness the power output of an entire star – then the waste heat could be detectable. Despite searches, no unambiguous Dyson sphere has been found (Carrigan Jr, 2009). Alternatively, if extraterrestrial civilizations construct large, non-circular structures in orbit around their host stars, then these may be detectable in unusual transits. In fact, a recent analysis of *Kepler* data revealed strange, aperiodic transits that were speculated to be a possible extraterrestrial megastructure (Wright *et al.*, 2015); however, the break-up of an exocomet turns out to be a more plausible hypothesis for a variety of reasons, including the high orbital inclination of the transiting material (Boyajian *et al.*, 2016).

Given the size of the cosmic haystack, it is unsurprising that SETI has not found unambiguous extraterrestrial broadcasts. There have, however, been a small number of ambiguous detections. The most intriguing of these is the appropriately named "Wow! signal", which was a narrowband radio signal detected in 1971 by the Big Ear radio telescope in Ohio. The signal was in the 'waterhole' region, a frequency that terrestrial broadcasters are prohibited from using. The location of the signal in the sky does not fit with the position of any known satellite or asteroid at the time, and is unlikely to be a reflection off space debris because the debris would have had to be perfectly stationary and well beyond low Earth orbit (Ehman, 2010). The signal was recorded for 72 s, but has never been detected again in follow-up observations. The source of the signal remains a mystery.

The physicist Enrico Fermi formulated a thought experiment related to SETI. In this so-called *Fermi paradox* we suppose that a technological civilization emerged somewhere in the galaxy. If this civilization decided to colonize other stellar systems, then the time required to colonize the entire galaxy is very short compared to the age of the universe – this is true even if we assume colonization of a new star system takes a long time and traveling between the stars is slow. Perhaps not every civilization in the galaxy develops the technology or has the desire to do this, but it only takes one sufficiently motivated civilization to colonize the galaxy. The question posed by Fermi is that if intelligent life is common in the cosmos, then why aren't they already here? Of course it is easy to dream up speculative solutions to the Fermi paradox that don't preclude the existence of extraterrestrial civilizations (Webb, 2015): perhaps interstellar travel is prohibitively resource-intensive, perhaps the Earth is the equivalent of a cosmic zoo, or perhaps spacecraft have visited the Earth at some point in its 4.57 billion year history! Naturally, it is hard to draw any firm conclusions about the Fermi paradox.

Conventional SETI targets may be too restrictive if extraterrestrial intelligence is not biological. Most experts in the field of artificial intelligence (AI) research believe that AI will surpass human-level intelligence by the end of the century

(Baum *et al.*, 2011). If this is true, then the length of time that biological intelligence is the dominant form of intelligence on Earth will be brief, and so any intelligent lifeforms we detect are likely to be machines (Schneider, 2016; Shostak, 2015). Machine intelligence may not need environments required by biological life, i.e. terrestrial planets in the habitable zone with surface liquid water. For example, machines might choose to live on airless planets close to their host stars to maximize solar energy. On the other hand, maintaining complex machinery may demand similar environments to biological life: abundant liquid water for manufacturing parts and a thick atmosphere to protect sensitive electronics from cosmic rays.

Despite it compelling nature, SETI has largely lagged behind other growth areas of astrobiology such as exoplanet research. The lack of government funding is a key reason. In the United States, Congress defunded NASA's SETI program in 1993, which was an arbitrary decision given that NASA spends a significant amount of money on searching for non-intelligent life. There is no reason to exclude intelligent life, which, after all, is the most interesting kind. Perhaps very few biospheres evolve intelligence, perhaps the origin of life itself is improbable, or perhaps intelligence civilizations are quite common. Presently, all of the last four terms in the Drake equation are unknown. For that reason, we should not handicap ourselves in the search for life beyond the Earth by ignoring intelligent life.

Conclusion

How likely is it that we will find life beyond Earth in the coming decades? *Hubble's* successor, the *James Webb Space Telescope* (*JWST*), is scheduled for launch in 2018. This telescope is not designed specifically to study exoplanets, and so it will probably observe only one to two potentially habitable worlds for signs of life, assuming that nearby targets can be found.[1] If life is ubiquitous then we might get lucky and detect atmospheric biosignatures with *JWST* in the early 2020s. Upcoming ground-based telescopes such as the European Extremely Large Telescope or Giant Magellan Telescope will also begin to directly image Earth-sized planets around the nearest stars in the mid-2020s. Proxima Centauri b – a habitable zone planet around our nearest star – will be a prime target for these telescopes. However, a null result with *JWST* and ground-based observations will mean little because they will only sample a handful of planets.

A likely outcome is that we will have to wait for a dedicated direct imaging mission sometime after 2025. A large space-telescope with starlight suppression technology could survey the nearest 1000 stars for planets with signs of life (Postman *et al.*, 2008). *Kepler* data suggest that approximately 5%–10% of Sun-like stars have planets in the habitable zone (Silburt *et al.*, 2015), and so this future telescope could survey around 50–100 habitable planets for biosignatures. If life is common in the universe then we will find it with such a telescope. Conversely, if we find no signs of life, the inescapable conclusion is that f_l, the fraction of habitable planets upon which life emerges, is very low. In fact, if we discovered that all the nearest planets were devoid of life then we might begin to suspect that the origin of life is hard.

What about the prospects for finding intelligent life in the coming decades? The first three terms in the Drake equation are known to be approximately $R = 7$ stars/year (Diehl et al., 2006), $f_p = 1$ (Cassan et al., 2012), and $n_e = 0.1$ (Silburt et al., 2015).[2] Let us suppose for the moment that life is common, perhaps $f_l = 0.5$. Additionally, let us optimistically assume that both intelligent life and communicating intelligent life are inevitable wherever life emerges (i.e. $f_c = f_i = 1$). For the final term in the Drake equation, the average lifetime of communicating civilizations, let us optimistically assume 1 million years, which is roughly the average lifetime of an animal species in the fossil record. Then it follows that the number of communicating civilizations that exist in the galaxy today is: $N = 7 \times 1 \times 0.1 \times 0.5 \times 1 \times 1 \times 1\,000\,000 = 350\,000$. This may seem like a large number, but there are around 300 billion stars in our galaxy, meaning that even in this extremely optimistic scenario only one in a million stars hosts an intelligent, communicating civilization. Current SETI searches have surveyed the nearest few thousand stars, not nearly enough to rule out even this best-case scenario. However, the Square Kilometer Array, a huge radio telescope scheduled to begin operation in 2020, will perhaps survey up to a million nearby stars for radio broadcasts (Siemion et al., 2014). If intelligent life is common in the galaxy, there is a good chance we will find it in the coming decades. If we find nothing, then the absence of evidence will start to provide evidence of absence.

Whatever we find, future generations will have one less cosmic question to ponder when they look up at the night sky. We may find that the universe is a desolate and sterile place. This knowledge will give the stars a very different character. Despite the incomprehensible vastness of the cosmos, we will know that Earth is exceptional, that terrestrial life is unique, and that the future of all life is our responsibility. But perhaps, we will instead discover exoplanets teeming with life. If so, this will become another fact to tell children about the night sky: the stars are distant suns, they all have planets just like our Solar System, and millions of them have life just like Earth. It may even be possible to point to an individual star in the sky and say *that one*, that star has a planet just like Earth. It is hard to imagine another scientific discovery that would instill more wonder than this.

Notes

1 *JWST* can only characterize *transiting* Earth-like planets because its starlight suppressing coronagraph will be too inefficient to image habitable planets. Therefore it will be necessary to find a nearby, transiting habitable planet in the right position in the sky to be frequently observed

with *JWST*, such as the TRAPPIST-1 system. The upcoming *Transiting Exoplanet Survey Satellite (TESS)* mission scheduled to launch in 2017 will search nearby stars for further *JWST* targets.

2 Strictly speaking, n_e is not equal to the fraction of Sun-like stars with habitable planets, which is the quantity Silburt *et al.* (2015) estimate. However, for the purposes of this approximate calculation they can be assumed to be equivalent.

References

Anglada-Escudé, G., Amado, P. J., Barnes, J. *et al.* (2016). A terrestrial planet candidate in a temperate orbit around Proxima Centauri. *Nature*, **536**(7617), 437–40.

Arnold, L., Gillet, S., Lardière, O., Riaud, P., & Schneider, J. (2002). A test for the search for life on extrasolar planets – Looking for the terrestrial vegetation signature in the Earthshine spectrum. *Astronomy & Astrophysics*, **392**(1), 231–37.

Barnes, R., Deitrick, R., Luger, R. *et al.* (2016). The habitability of Proxima Centauri b I: Evolutionary scenarios. *arXiv preprint arXiv:1608.06919*.

Baross, J., Benner, S., Cody, G. *et al.* (2007). The limits of organic life in planetary systems *Committee on the origins and evolution of life* (Vol. 38, pp. 1070): National Reseach Council.

Baum, S. D., Goertzel, B., & Goertzel, T. G. (2011). How long until human-level AI? Results from an expert assessment. *Technological Forecasting and Social Change*, **78**(1), 185–95.

Bessel, F. (1838). On the parallax of 61 Cygni. *Monthly Notices of the Royal Astronomical Society*, **4**, 152–61.

Boyajian, T., LaCourse, D., Rappaport, S. *et al.* (2016). Planet Hunters X. KIC 8462852 – Where's the flux? *Monthly Notices of the Royal Astronomical Society*, **457**(4), 3988–4004.

Burke, C. J., Christiansen, J. L., Mullally, F. *et al.* (2015). Terrestrial planet occurrence rates for the Kepler GK Dwarf sample. *The Astrophysical Journal*, **809**(1), 8.

Carrigan Jr, R. A. (2009). IRAS-based whole-sky upper limit on Dyson spheres. *The Astrophysical Journal*, **698**(2), 2075.

Cassan, A., Kubas, D., Beaulieu, J.-P. *et al.* (2012). One or more bound planets per Milky Way star from microlensing observations. *Nature*, **481**(7380), 167–9.

Catling, D. C. (2014). The Great Oxidation Event Transition. In H. D. H. a. K. K. Turekian, ed., *Treatise on Geochemistry*, 2nd edn, Vol. 6, Oxford: Elsevier, pp. 177–195.

Catling, D. C. & Kasting, J. (2017). *Atmospheric Evolution on Inhabited and Lifeless Worlds*, Cambridge: Cambridge University Press.

Crossfield, I. J. (2015). Observations of exoplanet atmospheres. *Publications of the Astronomical Society of the Pacific*, **127**(956), 941.

Dick, S. J. (1993). The search for extraterrestrial intelligence and the NASA High Resolution Microwave Survey (HRMS): historical perspectives. *Space science reviews*, **64**(1–2), 93–139.

Diehl, R., Halloin, H., Kretschmer, K. *et al.* (2006). Radioactive 26Al from massive stars in the Galaxy. *Nature*, **439**(7072), 45–7.

Domagal-Goldman, S. D., Meadows, V. S., Claire, M. W., & Kasting, J. F. (2011). Using biogenic sulfur gases as remotely detectable biosignatures on anoxic planets. *Astrobiology*, **11**(5), 419–41.

Ehman, J. (2010). The Big Ear Wow! signal (30th Anniversary Report). *Big Ear Radio Observatory*. Available online at http://www.bigear.org/Wow30th/wow30th.htm.

Einstein, A. (1936). Lens-like action of a star by the deviation of light in the gravitational field. *Science*, **84**(2188), 506–7.

Gillon, M., Triaud, A. H., Demory, B. O. et al. (2017). Seven temperate terrestrial planets around the nearby ultracool dwarf star TRAPPIST-1. *Nature*, **542**(7642), 456–60.

Harman, C., Schwieterman, E., Schottelkotte, J., & Kasting, J. (2015). Abiotic O_2 levels on planets around F, G, K, and M stars: Possible false positives for life? *The Astrophysical Journal*, **812**(2), 137.

Hart, M. H. (1978). The evolution of the atmosphere of the Earth. *Icarus*, **33**(1), 23–39.

Howard, A. W. (2013). Observed properties of extrasolar planets. *Science*, **340**(6132), 572–6.

Hubble, E. P. (1925). Cepheids in spiral nebulae. *The Observatory*, **48**, 139–42.

Jeans, J. (1942). Is there life on the other worlds? *Science*, **95**(2476), 589–92.

Kasting, J. F. (2013). What caused the rise of atmospheric O_2? *Chemical Geology*, **362**, 13–25.

Kawahara, H., & Fujii, Y. (2010). Global mapping of Earth-like exoplanets from scattered light curves. *The Astrophysical Journal*, **720**(2), 1333.

Kirchner, J. W. (2003). The Gaia hypothesis: conjectures and refutations. *Climatic Change*, **58**(1–2), 21–45.

Knoll, A. H. (2008). Cyanobacteria and Earth history. *The Cyanobacteria: Molecular Biology, Genomics, and Evolution*, **484**.

Kopparapu, R. K., Ramirez, R., Kasting, J. F. et al. (2013). Habitable zones around main-sequence stars: new estimates. *The Astrophysical Journal*, **765**(2), 131.

Krissansen-Totton, J., Bergsman, D. S., & Catling, D. C. (2016). On detecting biospheres from chemical thermodynamic disequilibrium in planetary atmospheres. *Astrobiology*, **16**(1), 39–67. doi: 10.1089/ast.2015.1327

Kuhn, J. R. & Berdyugina, S. V. (2015). Global warming as a detectable thermodynamic marker of Earth-like extrasolar civilizations: the case for a telescope like Colossus. *International Journal of Astrobiology*, **14**(03), 401–10.

Labeyrie, A. (1999). Snapshots of alien worlds – the future of interferometry. *Science*, **285**(5435), 1864.

Lane, N. (2002). *Oxygen: the Molecule that Made the World*, Oxford: Oxford University Press.

Léger, A., Fontecave, M., Labeyrie, A. et al. (2011). Is the presence of oxygen on an exoplanet a reliable biosignature? *Astrobiology*, **11**(4), 335–41.

Long, K., Obousy, R., Tziolas, A. et al. (2010). PROJECT ICARUS: Son of Daedalus, Flying Closer to Another Star. *arXiv preprint arXiv:1005.3833*.

Lovelock, J. E. (1965). A physical basis for life detection experiments. *Nature*, **207**(997), 568–70.

Lovelock, J. E. (1975). Thermodynamics and the recognition of alien biospheres. *Proceedings of the Royal Society of London B: Biological Sciences*, **189**(1095), 167–81.

Lovelock, J. E. & Margulis, L. (1974). Atmospheric homeostasis by and for the biosphere: the Gaia hypothesis. *Tellus*, **26**, 1–2.

Luger, R. & Barnes, R. (2015). Extreme water loss and abiotic O_2 buildup on planets throughout the habitable zones of M dwarfs. *Astrobiology*, **15**(2), 119–43.

Lyons, T. W., Reinhard, C. T., & Planavsky, N. J. (2014). The rise of oxygen in Earth's early ocean and atmosphere. *Nature*, **506**(7488), 307–15.

Marois, C., Macintosh, B., Barman, T. et al. (2008). Direct imaging of multiple planets orbiting the star HR 8799. *Science*, **322**(5906), 1348–52.

Martin, A. R. (1978). *Project Daedalus: The Final Report on the BIS Starship Study*: British Interplanetary Soc.

Mayor, M. & Queloz, D. (1995). A Jupiter-mass companion to a solar-type star. *Nature*, **378**(6555), 355–9.

Oliver, B. (1979). Rationale for the water hole. *Communication with Extraterrestrial Intelligence*, **6**(1–2), 71.

Pepe, F., Molaro, P., Cristiani, S. *et al.* (2014). ESPRESSO: The next European exoplanet hunter. *Astronomische Nachrichten*, **335**(1), 8–20.

Pohorille, A. & Pratt, L. (2012). Is water the universal solvent for life? *Origins of Life and Evolution of Biospheres*, **42**(5), 405–9.

Postman, M., Brown, T., Koekemoer, A. *et al.* (2008). *Science with an 8-meter to 16-meter optical/UV space telescope*. Paper presented at the SPIE Astronomical Telescopes+Instrumentation.

Preus, A. (2001). *Essays in Ancient Greek Philosophy VI: Before Plato* (Vol. 6), SUNY Press.

Ribas, I., Bolmont, E., Selsis, F. *et al.* (2016). The habitability of Proxima Centauri b. I. Irradiation, rotation and volatile inventory from formation to the present. *Astronomy & Astrophysics*, **596**, A111.

Robinson, T. D., Ennico, K., Meadows, V. S. *et al.* (2014). Detection of ocean glint and ozone absorption using LCROSS Earth observations. *The Astrophysical Journal*, **787**(2), 171.

Robinson, T. D., Meadows, V. S., & Crisp, D. (2010). Detecting oceans on extrasolar planets using the glint effect. *The Astrophysical Journal Letters*, **721**(1), L67.

Rodler, F. & López-Morales, M. (2014). Feasibility studies for the detection of O_2 in an Earth-like exoplanet. *The Astrophysical Journal*, **781**(1), 54.

Rubin, V. C., Ford, W. K., & Thonnard, N. (1980). Rotational properties of 21 SC galaxies with a large range of luminosities and radii, from NGC 4605/R= 4kpc/to UGC 2885/R= 122 kpc. *The Astrophysical Journal*, **238**, 471–87.

Sagan, C. (1994). *Pale Blue Dot: A Vision of the Human Future in Space*, Random House.

Sagan, C., Thompson, W. R., Carlson, R., Gurnett, D., & Hord, C. (1993). A search for life on Earth from the Galileo spacecraft. *Nature*, **365**(6448), 715–21.

Schneider, S. (2016). Superintelligent AI and the postbiological cosmos approach. In A. Losch. ed., *What is Life? On Earth and Beyond*, Cambridge: Cambridge University Press.

Schwieterman, E., Cockell, C., & Meadows, V. (2015). Nonphotosynthetic pigments as potential biosignatures. *Astrobiology*, **15**(5), 341–61.

Seager, S., Bains, W., & Hu, R. (2013). Biosignature gases in H_2-dominated atmospheres on rocky exoplanets. *The Astrophysical Journal*, **777**(2), 95.

Seager, S., & Sasselov, D. (2000). Theoretical transmission spectra during extrasolar giant planet transits. *The Astrophysical Journal*, **537**(2), 916.

Shapley, H. (1918). Studies based on the colors and magnitudes in stellar clusters. VII. The distances, distribution in space, and dimensions of 69 globular clusters. *The Astrophysical Journal*, **48**, 154–81.

Shklovskii, I. S. & Sagan, C. (1966). *Intelligent Life in the Universe*, San Francisco, CA: Holden Day.

Shostak, S. (2015). Searching for clever life. *Astrobiology*, **15**(11), 949–50.

Siemion, A. P., Benford, J., Cheng-Jin, J. *et al.* (2014). Searching for extraterrestrial intelligence with the Square Kilometre Array. *arXiv preprint arXiv:1412.4867*.

Silburt, A., Gaidos, E., & Wu, Y. (2015). A statistical reconstruction of the planet population around Kepler solar-type stars. *The Astrophysical Journal*, **799**(2), 180.

Snellen, I., de Kok, R., Birkby, J. *et al.* (2015). Combining high-dispersion spectroscopy with high contrast imaging: Probing rocky planets around our nearest neighbors. *Astronomy & Astrophysics*, **576**, A59.

Stark, C. C., Cady, E. J., Clampin, M. et al. (2016). *A direct comparison of exoEarth yields for starshades and coronagraphs*. Paper presented at the SPIE Astronomical Telescopes+ Instrumentation.

Tarter, J. (2001). The search for extraterrestrial intelligence (SETI). *Annual Review of Astronomy and Astrophysics*, **39**(1), 511–48.

Vernadsky, V. I. (1926). *The Biosphere*, New York: Copernicus Springer-Verlag (English translation of Vernadsky, VI, 1926).

Walker, J. C., Hays, P., & Kasting, J. (1981). A negative feedback mechanism for the long-term stabilization of the Earth's surface temperature. *Journal of Geophysical Research*, **86**(C10), 9776–82.

Webb, S. (2015). *If the Universe is Teeming with Aliens... Where is Everybody?: Fifty Solutions to the Fermi Paradox and the Problem of Extraterrestrial Life* (Second Edition), Springer Science & Business Media.

Wright, J. T., Cartier, K., Zhao, M., Jontof-Hutter, D., & Ford, E. B. (2015). The $\hat{G}$ search for extraterrestrial civilizations with large energy supplies. IV. The signatures and information content of transiting megastructures. *The Astrophysical Journal*, **816**(1), 17.

3

The Shape of Life: Morphological Signatures of Ancient Microbial Life in Rocks

BEDA A. HOFMANN

Introduction

As the reader will know, the remains of organisms conserved in rocks are called fossils (Fig. 3.1). Their recognition as remains of former life goes back to antiquity, but was also questioned up until the seventeenth century. With the advent of modern biology and geology in the nineteenth century, fossils rapidly became one of the most important study objects in geology. It was recognized that strata of different age contain very different organisms. This method of using index fossils, i.e., fossils restricted in occurrence to a limited range of strata and thus time, has since evolved and is calibrated with absolute age determinations of suitable rocks in fossil-containing sequences, typically volcanic rocks containing uranium-rich zircons (datable with the uranium–lead system) or potassium-rich minerals (datable with the potassium–argon system). Numerous types of fossils are used for age determination. These range from vertebrates to invertebrates and eucaryotic unicellular organisms such as foraminifera and radiolaria. Obviously, knowledge about the fossil counterparts of living organisms was distinctly lacking in the early days of "modern" biology and palaeontology, but this gap was rapidly closed. After the discovery of the tiniest organisms (prokaryotes, consisting of the domains of the archaea and bacteria), it was only natural to search for their fossil remains too. The detection of true microbial fossils was not made until the 1950s (Tyler & Barghoorn, 1954). A particular structure in sedimentary rocks called "stromatolites" was first recognized in Mesozoic rocks in Germany (Kalkowski, 1908), but later found most commonly in rocks of Precambrian age. Stromatolites are now considered to represent a sedimentary structure resulting from sediment accumulation on surfaces made "sticky" by the slime (or EPS: extracellular polymeric substances) of microorganisms. Stromatolites are basically structures made by microorganisms, similar to reefs made by corals. Stromatolites belong, today, to the oldest recognized features of life on Earth, with 3.4-billion-year-old examples found in Australia (Allwood *et al.*, 2007),

Fig. 3.1 Example of eukaryotic macrofossils. Ammonites from the Jurassic (early Callovian, ~166 million years old) found in the Swiss Jura. No organic matter is preserved, all information is based on morphology. Sample width 21 cm. (A black and white version of this figure will appear in some formats. For the colour version, please refer to the plate section.)

while living examples of stromatolites can be observed in various places, including Shark Bay, Australia (Suosaari *et al.*, 2016).

With increasing knowledge about life in extreme environments on Earth and the potential for life to exist on other planetary bodies, the interest in morphological biosignatures as a possible means for recognizing extraterrestrial life has been rejuvenated. The idea of searching for extraterrestrial life by searching for fossils in extraterrestrial matter is obvious and quite old. Repeatedly, purported life forms have been reported in meteorites (Hahn, 1880; Claus & Nagy, 1961; McKay *et al.*, 1996), but none of these claims has been substantiated by follow-up studies. Still, even today some meteorites (the ones from Mars) are the only extraterrestrial matter available from a source with a potential for life. With the newly intensified exploration of Mars since 1997 (*Mars Pathfinder* lander followed by four landers and five orbiters up to 2015) and several lander/rover missions planned with the specific aim of searching for ancient life, research on signatures of life has been intensified in recent years. These signatures comprise all manifestations of life including chemical, morphological, mineralogical, and isotopic features. While morphology is a classical method in palaeontology, its usefulness in recognizing the remains of prokaryotic microbial life is debated (Cady *et al.*, 2003; Grosch & McLoughlin, 2014). Nevertheless, life has specific characteristics that result in specific morphological remains in rocks, even in the case of seemingly uncharacteristic "slimy" microorganisms.

Habitats and What Life Really Needs

From the human point of view, an environment of 40 °C appears extreme but in the case of microbial life, the term "extreme" is rather used for conditions under which life appears unlikely or even impossible at first glance: sub-zero temperatures or pressurized water up to 120 °C, highly acidic or alkaline conditions, small cavities in rocks several kilometres below the Earth's surface, concentrated salt lakes under intense solar irradiation, and sites of extremely high radiation doses. In fact, the environments in which prokaryotic life is possible have few strict limitations. Still, such limitations exist: no life has yet been found that would reproduce at temperatures above 120 °C, and some free water needs to be available, putting limits to extremely low temperatures (tens of degrees below freezing temperature) and extremely saline conditions (e.g., saturated solutions of magnesium chloride). However, there is no need for free oxygen for many groups of microorganisms.

What life really needs are a few things: (i) sources of its material constituents: carbon, hydrogen, nitrogen, oxygen, phosphorous, sulphur, and a number of minor and trace elements; (ii) sources of energy (sunlight or chemical energy); (iii) environments consistent with the limits of life (approximately –10 to 120 °C, water available); and (iv) some space: even the tiniest microbe cannot exist inside a solid without pores. All of these conditions are met on the Earth's surface, in the oceans, in the lower atmosphere, but also in porous sediments and rocks to depths of several kilometres. The material constituents are available nearly anywhere on Earth. The main primary source of energy on the Earth's surface is sunlight. Photosynthesis is the main source of organic material and of oxygen on the Earth's surface and in the oceans. Eukaryotic life forms depend on photosynthesis.

However, life also exists in environments where no photosynthesis is possible (inside rocks and sediments and very hot springs), and also in environments that are free of molecular oxygen and of organic substances. The trick of these organisms is to use chemical energy, i.e., pairs of chemicals that should react with each other and give off energy, yet they do not due to lack of a catalyst. An example is organic matter and sulphate (SO_4^{-2}). Only the presence of specific microbial organisms that act as catalysts allows the reaction producing carbon dioxide, hydrogen sulphide, and energy to take place. Such anoxygenic organisms are very common wherever oxygen is lacking, from the intestines of a cow to beach sands a few millimetres or centimetres below the surface. The point is that conditions suitable for life exist in numerous places, not just on the surface of the continents and in the oceans, but also in sediments and hard rocks. Let us assume that the 120 °C isotherm, at approximately 3–5 km depth in the Earth's crust, is a real limit, and no life can be present below this boundary.

Besides temperature, the availability of chemical energy is a critical limit. Under conditions of perfect chemical equilibrium, there is no energy for microbes, and no life. Chemical energy is available where ground-waters of different chemistry are mixing, or where reactive reduced substances are interacting with oxygen-bearing ground-waters, e.g. in oxidizing ore deposits. Any situation with enhanced or channelled fluid flow has the potential to favour such a situation.

The Earth's crust as a habitat of life has only been recognized widely since about 1990. A specific branch of environmental microbiology ("deep subsurface microbiology") has been investigating such environments since then (Ghiorse & Wilson, 1988; Phelps et al., 1989; Lovley & Chapelle, 1995; Amy & Haldeman, 1997; Pedersen, 2000; Teske, 2005; Kallmeyer et al., 2012). It is now generally accepted that microbial life is present in such deep environments. The fraction of total biomass represented by subsurface microbes is not known. Estimates range from around 1% to >50% (Reith, 2011).

Microbial Taphonomy

Studies on microbial taphonomy, i.e., the processes leading from a living organism to a fossil, have demonstrated that rapid fossilization by encrustation with minerals strongly favours the preservation of microbial fossils. Fortunately, the same processes that provide the basis for flourishing microbial life often also favour its fossilization, including fluid mixing, rapid cooling, and steep chemical gradients. Environments of rapid mineral precipitation in combination with abundant microbial life do occur in hot springs where strong chemical and temperature gradients exist. Chemical gradients, in particular of redox potential, allow energy harvesting by microorganisms (chemoautotrophs) and also lead to mineral precipitation (e.g., of iron hydroxides), while temperature gradients induce mineral precipitation, e.g., of opaline silica ($SiO_2 \cdot H_2O$). Some hot spring deposits provide very well-preserved microbial fossils. Therefore hot spring deposits are one of the targets in the search for life on Mars.

Characteristic Morphologies of Life

The recognition of fossils of multicellular life is largely based on comparisons with living analogues. Even in the case of extinct organisms such as ammonites

(Fig. 3.1), the general morphological similarity allows an unambiguous identification as a fossil. A particular characteristic is the existence of numerous specimens with identical complex morphology, evidently a result of genetic encoding. Such complex and characteristic morphologies are even present in eukaryotic single-celled organisms, some of which provide important index fossils (foraminifera, radiolaria). However, eukaryotes evolved relatively late in the history of life. Their oldest traces date back to ~1700 million years (Rasmussen et al., 2008), indicating that the first two billion years of life were prokaryotic. Also, eukaryotes all depend on oxygen and/or sunlight and are restricted to the more conventional environments. It is likely that the development of eukaryotes was a result of increased oxygen concentrations in the atmosphere from about 2500 million years ago.

The conspicuous life forms in conventional environments (Earth's surface, oceans) can be summarized as morphologically complex, allowing an unambiguous morphological identification as forms of life. Prokaryotic organisms (bacteria and archaea) have sizes typically in the range of a few micrometres, and their shape is most commonly spherical or coccoid. Such organisms have a low chance of ending up as easily recognizable fossils. Reports exist (Westall, 1999; Westall et al., 2006; Wacey et al., 2011), but there are numerous debates about the nature of these features. Some prokaryotic microorganisms provide additional chances for morphology: Some grow in filamentous forms and/or excrete a slime-like sticky substance called EPS (extracellular polymeric substances or extracellular polysaccharides/exopolysaccharides, commonly identified as "slime"). Both have important implications in providing morphologically distinctive remains of prokaryotic microorganisms.

Filamentous microbes can form colonies that are morphologically distinctive and even unique. Numerous filaments attached to each other result in "streamers" (Figs. 3.2 and 3.3), a feature that can be observed quite commonly in hot and cold springs containing hydrogen sulphide. Such streamers can be fossilized by encrustation with calcium carbonate, silica or even by elementary sulphur resulting from partial oxidation of sulphide (Fig. 3.2). These filamentous forms can thus produce macroscopic build-ups with a characteristic texture, called biofabrics. Iron-oxidizing microbes can build up similar biofabrics consisting of iron hydroxides. In contrast to fossils, the texture of a biofabric results from a whole colony of organisms and can be distinctive even where the single organisms are not preserved. Such biofabrics can be regarded as analogues of massive reefs, where the individual organisms are coral or one of a number of other reef-builders.

The excretion of EPS is crucial in microbial colonies as it acts to bind together microbial cells, provides protection, binds nutrients, and also attaches sediment particles and in some cases microbial precipitates such as iron hydroxides. The characteristic morphology of stromatolites with steep-sided mounds is actually a

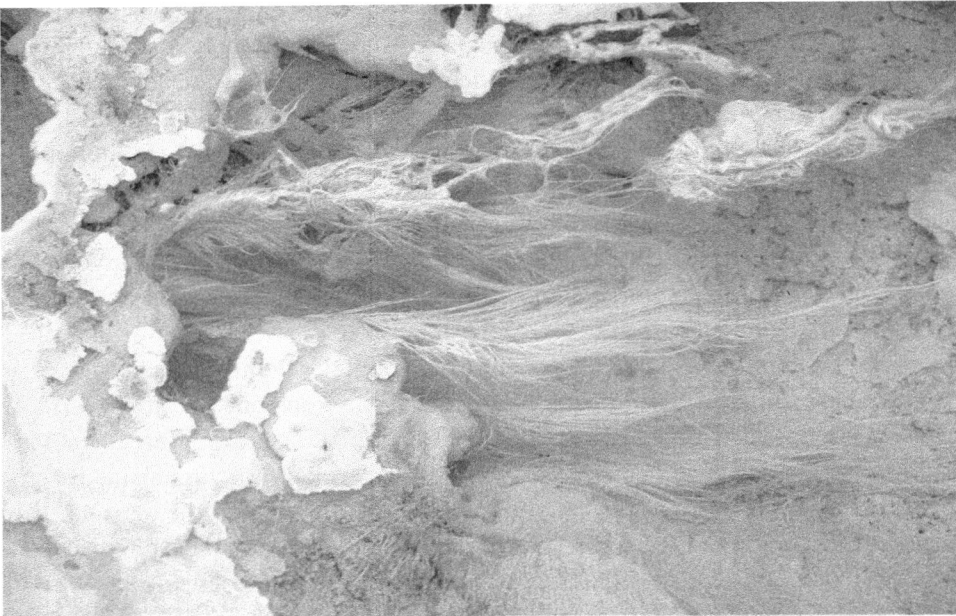

Fig. 3.2 Filamentous sulphur-oxidizing bacteria in a hot spring (Yellowstone Park, USA). Even though prokaryotic and individual cells are microscopic, the colony is visually conspicuous. Image width approx. 50 cm. (A black and white version of this figure will appear in some formats. For the colour version, please refer to the plate section.)

result of the sediment-binding capacity of EPS produced by microbial mats. In the following section, common types of morphologies preserved in rocks and resulting from prokaryotic microbial life are discussed, with a special focus on biologically induced textures from ancient subsurface environments.

Cellular Microfossils and Extracellular Excretions

Fossilized single microbial cells are just as small as living microbes, in the order of a micrometre or up to a few tens of micrometres. Single coccoid or rod-shaped fossil microbes are difficult to identify because the already rather unspecific shapes are typically not well preserved. If cell-like structures do occur in large numbers, it may be an indicative sign of their biological origin. More specific shapes such as spirochetes (corkscrew-like microbial cell forms) or bundles of filaments are more distinctive but require good preservation. There is a vast body of literature dealing with the requirements for identification of morphological structures as fossil microbes. Very generally, the better the preservation, the more specific the shape, and the higher the abundance of similar features, the higher the credibility. Besides their nature as the remains of a living organism, it is also important to prove that any

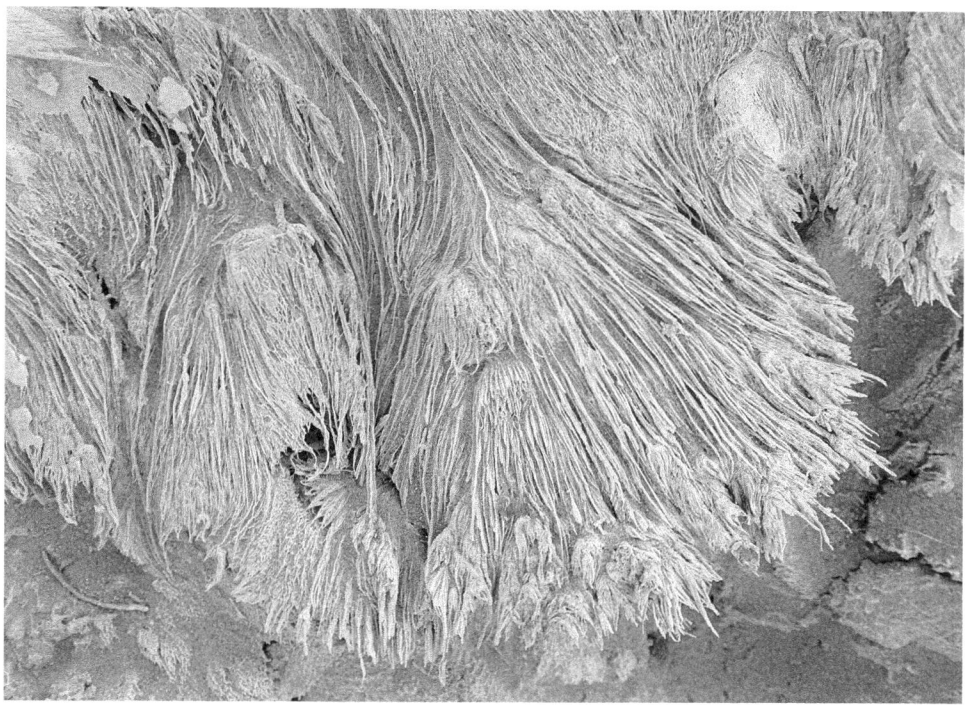

Fig. 3.3 Streamers of dried sulphur-oxidizing bacteria near the ground-water well Abu Tabul, Oman. The streamers consist mainly of elementary sulphur. Image width approximately 10 cm. (A black and white version of this figure will appear in some formats. For the colour version, please refer to the plate section.)

reported microbial fossils are syngenetic, i.e., they must be found in clear context embedded in the rock. Recent microbial colonization of fine fractures and pores of rocks may lead to false detections of cellular structures, as shown by Westall and Folk (2003). In general, the shape of single-cell microbial fossils is often not specific, and high magnifications (optical microscopes or electron microscopes) are required for their detection. Reports of single-cell microbial fossils are dominated by examples from sedimentary rocks, but there are some reports indicating the potential for single-cell fossilization in subsurface environments, e.g., basaltic aquifers (McKinley *et al.*, 2000).

A number of microorganisms produce extracellular precipitates with very specific morphologies. Examples are the iron-oxidizing bacteria *Gallionella ferruginea* and *Mariprofundus ferrooxidans* that produce twisted stalks of iron hydroxides of characteristic morphology (Chan *et al.*, 2011). Such morphologies can be preserved in geological materials under favourable conditions (Fig. 3.4) and provide a good biosignature. Reports of inorganically formed precipitates of similar

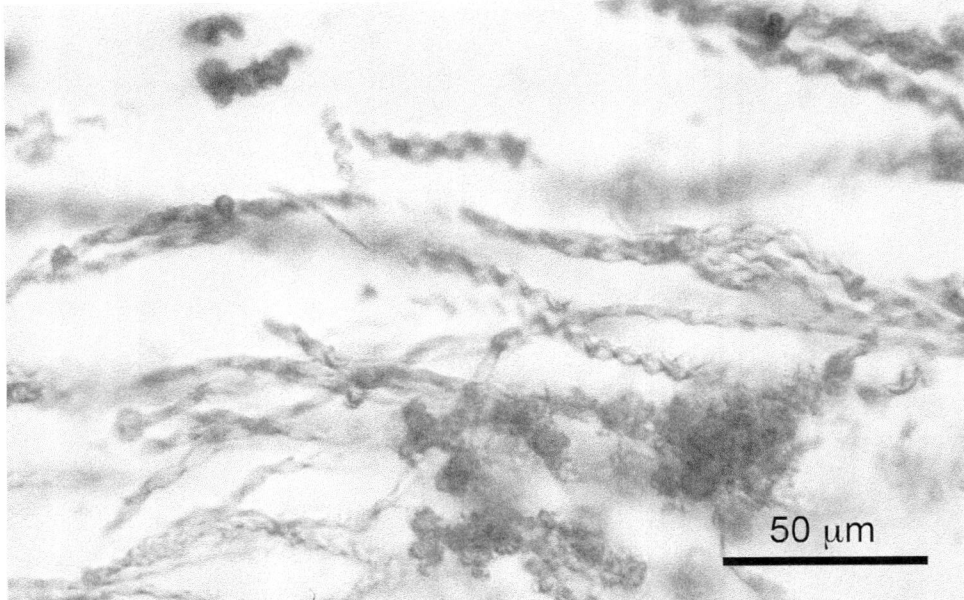

Fig. 3.4 Twisted stalks of iron hydroxides enclosed in a matrix of microcrystalline quartz. The stalks are products of microbes of the genera *Gallionella* or *Mariprofundus*. This sample is derived from a former cavity in plateau basalts in Breiðdalur, Eastern Iceland, and represents a site of microbial iron oxidation in a subsurface environment. Preserved by cementation with later quartz. (A black and white version of this figure will appear in some formats. For the colour version, please refer to the plate section.)

morphology (Garcia-Ruiz *et al.*, 2003) are based on experiments with chemical systems that are highly unlikely to be present in natural systems, and the resulting shapes lack the uniform size characteristic of biogenic precipitates.

Stromatolites

Stromatolites are layered sedimentary rocks with curved to domal, finely laminated strata corresponding to former sediment surfaces (Fig. 3.5). They are the result of the preservation of a layered succession of microbial mats at a specific locality over longer periods of time. The characteristic surface shape is attributed to a combination of phototaxis and EPS production. High microbial activity at the sites most exposed to light (elevated areas) is favoured by the ability of microbes to glide to their favoured environment. This leads to high bio-productivity, also of EPS, leading to stickiness of high-productivity areas and the resulting preferred attachment of sediment particles at these sites. Therefore, the highest sedimentation rates result at the sites of highest productivity, which are elevated areas. This auto-catalytic

Fig. 3.5 Cut surface of a stromatolite showing the characteristic domal morphology with steep slopes of the sedimentary layering. Sample from the Archean of the Pilbara craton, Australia, approx. 3.4 billion years old. Image width 15 cm. (A black and white version of this figure will appear in some formats. For the colour version, please refer to the plate section.)

process results in the formation of the characteristic domal shapes of stromatolites. In a lifeless environment without EPS, sediment particles are not sticky and no steep slopes as in the case of stromatolites can form. This is a very short and incomplete summary of a vast field of research. A good introduction to the details of stromatolite genesis can be found, e.g., in Awramik and Grey (2005) and Riding (2011). Stromatolites are macroscopic features with sizes of single domal structures ranging from a few millimeters to several meters. Some exceptionally large stromatolites are even visible on satellite imagery, e.g., on Tukarak Island, Canada (56.290° N 78.738° W; Hofmann, 1998). While the characteristic features of stromatolites in clear synsedimentary context are generally taken as good evidence of life, care must be taken, as non-biological pseudo-stromatolites do also exist (Grotzinger & Rothman, 1996; McLoughlin et al., 2008). Such pseudo-stromatolites can form by a number of processes, including chemical precipitation of minerals at sediment surfaces in fractures of rocks.

Microbial Mats

In contrast to stromatolites, microbial mats can form in a short time during single events of microbial activity on a substrate. Owing to the presence of EPS, such a single layer of microbes behaves very differently when compared with a sediment layer of similar thickness lacking biological colonization. Due to the stickiness, material may be attached on steep slopes, and, upon drying, microbial mats show typical roll-up features that have been preserved in the fossil record and present some of the earliest evidence of life (Noffke *et al.*, 2013). Wrinkle structures are decorations on the surface of sediment bedding planes that are thought to arise from the special binding provided by microbial colonization, including EPS production. However, wrinkle structures are similar to a number of non-biological decorations and are considered to be a weak biosignature (Porada & Bouougri, 2007; Mata & Bottjer, 2009).

Streamers of Microbial Filaments

Streamers form in dense colonies of filamentous microbes in flowing water in hot or cold springs or in streams. A typical recent example is shown in Fig. 3.3. The bio-productivity must be rather high to allow the development of a high density of cells. To allow fossilization, mineralization must be fast in order to preserve these textures. Figure 3.3 shows a "pre-fossil" stage where colonies are dried and preserved mainly as native sulphur. Such streamers have been documented in fossil hot spring environments (Walter *et al.*, 1998), but also in situations of microbial fabrics from subsurface environments, where streamers have formed in fractures of rock with flowing water (Hofmann *et al.*, 2008).

Subsurface Filamentous Fabrics (SFF)

The traditionally recognized morphologies of prokaryotic life in the geological record are vastly dominated by the types described previously: cellular remains, stromatolites, microbial mats, and streamers. Here I place special emphasis on a type of biofabric that is actually very common in occurrence, but not often recognized. The previously described morphologies all represent life that was flourishing on the Earth's surface, and their fossil remains are almost uniquely bound to ancient sediment surfaces. Morphological remains of life that was active in subsurface environments, i.e., in fractures, pores, and cavities in rocks within the upper few kilometres of the Earth's crust, generally receive little attention. Let us bear in mind that a sedimentary rock, e.g., a layer of limestone of a few centimetres of thickness, may have formed within a few years. The evidence of surface-bound life potentially contained in this rock must all have formed within the few years it was in the surface. If the rock is of Precambrian age, thousands of millions of years followed while the rock was buried at some depth. During this long time period, microbial activity may have occurred within the rock and some of these microbes may have been preserved as fossils. Even if the probability of preservation of microbial fossils in the subsurface were lower by a factor of 10^9 as compared with the surface environment (which appears excessively low), the probability of such a rock to contain traces of subsurface life will still be in the same order of magnitude as to contain traces of life from the time of sedimentation.

Morphological forms reminiscent of filaments, microbial mats, and streamers do occur in a number of geological settings and rock types that clearly represent former subsurface environments: cavities in volcanic rocks, porous zones in the oxidation zone of ore deposits, palaeokarst, and even cavities within macrofossils, to name but a few. While such features have been known for a long time and have even been described as "fossil moss" and similar (e.g., Daubenton, 1782; Razumovsky, 1835), an organic origin was often questioned later. However, in a number of situations the likely biological origin of such features was evidenced (Kretzschmar, 1982; Trewin & Knoll, 1999; Schumann *et al.*, 2004) and a common origin of these "subsurface filamentous fabrics" (SFF) as mineralized filamentous microbes from former subsurface settings was proposed by Hofmann and Farmer (2000) and Hofmann *et al.* (2008). The main characteristics of SFF can be summarized as follows: (a) presence of filamentous cores with a primary inner diameter in the order of a few microns, heavily encrusted by minerals (iron hydroxides, silica, calcite, and may others); (b) association of single filaments to mat-like fabrics; and (c) evidence of flexibility of the filamentous cores prior to mineral encrustation, such as formation of U-loops and vertically arranged features resembling stalactites (also known as pool fingers in speleology). Depending on the degree of cementation, the final result can range from a delicate fabric in rock cavities, often vertically arranged (Fig. 3.6), to massive rocks in which the filamentous fabric is only visible by investigation under the microscope.

A particular feature of SFF is their macroscopic aspect, which they have in common with stromatolites and streamers. This allows for their relatively easy recognition. In particular, the pseudo-stalactitic morphologies (Fig. 3.6) are very conspicuous, but so are also the mat-like fabrics (Fig. 3.7).

Distinction of Biogenic and Non-Biogenic Morphologies

While eukaryotic fossils, both macro and micro, show numerous distinctive morphological features that allow an unambiguous identification, this is not the case for prokaryotes. The reasoning for the identification of morphologies as being of

Fig. 3.6 Pseudo-stalactitic subsurface filamentous fabric in a cavity of basalt. Original microbial filaments are overgrown by microcrystalline quartz (chalcedony), Ballina, New South Wales, Australia. Field of view is 30 mm. (A black and white version of this figure will appear in some formats. For the colour version, please refer to the plate section.)

biological origin is typically less certain and often debated. Still, the most important morphological characteristics of fossil prokaryotic life can be summarized as follows. (i) Textural evidence of "stickiness" or "slime", indicating the presence of EPS, for which no non-biological analogue with similar properties and resulting sedimentary textures is known (finely laminated sediments and stromatolites). (ii) Filaments with a consistent diameter and evidence of flexibility and mat-like fabrics consisting of filaments. Morphometric parameters may yield additional criteria for a distinction from non-biological fibrous minerals (Hofmann *et al.*, 2008; Williams *et al.*, 2015). (iii) Cell-like morphologies (cocci, rods, etc.) present in large numbers in similar size and shape. (iv) Morphologies known as extracellular excretions of microbes, e.g., twisted stalks made up of iron hydroxides. None of these morphologies is 100% clearly biological. Providing context information is supportive if it can be shown that the necessary conditions for microbial activity were present (e.g., in terms of temperature).

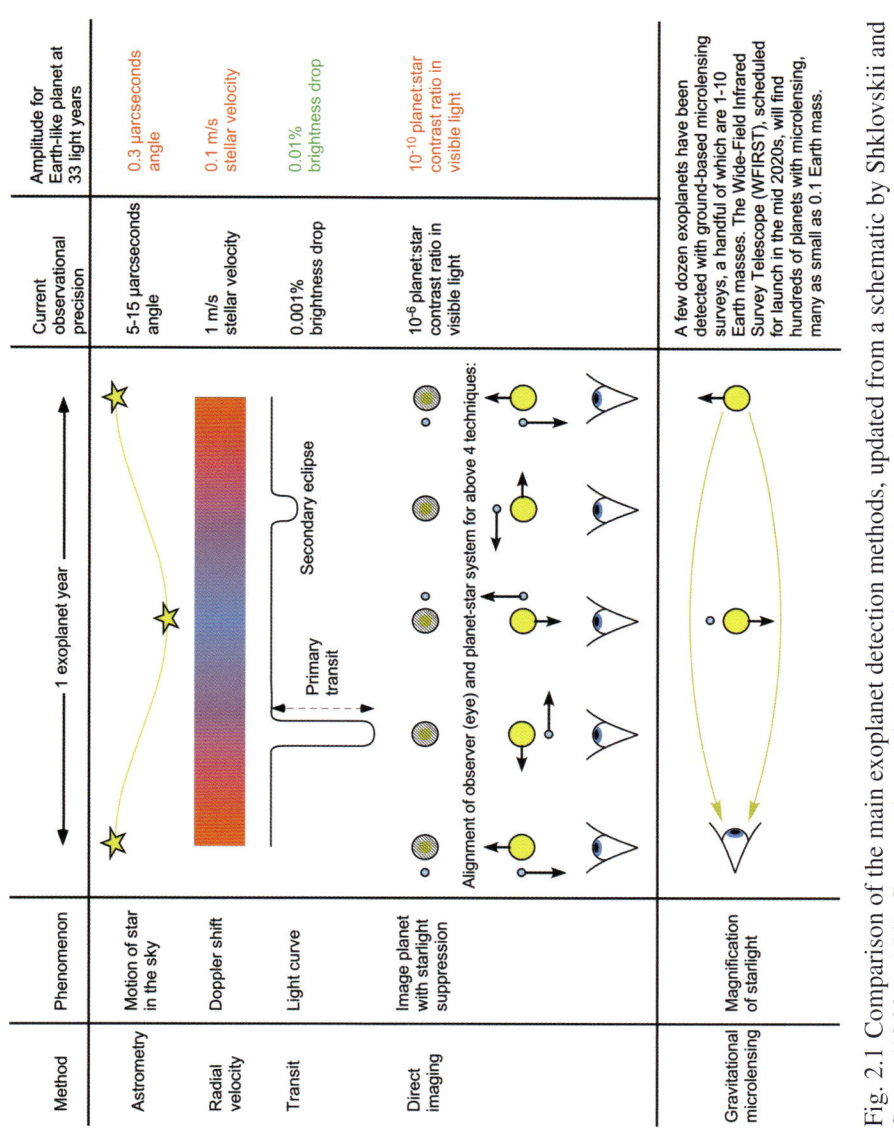

Fig. 2.1 Comparison of the main exoplanet detection methods, updated from a schematic by Shklovskii and Sagan (1966, p. 154). Current observational precision and the precision required to detect an Earth-like planet around a Sun-like star at 33 light years is contrasted. Techniques currently capable of detecting Earth-like planets are highlighted in green, whereas those not yet sufficiently precise to detect Earth-like planets are in red. The obscured stars in direct imaging represent starlight suppression. Gravitational microlensing is separated from the other methods because it cannot be used to target individual nearby stars. Instead, it is necessary to survey a very large field of stars for exceptionally rare stellar alignments.

INPUT

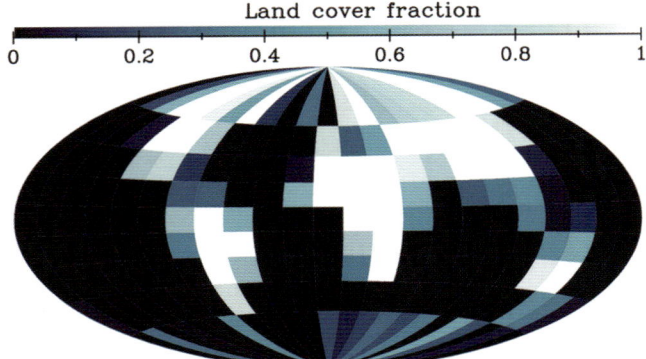

RECOVERED MAP : $\zeta=90.00°$, $\sigma=0.0\%$

Fig. 2.5 Surface maps like this one could be obtained for Earth-like exoplanets using only observations of planet brightness. The position of continents and oceans can be extracted from precise observations of brightness variations over a planet's rotation and orbit (Kawahara & Fujii, 2010). Reproduced with permission from IOP publishing.

Fig. 2.6 In the more distant future, a flotilla of space telescopes could obtain images of Earth-like exoplanets, revealing continents, oceans, clouds, and vegetation. This image shows how the Earth would appear using such a telescope from a distance of 10 light years (Labeyrie, 1999). Reproduced with permission from AAAS.

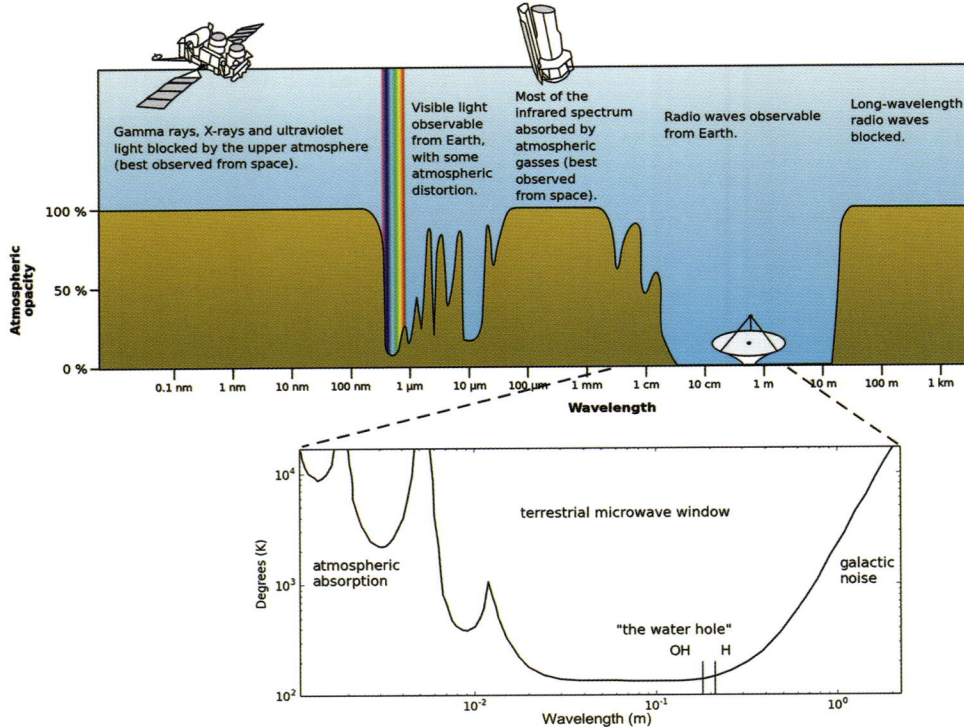

Fig. 2.8 Most of the electromagnetic spectrum does not make it through the Earth's atmosphere and so is unsuitable for ground-based SETI. Radio wavelengths and visible light are ideal for SETI since both penetrate the atmosphere. Within the radio spectrum there is a region around 10 cm where natural background radio noise is minimal. The "water hole" lies within this window. Figures adapted from NASA and Wikimedia commons.

Fig. 3.1 Example of eukaryotic macrofossils. Ammonites from the Jurassic (early Callovian, ~166 million years old) found in the Swiss Jura. No organic matter is preserved, all information is based on morphology. Sample width 21 cm.

Fig. 3.2 Filamentous sulphur-oxidizing bacteria in a hot spring (Yellowstone Park, USA). Even though prokaryotic and individual cells are microscopic, the colony is visually conspicuous. Image width approx. 50 cm.

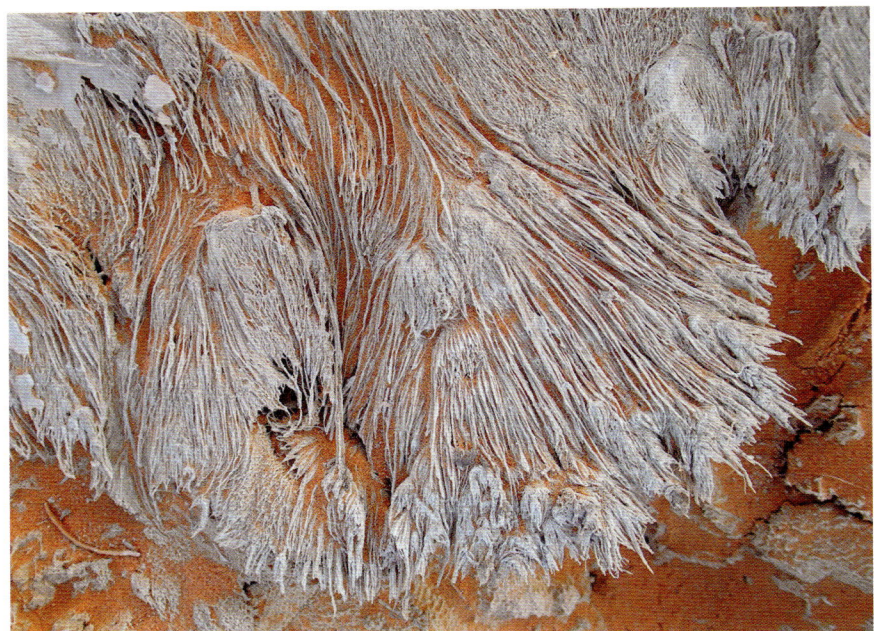

Fig. 3.3 Streamers of dried sulphur-oxidizing bacteria near the ground-water well Abu Tabul, Oman. The streamers consist mainly of elementary sulphur. Image width approximately 10 cm.

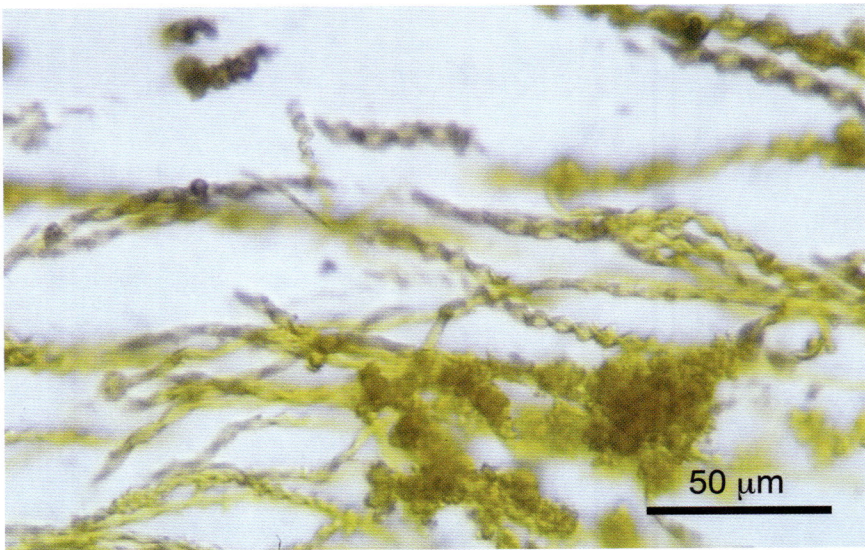

Fig. 3.4 Twisted stalks of iron hydroxides enclosed in a matrix of microcrystalline quartz. The stalks are products of microbes of the genera *Gallionella* or *Mariprofundus*. This sample is derived from a former cavity in plateau basalts in Breiðdalur, Eastern Iceland, and represents a site of microbial iron oxidation in a subsurface environment. Preserved by cementation with later quartz.

Fig. 3.5 Cut surface of a stromatolite showing the characteristic domal morphology with steep slopes of the sedimentary layering. Sample from the Archean of the Pilbara craton, Australia, approx. 3.4 billion years old. Image width 15 cm.

Fig. 3.6 Pseudo-stalactitic subsurface filamentous fabric in a cavity of basalt. Original microbial filaments are overgrown by microcrystalline quartz (chalcedony), Ballina, New South Wales, Australia. Field of view is 30 mm.

Fig. 3.7 Subsurface filamentous fabric in the form of mineralized microbial mats enclosed in solid quartz (chalcedony). Formed in a cavity in basaltic volcanic rock, Arz Bogd, Mongolia. Field of view 15 mm.

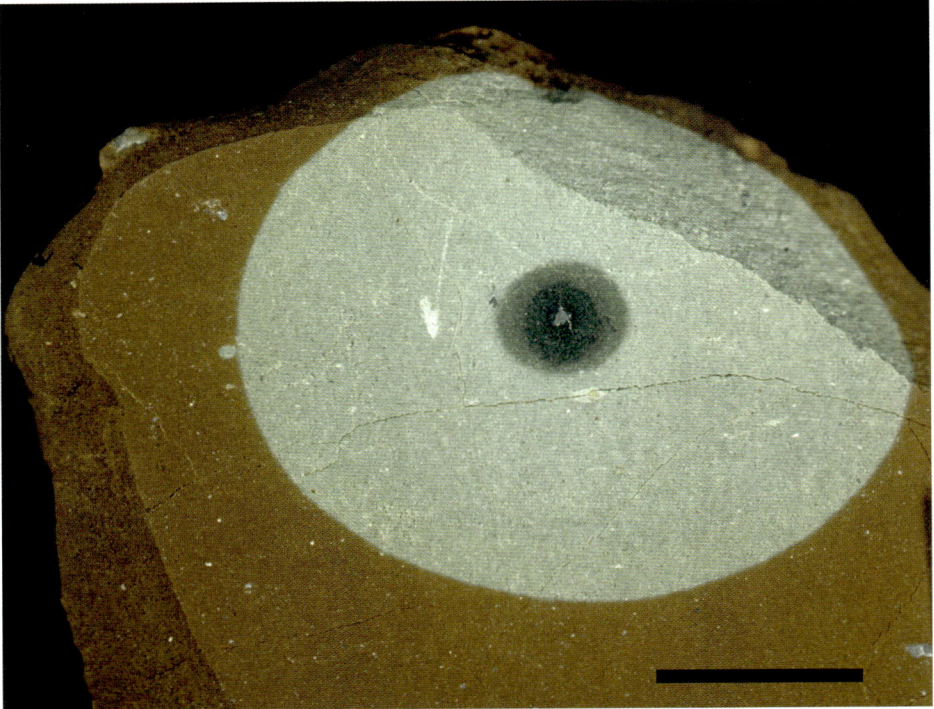

Fig. 3.8 Reduction spot, evidence of localized redox reactions in a sedimentary rock, likely to have been catalysed by microorganisms. Permian red beds, Kaisten well, Switzerland. Scale bar is 1 cm.

Fig. 3.7 Subsurface filamentous fabric in the form of mineralized microbial mats enclosed in solid quartz (chalcedony). Formed in a cavity in basaltic volcanic rock, Arz Bogd, Mongolia. Field of view 15 mm. (A black and white version of this figure will appear in some formats. For the colour version, please refer to the plate section.)

Other Visually Discernible Signatures of Life

Besides morphological remains of microbial activity in the shape of individual microbes or colonies of microbes, there are a number of other signatures of life that may be used to infer the former (in geological time) presence and activity of microbial life in rocks. A number of these signatures are of chemical or isotopic nature and can only be deciphered with the use of sophisticated analytical instruments. But there are also optically visible signatures of life that are not in the shape of microbial cells or colonies. Such morphologies are due to chemical transformations induced by microbes, in their function as catalysts of chemical reactions. These chemical transformations induce chemical reactions that can lead to visible changes in rocks. Gradients of redox potential in the pore-water of rocks induce characteristic colour changes, often due to mineralogical transformations of iron phases. An example of such evidence of redox reactions in rocks are reduction spots (or reduction spheroids), which occur worldwide in red-bed sediments, i.e., sedimentary rocks containing a red iron oxide pigment indicating oxidizing conditions (Fig. 3.8). Local redox reactions result in the reduction of iron, vanadium, and a series of rare elements. The need for localized catalysts in the rocks

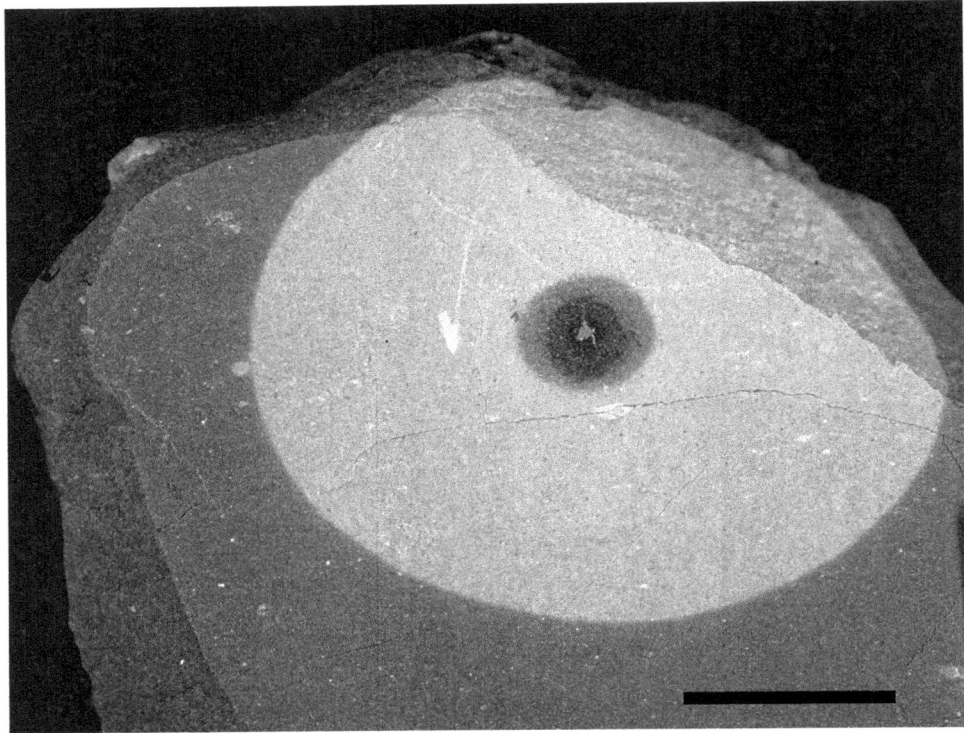

Fig. 3.8 Reduction spot, evidence of localized redox reactions in a sedimentary rock, likely to have been catalysed by microorganisms. Permian red beds, Kaisten well, Switzerland. Scale bar is 1 cm. (A black and white version of this figure will appear in some formats. For the colour version, please refer to the plate section.)

can be explained by local microbial activity (Hofmann, 1990, 1991, 2011; Spinks et al., 2010; Parnell et al., 2016). Reduction spots and other mineralogical changes due to redox reactions can be preserved over geological time. Since many redox reactions in the pore-water of sediments and rocks are microbially mediated, such colour changes are in fact signatures of life. Also the presence of pyrite (iron sulfide, FeS_2) in un-metamorphosed sedimentary rocks, a mineral that is often easily identified with the naked eye, is a strict indicator of microbial activity: the reduction of abundant sulphate in sea water to sulphide is a process that, under conditions of sedimentary rock formation, only works by means of microbial catalysis.

Application During Space Missions

The search for another example of life (outside Earth), extant or extinct, is a major scientific goal. Morphological expressions of life are of small to medium scale and therefore can only be searched for during lander missions. The application of morphological signatures of life in the near and medium future is restricted to the exploration of Mars. This is the only planetary body with a potential for life that is relatively easily accessible, with several surface missions planned. In principle, there is also a potential in the subsurface of the icy moons of Jupiter, but access is much more difficult. In practice, the search for morphological signatures of life during surface missions is focused on Mars. The currently active NASA rovers on Mars (*Opportunity*, active since 2004, and *Curiosity*, active since 2012) have already sent back thousands of images at resolutions that would, in principle, allow the recognition of a series of biologically induced textures in rocks. Although there have been some interpretations going in this direction (Noffke, 2015), no features have been detected so far that would allow classification as hot candidates. Additional opportunities will come up with the *ExoMars* rover mission of ESA in 2020 and the *Mars 2020* mission of NASA. The current and future missions are equipped with cameras that allow the visualization of the geological environment at scales comparable with the visual impressions gained by a geologist during field-work, including the use of a hand lens (close-up cameras on the Mars rovers). It will be important to investigate all potential signatures including those in non-sedimentary rocks. Besides potential remains of surface-bound life, the signature of ancient subsurface life might be an equally likely target.

However, no technique is available during the coming Mars missions that would be comparable to the use of thin sections and a petrographic microscope, standard techniques in geological laboratory work. If features are detected on Mars that resemble biologically induced morphologies on Earth, such evidence will hardly be accepted as sole proof for the former presence of life. However, morphological signatures of life would importantly contribute, together with chemical and isotopic signatures, to a comprehensive set of arguments. Different types of biosignatures will be searched for during the *ExoMars* rover mission planned for 2020. This rover will be equipped with close-up cameras, a drilling system able to retrieve cores from up to two meters depth, and a sophisticated laboratory.

Conclusions

Microbial life is able to produce macroscopic textures with characteristic morphologies. Because microbial life is active not only on the Earth's surface but also in the upper few kilometres of the crust, morphological signatures of ancient life in the form of microscopic cellular remains as well as macroscopic textures resulting from the activity of microbial colonies occur not only in sedimentary rocks but in any rock type that has provided suitable conditions for microbial activity at any time since the formation of the rock. Therefore, the search for ancient life on Earth and also on other planetary bodies such as Mars, must include, but not be restricted to, sedimentary rocks.

References

Allwood, A. C., Walter, M. R., Burch, I. W. & Kamber, B. S. (2007). 3.43 billion-year-old stromatolite reef from the Pilbara Craton of Western Australia: Ecosystem-scale insights to early life on Earth. *Precambrian Research*, **158**, 198–227.

Amy, P. S. & Haldeman, D. L. (1997). *The Microbiology of the Terrestrial Deep Subsurface*, Boca Raton: Lewis.

Awramik, S. M. & Grey, K. (2005). Stromatolites: biogenicity, biosignatures, and bioconfusion. In R. B. Hoover, G. V. Levin, A. Y. Rozanov, and G. R. Gladstone, eds., *Astrobiology and Planetary Missions, Proceedings of SPIE 5906*, 1–9.

Cady, S. L., Farmer, J. D., Grotzinger, J. P., Schopf, J. W. & Steele, A. (2003). Morphological biosignatures and the search for life on Mars. *Astrobiology*, **3**, 351–68.

Chan, C. S., Fakra, S. C., Emerson, D., Fleming, E. J. & Edwards, K. J. (2011). Lithotrophic iron-oxidizing bacteria produce organic stalks to control mineral growth: implications for biosignature formation. *The ISME Journal*, **5**, 717–27.

Claus, G. & Nagy, B. (1961). A microbiological examination of some carbonaceous chondrites. *Nature*, **192**, 594.

Daubenton, L. J. M. (1782). Sur les causes qui produisent trois sortes d'herborisations dans les pierres. *Memoires Acad. Royale*, 667–73.

Garcia-Ruiz, J. M., Hyde, S. T., Carnerup, A. M. *et al.* (2003). Self-assembled silica-carbonate structures and detection of ancient microfossils. *Science*, **302**, 1194–7.

Ghiorse, W. C. & Wilson, J. T. (1988). Microbial ecology of the terrestrial subsurface. *Advances in Applied Microbiology*, **33**, 107–72.

Grosch, E. G. & McLoughlin, N. (2014). Reassessing the biogenicity of Earth's oldest trace fossil with implications for biosignatures in the search for early life. *Proceedings of the National Academy of Sciences*, **111**(23), 8380–5.

Grotzinger, J. P. & Rothman, D. H. (1996). An abiotic model for stromatolite morphogenesis. *Nature*, **383**, 423–5.

Hahn, O. (1880). *Die Meteorite (Chondrite) und ihre Organismen*, Tübingen: Verlag der H. Laupp'schen Buchhandlung.

Hofmann, B. (1990). Reduction spheroids from northern Switzerland: Mineralogy, geochemistry and genetic models. *Chemical Geology*, **81**, 55–81.

Hofmann, B. (1991). Mineralogy and geochemistry of reduction spheroids in red beds. *Mineralogy and Petrology*, **44**, 107–24.

Hofmann, B. A. & Farmer, J. D. (2000). Filamentous fabrics in low-temperature mineral assemblages: Are they fossil biomarkers? Implications for the search for a subsurface fossil record on the early Earth and Mars. *Planetary and Space Science*, **48**, 1077–86.

Hofmann, B. A., Farmer, J. D., von Blanckenburg, F. & Fallick, A. E. (2008). Subsurface filamentous fabrics: An evaluation of possible modes of origins based on morphological and geochemical criteria, with implications for exopalaeontology. *Astrobiology*, **8**, 87–117.

Hofmann, B. A. (2011). Reduction spheroids. In J. Reitner and V. Thiel, eds., *Encyclopedia of Geobiology*, Dordrecht: Springer, pp. 761–62.

Hofmann, H. J. (1998). Synopsis of Precambrian fossil occurrences in North America. In S. B. Lucas and M. R. St-Onge, eds., *Geology of the Precambrian Superior and*

Grenville Provinces and Precambrian Fossils in North America, Ottawa: Geological Survey of Canada, pp. 271–376.

Kalkowski, E. (1908). Oolith und Stromatolith im norddeutschen Buntsandstein. *Zeitschrift der deutschen geologischen Gesellschaft*, **60**, 68–125.

Kallmeyer, J., Pockalny, R., Adhikaria, R. R., Smith, D. C. & D'Hondt, S. (2012). Global distribution of microbial abundance and biomass in subseafloor sediment. *Proceedings of the National Academy of Sciences*, **109**(40), 16 213–16.

Kretzschmar, M. (1982). Fossile Pilze in Eisen-Stromatolithen von Warstein (Rheinisches Schiefergebirge). *Facies*, **7**, 237–60.

Lovley, D. R. & Chapelle, F. H. (1995). Deep subsurface microbial processes. *Reviews of Geophysics*, **33**, 365–81.

Mata, S. A. & Bottjer, D. J. (2009). Development of lower Triassic wrinkle structures: Implications for the search for life on other planets. *Astrobiology*, **9**, 895–906.

McKay, D. S., Gibson, E. K., Thomas-Keprta, K. L. et al. (1996). Search for past life on Mars: Possible relic biogenic activity in Martian meteorite ALH84001. *Science*, **273**, 924–30.

McKinley, J. P., Stevens, T. O. & Westall, F. (2000). Microfossils and paleoenvironments in deep subsurface basalt samples. *Geomicrobiology Journal*, **17**, 43–54.

McLoughlin, N., Wilson, L. A. & Brasier, M. D. (2008). Growth of synthetic stromatolites and wrinkle structures in the absence of microbes – implications for the early fossil record. *Geobiology*, **6**, 95–105.

Noffke, N. (2015). Ancient sedimentary structures in the < 3.7 Ga Gillespie Lake Member, Mars, that resemble macroscopic morphology, spatial associations, and temporal succession in terrestrial microbialites. *Astrobiology*, **15**, 1–24.

Noffke, N., Christian, D., Wacey, D. & Hazen, R. M. (2013). Microbially induced sedimentary structures recording an ancient ecosystem in the ca. 3.48 billion-year-old Dresser formation, Pilbara, Western Australia. *Astrobiology*, **13**, 1103–24.

Parnell, J., Brolly, C., Spinks, S. & Bowden, S. (2016). Metalliferous biosignatures for deep subsurface microbial activity. *Origins of Life and Evolution of Biospheres*, **46**, 107–18.

Pedersen, K. (2000). Exploration of deep intraterrestrial microbial life: current perspectives. *FEMS Microbiology Letters*, **185**, 9–16.

Phelps, T. J., Raione, E. G., White, D. C. & Fliermans, C. B. (1989). Microbial activities in deep subsurface environments. *Geomicrobiology Journal*, **7**, 79–92.

Porada, H. & Bouougri, E. (2007). 'Wrinkle structures' – a critical review. In J. Schieber, P. K. Bose, P. G. Eriksson et al., eds., *Atlas of Microbial Mat Features Preserved Within the Clastic Rock Record*, Amsterdam: Elsevier, pp. 135–44.

Rasmussen, B., Fletcher, I. R., Brocks, J. J. & Kilburn, M. R. (2008). Reassessing the first appearance of eukaryotes and cyanobacteria. *Nature*, **455**, 1101–5.

Razumovsky, G. (1835). Les agates mousseuses. *Bulletin de la société géologique de France*, **6**, 165–8.

Reith, F. (2011). Life in the deep subsurface. *Geology*, **39**, 287–8.

Riding, R. (2011). The nature of stromatolites: 3,500 million years of history and a century of research. In J. Reitner, N.-V. Quéric and G. Arp, eds., *Advances in Stromatolite Geobiology, Lecture Notes in Earth Sciences 131*, Berlin: Springer, pp. 29–74.

Schumann, G., Manz, W., Reitner, J. & Lustrino, M. (2004). Ancient fungal life in north Pacific Eocene oceanic crust. *Geomicrobiology Journal*, **21**, 241–6.

Spinks, S. C., Parnell, J. & Bowden, S. A. (2010). Reduction spots in the Mesoproterozoic age: implications for life in the early terrestrial record. *International Journal of Astrobiology*, **9**, 209–16.

Suosaari, E. P., Reid, R. P., Playford, P. E. *et al.* (2016). New multi-scale perspectives on the stromatolites of Shark Bay, Western Australia. *Scientific Reports*, **6**, 20557.

Teske, A. P. (2005). The deep subsurface biosphere is alive and well. *Trends in Microbiology*, **13**, 402–4.

Trewin, N. H. & Knoll, A. H. (1999). Preservation of Devonian chemotrophic filamentous bacteria in calcite veins. *Palaios*, **14**, 288–94.

Tyler, S. A. & Barghoorn, E. S. (1954). Occurrence of structurally preserved plants in Precambrian rocks of the Canadian Shield. *Science*, **119**, 606–8.

Wacey, D., Kilburn, M. R., Saunders, M., Cliff, J. & Brasier, M. D. (2011). Microfossils of sulphur-metabolizing cells in 3.4-billion-year-old rocks of Western Australia. *Nature Geoscience*, **4**, 698–702.

Walter, M. R., McLoughlin, S., Drinnan, A. N. & Farmer, J. D. (1998). Paleontology of Devonian thermal spring deposits, Drummond Basin, Australia. *Alcheringa*, **22**, 285–314.

Westall, F. (1999). The nature of fossil bacteria: A guide to the search for extraterrestrial life. *Journal of Geophysical Research*, **104**, 16 437–51.

Westall, F. & Folk, R. L. (2003). Exogeneous carbonaceous microstructures in early Archaean cherts and BIFs from the Isua greenstone belt: Implications for the search for life in ancient rocks. *Precambrian Research*, **126**, 313–30.

Westall, F., Vries, S. T. d., Nijman, W. *et al.* (2006). The 3.466 Ga "Kitty's Gap Chert," an early Archean microbial ecosystem. *Geological Society of America Special Paper*, **405**, 105–31.

Williams, A. J., Sumner, D. Y., Alpers, C. N., Karunatillake, S. & Hofmann, B. A. (2015). Preserved filamentous microbial biosignatures in the Brick Flat Gossan, Iron Mountain, California. *Astrobiology*, **15**, 337–668.

4

Precellular Evolution and the Origin of Life: Some Notes on Reductionism, Complexity and Historical Contingency

ANTONIO LAZCANO

The secular description of life phenomena that has shaped contemporary biology is one of the most remarkable accomplishments of the freethinking atmosphere of the Enlightenment. It is a major intellectual watershed that permeates the scientific and philosophical treatises written from the start of the nineteenth century onwards, and is echoed in a masterly fashion in *Frankenstein or, the Modern Prometheus*, published anonymously in 1818 by the young Mary Wollstonecraft Shelley. In some of the most gripping pages of the book, Dr. Viktor Frankenstein exposes the dead body of the creature he has completed by sewing up bits and pieces of corpses illegally obtained, to the thunderbolts that break up the night sky during a tempest. Soon a lightning strike hits the conductors and, as the electricity travels downwards and reaches the inert body, it starts to move and becomes alive. The rabbi Loew ben Bezalel animated the Golem in Prague using religious invocations, but Dr. Frankenstein achieved the forbidden by bringing back to life the dead by making use of electricity, a purely physical force.

Few novels reflect in such an extraordinary manner the scientific developments of their times. After many years of experimentation on the effects of electricity on severed frog legs in Bologna performed with the help of his wife Lucia Galeazzi, in 1791 Luigi Galvani published his *Commentary on the Effects of Electricity on Muscular Motion*, summarizing the observations that had led him to hypothesize on the existence of an animal electric fluid that originated in the brain and traveled through nerves and muscles. Galvani was not a necromancer but a child of the Enlightenment. Like many of his contemporaries, he avidly explored philosophy, mathematics and science, and the fascination that his observations awoke both among his colleagues and the lay public is a clear indication of the gradual but inexorable abandonment of metaphysical causes as naturalists and philosophers alike turned to physical explanations that shook the idea of a vital force (Lazcano, 2008).

Galvani's nephew Giovanni Aldini became the main promoter of his ideas, and gained considerable notoriety as he crisscrossed European theaters and salons applying electric discharges to twitching human and animal corpses. Very soon galvanic currents were considered as therapeutic agents against a wide range of illnesses (Fara, 2002; Montillo, 2013). As stated in the preface of *Frankenstein*, which may have been written by Percy Shelley, "... [T]he event on which this fiction is founded has been supposed, by Dr. Darwin, and some of the physiological writers of Germany, as not of impossible occurrence". The said Dr. Darwin was none other than Erasmus Darwin, the famous physician, poet, naturalist and libertarian whose grandson, Charles R. Darwin, would eventually become one of the most influential scientists of all times.

Mary Wollstonecraft Shelley's father had been a good friend of Humphry Davy, one of the founders of electrochemistry, and her appreciation of Erasmus Darwin is not altogether surprising. Always in the forefront of medical sciences, he was a keen follower of the ideas and practices of French physicians and of the philosophical trends in Europe, and had been one of the first to promote in England the use of electricity as a therapeutic agent to treat a number of ailments, including melancholy and other mental disturbances. The expectations were often higher than the outcomes, but rumor has it that he was asked to treat King George II from the incapacitating disease that so much affected the politics of the realm. The dynasty may have been spared by Dr. Darwin's refusal, but as Fara has written, the fascination that he and others had with galvanic currents reflects that "electricity was the greatest scientific invention of the Enlightenment, . . . [and] by the middle of the eighteenth century, electrical experiments were being performed all over Europe with new, powerful instruments that could produce, store and discharge static electricity" (Fara, 2002, p. 2).

Like Erasmus Darwin and many of his contemporaries, the Swedish naturalist Jöns Jacob Berzelius was attracted at first to the medical applications of galvanic currents. After studying to no effect their role in a number of illnesses, Berzelius began working on what would become the field of electrochemistry, and in 1813 proposed a dual classification of substances depending on their behavior during electrolysis, dividing them into electropositive and electronegative (Wisniak, 2000). As early as 1780 Linneus' former student Torbern Bergman had noted that the compounds extracted from plants and animals, i.e., "organized beings", were more reactive, and started to speak of the differences between organic and inorganic substances. Berzelius became rapidly convinced that the problems with the electrical breakdown of compounds of biological origin evidenced their special nature, and coined the term "organic chemistry" to set them apart. He carried his arguments further on, and wrote that "[W]e can regard the whole animal body as a machine which gathers materials for ceaseless chemical processes out of the food it receives . . . but the cause of most of the phenomenon in the animal body is so deeply hidden from our understanding that we will never discover it. We call this hidden cause the vital force..." (cf. Friedmann, 1997, p. 72). Berzelius may have not been the first to use the term, but his academic prestige helped its promotion, and he soon concluded that the separation between organic and inorganic chemistry was impassable. Such a gap did not last for long. Although in 1827 Berzelius had stated that "art cannot combine the elements of inorganic matter in the manner of living nature", only a year later his former student Friedrich Wöhler demonstrated that urea could be formed in high yield by heating ammonium cyanate "without the need of an animal kidney" (Leicester, 1974). The seemingly deep chasm separating the living from the non-living had started to close, but the term organic chemistry had taken hold.

Wöhler seems not have been aware in full of the philosophical implications of his results, but his work ushered a new epoch in chemical research. In 1850 Adolph Strecker synthesized alanine from a mixture of an aldehyde, potassium cyanide and ammonium chloride. This was followed by the laboratory synthesis of hydrocarbons achieved by Marcellin Berthelot, and the formation of a wide array of sugars from an aqueous solution of formaldehyde under slightly basic conditions reported by Alexander Butlerow. The breach separating organic compounds from inorganic substances was weakened over and over again. The laboratory synthesis of organic molecules was based on increasingly complex experimental settings, and by the end of the nineteenth century it had become a vigorous, well-established field that had achieved the formation of a bounty of biologically active compounds.

Berzelius was an accomplished mineralogist, and in the early 1830s he undertook the study of several meteorites. His analysis of the Alais meteorite, which had fallen in France in 1806, led to the identification of clays, water, carbon dioxide, an ammonia-containing salt and, most importantly, a complex, insoluble carbon-rich bitumen-like substance. As reviewed by Burke, Berzelius concluded that "the presence of a carbonaceous substance in the meteoritic earth has an analogy with the humus contained in terrestrial earth, but it is likely that it was added in a different manner, that it has different properties, and that it does not justify the conjecture that it has an analysis analogous to the carbonaceous matter in terrestrial earth" (Burke, 1991, p. 167), adding that "the carbonaceous substance, which is intermixed with this earth, appears not to justify the conclusion that organic nature is present at the place of this earth's origin" (Burke, 1991, p. 168).

A number of brilliant nineteenth-century chemists, including Berthelot, Wöhler, Clöez and others, followed suit and analyzed the Kaba, Orgeuil and other meteorites. Although several of them reported the presence of mixtures of organic polymers (Nagy, 1975; Burke, 1991), none suggested a possible evolutionary connection with organisms. Chemistry had been left out of the evolutionary world view that permeated natural and social sciences during the nineteenth century (Lewontin

& Levin, 2007), and no one saw abiotic organic syntheses as evidence of a continuity between chemistry and biology.

For all we know, Jean-Baptiste de Lamarck was the first to assume that spontaneous generation was not only a means of non-sexual reproduction of insects, worms and microbes, but also explained the emergence of the first life forms from which the Animal and Plant kingdoms arose (Farley, 1977). Lamarck wrote that "[N]ow light is known to generate heat, and heat has been justly regarded as the mother of all generations" (Lamarck, 1914, p. 245), but like many of his contemporaries he felt attracted to the life-giving properties of galvanic currents and became convinced that together with other environmental factors, electricity had played a key role in the appearance of the first organisms. As he concluded in 1809 in his *Zoological Philosophy*, "Nature, by means of heat, light, electricity and moisture, forms direct or spontaneous generation at the extremity of each kingdom of living bodies, where the simplest of these bodies are found" (Lamarck, 1914, p. 237).

Charles Darwin took a somewhat similar stand. He did not share Shelley's morbid fascination with putrefaction, nor did he accept the possibility that the decomposition of preexisting organic compounds could lead to the appearance of microbes. Nonetheless, he was convinced that life was the outcome of natural processes, and as early as 1837 he wrote in his notebooks that "[t]he intimate relation of Life with laws of chemical combination, & the universality of latter render spontaneous generation not improbable" (Darwin *et al.*, 1987, p. 269). It was not until 1863 that he mentioned his views on the origin of life in an open letter sent to a London periodical in response to the criticisms that Robert Owen had raised against his theory. In it Darwin stated that "[t]here must have been a time when inorganic elements alone existed on our planet: let any assumptions be made, such as that the reeking atmosphere was charged with carbonic acid, nitrogenized compounds, phosphorus, &c. Now is there a fact, or a shadow of a fact, supporting the belief that these elements, without the presence of any organic compounds, and acted on only by known forces, could produce a living creature?" (Darwin, 1863, p. 554).

A few years later Darwin restated his ideas using the "warm little pond" metaphor in a letter mailed on February 1st, 1871, to his friend and confidant Joseph Dalton Hooker (Peretó *et al.*, 2009). In it Darwin acknowledged, like his grandfather Erasmus had done before him, the role of electricity in life phenomena and, like Lamarck, hypothesized on the role of heat and galvanic currents on the origin of life.

"It is often said that all the conditions for the first production of a living organism are now present, which could ever have been present", wrote Darwin to Hooker, adding that "But if (and oh what a big if) we could conceive in some warm little pond with all sorts of ammonia and phosphoric salts, – light, heat, electricity &c. present, that a protein compound was chemically formed, ready to undergo still more complex changes, at the present day such matter would be instantly devoured, or absorbed, which would not have been the case before living creatures were formed." *(Darwin, 1887, p. 18)*

Darwin had a good understanding of chemistry, and his "big if" is a reminder that he was keenly aware of the lack of evidence for the scheme he had outlined. How life appeared is still not known but, as argued below, an evolutionary framework derived from Darwin's perspective has led to the most fruitful approaches to investigate this problem.

Darwin's sharp rebuttal against Owen appears to be the only time in which he publicly addressed the origins of life. In fact, his silence on this issue led Ernst Haeckel, his foremost champion, to openly criticize him in a short footnote of small book published in 1862, in which he wrote that "[T]he chief defect of the Darwinian theory is that it throws no light on the origin of the primitive organism – probably a simple cell – from which all the others have descended. When Darwin assumes a special creative act for this first species, he is not consistent, and, I think, not quite sincere..." (Haeckel, 1862, p. 232 (ann. 1)).

Haeckel did not shy away from this issue, and devoted a whole chapter to the origins of life in his popular 1868 book *Natürliche Schöpfungsgeschichte* (engl. edition *The History of Creation* 1876). In it he underlined the differences between Lamarck and Darwin. While the former had accepted the possibility of spontaneous generation, wrote Haeckel, "...Darwin passes over and avoids this subject, as he expressly remarks that he has 'nothing to do with the origin of the soul, nor with that of life itself'. At the conclusion of his work he expresses himself more distinctly in the following words: 'I imagine that probably all organic beings which ever lived on this earth descended from some primitive form, which was first called into life by the Creator'" (Haeckel, 1876, p. 371).

The influence that Haeckel's ideas exerted on a number of scientists, including Stephane Leduc, Jerome Alexander and Alfonso L. Herrera, on the one hand, and on Alexandr I. Oparin and other Russian researchers, on the other, merits a detailed examination of his writings. Haeckel was convinced that the origin of life, which he associated with albumins, i.e., proteins, should be understood as part of a grandiose secular scheme of cosmic evolution that started with the formation of the Earth and the condensation of the primitive oceans, although he never mentioned the formation of organic compounds. As he stated, "[I]t was not till the earth's crust had so far cooled that the water had condensed into a fluid form, it was not till the hitherto dry crust of the earth had for the first time become covered with liquid water, that the origin of the first organisms could take place. For all animals and all plants – in fact, all organisms – consist in great measure of fluid water, which combines in a peculiar manner with other substances, and brings them into a semi-fluid state of aggregation. We can therefore, from these general outlines of the inorganic history of the earth's crust, deduce the important fact, that at a certain definite time life had its beginning on earth, and that terrestrial organisms did not exist from eternity, but at a certain period came into existence for the first time" (Haeckel, 1876, p. 327).

"Now, how are we to conceive of this origin of the first organisms?", asked Haeckel, adding that "This is the point at which most naturalists, even at the present day, are inclined to give up the attempt at natural explanation, and take refuge in the miracle of an inconceivable creation. In doing so, as has already been remarked, they quit the domain of scientific knowledge, and renounce all further insight into the eternal laws which have determined nature's history. But before despondingly taking such a step, and before we despair of the possibility of any knowledge of this important process, we may at least make an attempt to understand it. Let us see if in reality the origin of a first organism out of inorganic matter, the origin of a living body out of lifeless matter, is so utterly inconceivable and beyond all experience. In one word, let us examine the question of *spontaneous generation, or archigony*. In so doing, it is above all things necessary to form a clear idea of the principal properties of the two chief groups of natural bodies, the so-called inanimate or inorganic, and the animate or organic bodies, and then establish what is common to, and what are the differences between, the two groups"(Haeckel, 1876, pp. 327–8), and continued "... to me the idea that the Creator should have in this one point arbitrarily interfered with the regular process of development of matter, which in all other cases proceeds entirely without his interposition, seems to be just as unsatisfactory to a believing mind as to a scientific intellect. If, on the other hand", added Haeckel, "we assume the hypothesis of spontaneous generation for the origin of the first organisms, which in consequence of reasons mentioned above, and especially in consequence of the discovery of the Monera, has lost its former difficulty, then we arrive at the establishment of an uninterrupted natural connection between the development of the earth and the organisms produced on it, and, in this last remaining lurking-place of obscurity, we can proclaim the *unity of all Nature, and the unity of her laws of Development*" (Haeckel, 1876, p. 349).

For Haeckel the "unity of all Nature" meant the evolutionary continuity between the inorganic world and living entities. As he concluded in *The Natural History of Creation*, "[t]he differences which exist between the simplest organic individuals and inorganic crystals are determined by the solid state of aggregation of the latter, and by the semi-fluid state of the former. Beyond that the causes producing form are exactly the same in both. This conviction forces itself upon us most clearly, if we compare the exceedingly remarkable phenomena of growth, adaptation, and the 'correlation of parts' of developing crystals with the corresponding phenomena of the origin of the simplest organic individuals (Monera and cells). The analogy between the two is so great that, in reality, no accurate boundary can be drawn" (Haeckel, 1876, p. 338).

It is easy to see how the so-called chemical gardens and their microscopic equivalents, developed as models of cells by Alexander, Leduc and Herrera (Pereto & Catalá, 2007), fitted smugly in Haeckel's scheme. As shown by the attention that many had devoted since the seventeenth century to the so-called "tree of Diana", a metallic dendritic structure with plantlike morphologies formed from an amalgam of silver and mercury dissolved in nitric acid, there was a long interest, rooted in part in the alchemic traditions to which Newton was no stranger, that hailed biomorphic structures as part of a grand scale of things (cf. Lazcano, 2003).

In an attempt to understand the origin and nature of life from a strictly materialistic perspective, Herrera and others devoted the best part of their careers to the production of life-like structures from various combinations of inorganic compounds, crystals and different fluids, as part of the now largely forgotten field "plasmogeny". Although Marcel Florkin mercilessly described these approaches as the "dark age of biocolloidology", the mesmerizing reports of life-like behavior in microscopic droplets of different compositions led many to the conclusion that these structures provided direct insights on the nature of cells and living processes, convinced that the resemblances of form and size sufficed to connect the living and the non-living worlds. A good example is Herrera's statement that "[S]ulfur alone sublimed on cold glass yields no end of cellular patterns by virtue of its molecular polymorphism and resulting allotropic states... sulfur is present in nearly all proteins and in all living organisms and thus merits special attention in any theory of the origins of life" (Herrera, 1942, p. 14).

The work of Herrera and his followers provided morphological analogs of cell-like features but, with few exceptions (Perezgasga *et al.*, 2003; Saladino *et al.*, 2016), their chemical precursors and the conditions under which they are formed are not truly relevant to our current understanding of the origins of life. As shown by the intense debates on the ultimate origin of some of the microstructures found in early Precambrian sediments (Garcia-Ruiz *et al.*, 2003) and in the Martian Alan Hills 84001 meteorite (McKay *et al.*, 1996), the nature of cell-like microscopic biomorphs is still a source of major debates. It is easy to understand the allure of the striking analogy between the form of mitotic apparatus and the aligned iron scrapings that reveal the orientation of a magnetic field, but one should avoid stretching the analogies beyond their superficial similarity. It may be difficult nowadays to understand the fascination for such physicochemical structures, which survive in increasingly rare children's chemical sets, and in attempts to examine them within the physics of complex systems (Steinbock *et al.*, 2016). However, one should acknowledge the passionate efforts of Herrera, Leduc and their colleagues as part of a failed attempt to describe and understand the basic properties of life based on purely physical phenomena that included magnetism, surface tension and radioactivity (Fox-Keller, 2002; Campos, 2015).

As shown by Thomas Mann's complex novel *Doktor Faustus*, the description of these startling microscopic structures eventually reached literary heights. "The vessel of crystallization was three-quarters full of slightly muddy water – that is, dilute water-glass – and from the sandy bottom there strove upwards a grotesque little landscape of variously coloured growths: a confused vegetation of blue, green, and brown shoots which reminded one of algae, mushrooms, attached polyps, also moss, then mussels, fruit pods, little trees or twigs from trees, here and there of limbs. It was the most remarkable sight I ever saw, and remarkable not so much for its appearance, strange and amazing though that was, as on account of its profoundly melancholy nature. For when Father Leverkühn asked us what we thought of it and we timidly answered him that they might be plants: 'No,' he replied, 'they are not, they only act that way. But do not think the less of them. Precisely because they do, because they try to as hard as they can, they are worthy of all respect'".

"It turned out that these growths were entirely unorganic in their origin, they existed by virtue of chemicals from the apothecary's shop, the 'Blessed Messengers'. Before pouring the water-glass, Jonathan had sprinkled the sand at the bottom with various crystals, if I mistake not potassium chromate and sulphate of copper. From this sowing, as the result of a physical process called 'osmotic pressure' there sprang the pathetic crop for which their producer at once and urgently claimed our sympathy. He showed us that these pathetic imitations of life were light-seeking, heliotropic, as science calls it. He exposed the aquarium to the sunlight, shading three sides against it, and behold, toward that one pane through which the light fell, thither straightway slanted the whole equivocal kith and kin, mushrooms, phallic polyp-stalks, little trees, algae, half-formed limbs. Indeed, they so yearned after warmth and joy that they actually clung to the pane and stuck fast there" (Mann, 1947, p. 19).

By the turn of the twentieth century the characterization of an increasingly large number of proteins led many to the idea that life could be associated with specific enzymes, and that submicroscopic aggregates or micelles could exhibit the properties of life. Many saw enzymes as colloidal catalysts, which appeared to support the possibility that entities smaller and simpler than protoplasm itself could be alive. This possibility was reinforced by the discovery of viruses, that were mistakenly considered by Muller and few others as naked genes, reinforcing the assumption that life could have started with "living" molecules (Fry, 2006; Lazcano, 2010).

The American physicist Leonard Troland championed this view in a series of articles published between 1914 and 1917, in which he argued that the first living entity had been nothing more than a self-replicating enzyme-like molecule that had appeared by chance in the primitive oceans. Troland spoke first of living molecules and then of a "genetic enzyme", which he eventually identified with nucleoproteins (Fry, 2006). It did not take long for Hermann J. Muller, who was part of Thomas Hunt Morgan's group in Columbia University where *Drosophila* genetics was born, to adapt Troland's hypothesis and propose that the ancestral molecule had not been an enzyme but a gene, although at the time these were purely formal objects and neither their actual nature nor their chemical composition were understood. Muller's appreciation of genetic mutation as the fundamental mechanism of evolutionary novelties (Carlson, 1981) led him to argue that the first living entities consisted of nothing more than a gene with self-replicative properties, the ability to modify its chemical environment and to evolve, i.e., life had originated as an autocatalytic genetic entity endowed with heterocatalysis and mutability (Muller, 1926).

Muller fancied himself as an evolutionist, but his explanation of the origin of life reveals a mutationist's attitude, not a Darwinian one. His approach is easy to understand. As documented by Bowler (1988), towards the end of the nineteenth century Darwinian explanations based on natural selection were increasingly criticized, and alternative proposals based on Mendelism and other mechanisms were on the rise. The situation was quite different in the Russian Empire, where Darwin's ideas had been introduced and popularized thanks to the tireless efforts of Kliment A. Tymiriazev, a renowned plant physiologist and agronomist. Darwinism spilled beyond the boundaries of academia and was rapidly transformed into a battle cry of those searching a political alternative to the autocratic Czarist regime. This situation continued well into the twentieth century, when the foundations of what would become a short-lived but brilliant school of genetics were being laid by a number of insightful researchers like Nikolai K. Koltzov, who proposed that genes were macromolecules whose structure and chemical composition could explain genetic phenomena (Soyfer, 2001). However, during those politically charged times many Russian scientists, writers and left-wing politicians committed to Darwin's ideas had the upper hand in their struggle against geneticists who were simplistically portrayed as opposing evolution by natural selection. Thus, well before the savage campaigns orchestrated by Lysenko and his followers against genetics, in Russia first and in the Soviet Union afterwards, many had fallen into the false disjunction of confronting Darwin's theory against Mendel's ideas.

In contrast to Muller, Alexandr I. Oparin approached the origins of life from a purely Darwinian perspective and from the standpoint of biochemistry and cell biology. As an undergraduate he became acquainted with Tymiriazev, who invited him to the weekly seminars and discussions on evolutionary biology he held in his Moscow flat. During the same years Oparin had joined the laboratory of Alexei N.

Bakh, where he received a strong training in biochemistry and became acquainted with the functional and structural intricacies of photosynthesis. Thus, by the time Oparin graduated from Moscow University in the fateful year of 1917, his academic background combined natural history, biochemistry and plant physiology, a knowledge acquired within a research tradition strongly committed to integral approaches in the analysis of natural phenomena. Equally important, he was not only familiar with nearly all the literature on evolution available in Russia at the time, but had also developed a deep understanding of the Darwinian method of comparative analysis and historical interpretation of life features (Lazcano, 2010).

From the very beginning Oparin opposed the possibility of a spontaneous photosynthetic origin of life, and based on the simplicity and ubiquity of fermentative reactions concluded that the first life forms had been anaerobic heterotrophic microorganisms. This assumption required the prebiotic synthesis of organic material prior to the origin of life itself, a possibility that Oparin argued was supported by the work of Wöhler, Butlerov and many others, as well as by the presence of organic molecules in meteorites.

Oparin faced some issues with the reviewers, but he finally published a small book titled *The Origin of Life* (Oparin, 1924). The writings that the geochemist Charles Lipman, the microbiologist R. B. Harvey and the geneticist John B. S. Haldane, a towering figure who became one of the founding fathers of neodarwinism, produced in the 1920s, demonstrate that some of his contemporaries were developing similar proposals, but none of these was as refined as Oparin's hypothesis (Bada & Lazcano, 2003). Twelve years later, Oparin published a second volume that was soon translated into English (Oparin, 1938). Although both books bear the same title, the second one is intellectually and scientifically much more sophisticated. In it, Oparin presented a much more documented picture of the primitive environment scheme, based on a painstakingly thorough analysis of data from widely different fields to support his assumption of a highly reducing primitive milieu in which the molecular precursors to life, including protein-like compounds, had formed abiotically and accumulated in droplets rich in organic polymers, from which the first heterotrophic anaerobic bacteria evolved.

Like many of his contemporaries, Oparin was convinced that coacervates, the microscopic organic colloidal droplets that had been popularized by Bundemberg de Jong as physicochemical models of protoplasm, may have existed in the primitive oceans as mimicking the surmised properties of precellular systems. In doing so, he joined a venerable scientific tradition that assigned life's essential properties to cells. In 1805 the German naturalist Lorenz Oken had written a small booklet titled *The Creation*, in which he stated that "all organic beings originate from and consist of vesicles of cells" (Oken, 1805, p. 74), a conclusion that appeared to be supported by the observations of the French microbiologist Felix Dujardin, who described a jellylike, water insoluble substance that he termed "gelée vivante" when he crushed protists under the microscope.

Several decades later such a gel-like substance was found inside all cells and was termed "protoplasm" by Johann E. Purkinje and Hugo von Mohl, who concluded that it was the basic physicochemical component of life. This was followed by Thomas Graham's 1861 proposal that the protoplasm was a colloid formed by a homogenous, proteinaceous substance, which was understood by many as implying, as Thomas Henry Huxley wrote a few years later, that the basic traits of life were defined by the chemical and physical properties of the molecules that made up protoplasm.

A year after Graham published his ideas, the chroniclers of the Parisian social and intellectual life Edmond and Jules de Goncourt registered in their *Journal* a short entry on the nature of life "Qu'est-ce que la vie? L'usufruit d'une agrégation de molecules" (de Goncourt & de Goncourt, 1904, p. 52). How protoplasm managed to reach French literary salons is unknown, but many would agree nowadays with this insightful summary that comes not from scientists but from literary critics.

The evolutionary scheme proposed by Oparin (1938) represents a pioneering attempt to undertake a Darwinian analysis of the traits and the molecular constituents of life and their interaction in the intracellular microenvironment. Even the simplest organisms, he argued, "... possesses a definitive dynamically stable structural organization which is founded upon a harmonious combination of strictly coordinated chemical reactions. It would be senseless to expect that such an organization could originate accidentally in a more or less brief span of time from simple solutions or infusions" (Oparin, 1938, p. 246). In other words, the first living entities were the outcome of a step-wise, non-teleological process of precellular evolution during which the exquisite functional properties of subcellular components were shaped, prior to the onset of Darwinian evolution.

As underlined by Morange (2012), Oparin was the first to recognize that "the specific peculiarity of living organisms is only that in them there have been collected and integrated an extremely complicated combination of a large number of properties and characteristics which are present in isolation in various dead, inorganic bodies" (Oparin, 1938, p. 32), i.e., life is not characterized by any special properties but by a definite, specific combination of these properties that is the outcome of a historical process. In other words, it cannot be defined on the basis of a single trait or substance, and the appearance of living systems can be understood as an outcome of the synchronic emergence and coevolution of their basic components.

Oparin's refusal to assume that life had started with a gene needs to be understood not only from his unwillingness to adopt a reductionist perspective, but has to be discussed within the framework of Cold War politics and his controversial affiliation with the Soviet establishment and with Lysenko. However unpalatable these associations were, there is no connection between them and his ideas on the origins of life, and they should not keep us from recognizing that Oparin's proposal played a major role in stimulating the scientific discussion on the emergence of life. Indeed, in spite of his shameful political associations during the harsh Stalinist period, there are a number of reasons to cherish Oparin's scientific contributions. As underlined by Kamminga (1988) and Sankaran (2012), Oparin's work helped to distinguish between the spontaneous generation of organisms, and the chemical and biochemical origins of life. His heterotrophic scheme set the question of the origin of life within a Darwinian framework and led to a multi- and interdisciplinary research program that allowed the reinterpretation of many isolated facts and observations within a non teleological evolutionary sequence. In sharp contrast to Berzelius, Wohler, Strecker and other nineteenth-century scientists, including his mentor Bakh, he was the first to interpret chemical data and phenomena within an evolutionary context while, at the same time, he argued that life was a property of membrane-bounded systems of interacting components and not of a "living molecule".

Muller and Oparin never met. Although at first Muller viewed Oparin's ideas with sympathy, they soon clashed in an intense debate characterized by sharp ideological and scientific exchanges that reverberate to this day (Falk & Lazcano, 2012). Like many of his contemporaries, Muller was attracted to the Soviet Union and moved there convinced that science could help to build a better society. He rapidly became disappointed with the dictatorial aspects of Soviet life and their influence in the scientific environment, and was forced to leave the country. His passionate disagreement with Oparin's views was the complex outcome of a conflict shaped by their opposing views on the nature of life, the processes that had preceded the first living entities, their different perspectives on heredity and evolution and, last but not least, by the atmosphere of the Cold War (Lazcano, 2010; Falk & Lazcano, 2012).

Contrary to what Muller and others assumed, Oparin's books were not commissioned by the Soviet Communist Party as part of an ideological conjure. However, it is true that following the Bolschevik Revolution many attempts were made in the USSR to develop science, art and culture in a cultural and scientific milieu shaped by dialectical materialism, and that Oparin and other Soviet scientists were sincerely persuaded that Marxism was not only an official state ideology but also provided a philosophical framework on which entire research programs could be based (Graham, 1972; Farley, 1977; Tagliagambe, 1978; Lazcano 2010).

Following the publication of his second book, Oparin concentrated his efforts in the study of coacervates and the refinement of the concept of precellular evolution, taking for granted the possibility of the synthesis and accumulation of organic compounds in the primitive environment. In 1952, however, Harold C. Urey published a seminal book titled *The Planets: Their Origin and Development*, in which he argued that the highly reducing environment of the primitive Earth would had led to the formation of organic molecules (Urey, 1952). Based on the atmospheric models developed by Urey and Oparin's theoretical scheme, in 1952 Stanley L. Miller, a young graduate student, simulated the ocean–atmosphere system on the primitive Earth by investigating the action of electric discharges on a highly reducing mixture of CH_4, NH_3, H_2 and H_2O, and was able to synthesize racemic mixtures of several protein amino acids, as well as hydroxy acids, urea and other organic molecules.

The extraordinary impact of 1953 Miller's report on the synthesis of amino acids and other organic compounds under possible Earth conditions (Miller, 1953) was a striking demonstration of the epistemological significance of Oparin's proposal. It is frequently forgotten, in fact, that only two weeks earlier Watson and Crick had published their double helix model of the DNA molecules (Watson & Crick, 1953). It is somewhat surprising to learn that it took Muller more than a year to find the time to read the Watson and Crick (1953) paper on the DNA double helix model, but once he did he immediately grasped its significance for his own ideas on the origin and nature of life.

Muller remained stubbornly attached to his reductionist view, updating his ideas by identifying primordial genes with replicating DNA sequences, which he argued had formed abiotically in the waters of the primitive Earth. Few would agree today with this highly reductionist definition of life, but following the *in vitro* enzymatic synthesis of strands of DNA reported by Arthur Kornberg and his associates (Lehman *et al.*, 1958), during the University of Chicago 1959 celebrations of the centennial of Darwin's *Origin of Species*, Muller famously declared that "those who define life as I do will admit that the most primitive forms of things that deserve to be called living had already been made in the test tube by A. Kornberg" (Muller, 1960, p. 93). Forty years after his 1926 paper, Muller still insisted that the essence of life could be found in the combination of autocatalysis, heterocatalysis and mutability. "The gene material alone, of all natural materials", he added, "possesses these faculties, and it is therefore legitimate to call it living material, the present-day representative of the first life" (Muller, 1966, p. 497).

An alternative to Muller's reductionism would be the recognition that the properties of genes and DNA are outcomes of a process of evolution. In other words, the first replicating genetic polymers did not arise spontaneously from an unorganized prebiotic organic broth due to an extremely improbable accident, nor was precellular evolution a continuous, unbroken chain of progressive transformations steadily proceeding to the first living entities. However, Muller never advocated this possibility, nor did he overcome his ideological rejection and bitterness against Lysenko and his associates, whom he blamed for the destruction of the Soviet schools of genetics. He included Oparin in this group, as shown by the asperity with which he described his proposal: "that protoplasm had originated first and had the capability of manufacturing not only the genetic material but also its own complex organization... The Russian Oparin has since the early 1930's espoused this view and has followed the official Communist Party line by giving the specific genetic materials a back seat" (Muller, 1961, p. 7).

Muller was not being fair. As noted above, while Haeckel presented a general scheme of evolution in which he claimed that spontaneous generation was the only secular mechanism to explain the origin of life, Oparin advocated a gradual, stepwise slow process of precellular evolution during which the basic traits of replication and metabolism had developed in membrane-bounded molecular networks. It is not surprising that Oparin's original proposals did not include genes because, as late as 1942, Julian Huxley, one of leading figures of neodarwinism, had stated that "... bacteria have no genes in the sense of accurately quantized portions of hereditary substances; and therefore have no need for accurate division of the genetic system which is accomplished by mitosis" (Huxley, 2010, p. 131).

It took some time before Oparin abandoned his pre-Mendelian genetics and accepted the role of nucleic acids in heredity, and he did so only half-heartedly. Like Haldane, Pirie, Bernal and many others, he became convinced that RNA molecules had preceded DNA genomes (Oparin, 1961). However, until the very end of his long scientific career he refused to adhere to Muller's reductionist proposal and wrote, in what may be the final, most refined statement of his views, that "... [O]ne must clearly imagine that there were no polynucleotides capable of replication or polypeptides having some specificity in their sequence under the influence of these polynucleotides which entered the selective process but rather as a whole system..." (Oparin, 1972, p. 10).

During the first half of the twentieth century many life scientists were blissfully ignorant of the ongoing debates on the origins and nature of life, and some were actually convinced that they could not even be understood. This was also true of most chemists and physicists, as epitomized by the 1932 Niels Bohr famous remark that "[t]he existence of life must be considered as an elementary fact that cannot be explained, but must be taken as a starting point in biology" (Bohr, 1933, p. 458).

As shown by the popularity of Schroedinger's *What is Life?*, not all physicists agreed with this somewhat negative perspective. Their interest as a group is also manifested in the many attempts to describe the emergence and nature of life in terms of non-linear interactions and non-equilibrium constraints, the thermodynamics of irreversible processes, pattern formation, chaos, attractors, fractals and, more recently, complexity theory (Fox-Keller, 2002).

Self-assembly of complex systems may be found in a wide variety of situations, including cellular automata, the flow patterns of different fluids and in cyclic chemical phenomena, such as the Belousov–Zhabotinsky reaction. There are indeed some common features among these systems, and complexity theorists claim that they follow general principles that are in fact equivalent to universal laws of nature that can also explain the origin and ultimate nature of life (Kaufmann, 1993). Is this really so? In addition to natural selection, it is obvious that there are other mechanisms of ordered complexity that operate, and there can be little doubt that self-organization phenomena played a role in the origin of life. The examples include the formation of micelles, liposomes and lipid vesicles from prebiotic amphiphiles, the self-assembly of nucleic acids and other base-bearing polymers, the formation mineral and organic compounds complexes, and autocatalytic synthetic reactions such as the formose reaction, among others (Lehn, 2002; Orgel, 2008; Budin & Szostak, 2010; Lazcano, 2010).

During the past few decades there have been a number of attempts to consider life in terms of somewhat ill-defined "emergent properties" or "self-organizing principles", by assuming that at first living beings consisted of a continuously renewing interactive system of biochemical pathways that lacked genetic polymers. The basic assumption underlying many of these proposals is that there are intrinsic phenotypic laws rooted in physical processes, i.e., that a set of a large number of interacting molecules could assist the replication of some of its components, eventually leading to autocatalytic replication. However, there is no chemical evidence that long sequences of reactions organize spontaneously. As underlined by Orgel (2008), these proposals are also hindered not only by the lack of empirical evidence, but also by a number of rather unrealistic assumptions about the properties of minerals and other catalysts required to spontaneously organize such sets of autocatalytic chemical reactions.

In fact, our current understanding of biology indicates that life could have not evolved in the absence of a genetic replicating mechanism insuring the stability and diversification of its basic components. It is precisely in this context that the significance of the catalytic activity of RNA molecules for the origins of life should be seen. The RNA World model does not imply that wriggling autocatalytic nucleic acid molecules were floating in the waters of the primitive oceans ready to be used as primordial genes, or that the RNA World sprang completely assembled from simple precursors present in the prebiotic soup (Lazcano, 2007). There is no formal definition of the RNA World, but it can be pictured as an early, perhaps primordial, stage during which RNA molecules played a much more conspicuous role in both heredity and metabolism.

There are many indications of the robustness of the RNA World hypothesis. Although the ultimate origin of ribozymes is still an open question, the catalytic, regulatory and structural properties of RNA molecules and ribonucleotides, combined with their ubiquity in cellular processes, are consistent with the proposal that they played a key role in early evolution and, perhaps, in the origin of life itself (Lazcano, 2014). Acceptance of the RNA World is by no means a mere reformulation of Muller's ideas of a living molecule, but rather the recognition that the properties of polyribonucleotides and ribozymes are the outcome of a step-wise process of chemical and precelullar evolution. This conclusion has important implications for understanding the emergence of living systems: if the origin of life is seen as the evolutionary transition between the non-living and the living, then it is also necessary to recognize it as a historical process shaped in part by the vagaries of contingent events. It also suggests that attempts to draw a strict line between these two worlds may be meaningless. In other words, the appearance of life on Earth should be seen as a non-progressive evolutionary continuum that seamlessly joins the prebiotic synthesis and accumulation of organic molecules in the primitive environment, with the emergence of self-sustaining, replicative chemical systems capable of undergoing Darwinian evolution.

It is of course impossible to demonstrate that this is the evolutionary pathway that led to the appearance of the first organisms, since the details of how it happened have been lost to time. However, the available evidence from widely different scientific fields is consistent with the possibility that it could have happened this way. Moreover, the recognition that life is the outcome of an evolutionary process can lead to the acceptance that the properties associated with living systems, such as replication, self-assemblage or catalysis are also found in non-living entities.

The spectacular developments in our understanding of the molecular basis that underlies biological phenomena have not led to a generally agreed definition of life, but this does not imply that biology is epistemologically weaker or less mature than others fields like physics or mathematics. As Mayr wrote some time ago, "biology is, like physics and chemistry, a science. But biology is not a science like physics and chemistry; it is rather an autonomous science on a par with the equally autonomous physical sciences." (Mayr, 1997, p. 32). One of the main differences between biology and these other disciplines is, of course, its historical nature. Although natural selection is consistent with physical laws, it cannot be deduced from them, and to understand the nature of life we must recognize both the limits imposed by physics and chemistry, as well as the role of history's contingency.

Life sciences are going through one of the most important and contested periods of their history, and every book, every article, every publication on cell biology, biochemistry and molecular genetics can be read as a secular paean to life phenomena. To complete this extraordinary accomplishment, non-teleological, non-progressive explanations are required to refine our understanding of the properties of the subcellular constituents of life and their interactions. Attempts to find a timeless definition of life may be a useless endeavor bound to fail but, so far, the intellectual drive to define it has been more intellectually enriching than the bleak results we have achieved so far.

Acknowledgements

I am indebted to Dr. Andreas Losch for his warm hospitality, intellectual support and extraordinary patience. This chapter is based in part on previously published analyses (Lazcano, 2008, 2010, 2012). Support from project UNAM-PAPIIT IN223916 is gratefully acknowledged. I thank Sara Islas and Ricardo Hernandez-Morales for help with this manuscript.

References

Bada, J. L. & Lazcano, A. (2003). Prebiotic soup: revisiting the Miller experiment. *Science*, **300**, 745–6.
Bohr, N. (1933). Light and life. *Nature*, **133**, 457–9.
Bowler, P. J. (1988). *The Non-Darwinian Revolution: Reinterpreting a Historical Myth*, Baltimore, MD: The Johns Hopkins University Press.
Budin, I. & Szostak, J. W. (2010). Expanding roles for diverse physical phenomena during the origin of life. *Annual Review of Biophysics*, **39**, 245–63.
Burke, J. G. (1991). *Cosmic Debris: Meteorites in History*, Berkeley, CA: California University Press.
Campos, L. A. (2015). *Radium and the Secret of Life*, Chicago, IL: Chicago University Press.
Carlson, E. A. (1981). *Genes, Radiation, and Society: The Life and Work of H J Muller*, Ithaca, NY: Cornell University Press.
Darwin, C. (1863). Doctrine of heterogeny and modification of species. *Athenaeum*, Apr. 25, 554–5.
Darwin, C., Barrett, P. H., Gautrey, P. J. et al. (1987). *Charles Darwin's Notebooks. 1836–1844*, London: British Museum (Natural History).
Darwin, F. (1887). *The Life and Letters of Charles Darwin*. Vol. III, London: J. Murray.
Falk, R. & Lazcano, A. (2012). The forgotten dispute: A. I. Oparin and H. J. Muller on the origin of life. *History and Philosophy of the Life Sciences*, **34**, 373–90.
Fara, P. (2002). *An Entertainment for Angels: Electricity in the Enlightenment*, New York, NY: Columbia University Press.
Farley, J. (1977). *The Spontaneous Generation Controversy from Descartes to Oparin*, Baltimore, MD: Johns Hopkins University Press.
Fox-Keller, E. (2002). *Making Sense of Life: Explaining Biological Development with Models, Metaphors, and Machines*, Cambridge, MA: Harvard University Press.
Friedmann, H. C. (1997). From Friedrich Wöhler's urine to Eduard Buchner's alcohol. In A. Cornish-Bowden, ed., *New Beer in an Old Bottle: Eduard Buchner and the Growth of Biochemical Knowledge*, Valencia: Universitat de Valencia, pp. 67–122.
Fry, I. (2006). The origins of research into the origins of life. Endeavour, 30, 24–8.
Garcia-Ruiz, J. M., Hyde, S. T., Carnerup, A. M. *et al.* (2003). Self-assembled silica-carbonate structures and detection of ancient microfossils. *Science*, **302**, 1194–7.

de Goncourt, E. & de Goncourt, J. (1904). *Journal des Goncourt–Mémoires de la vie littéraire ... : 1862–1865*, Paris: E. Fasquelle.

Graham, L. R. (1972). *Science and Philosophy in the Soviet Union*, New York, NY: Alfred A. Knopf.

Haeckel, E. (1862). *Die Radiolarien (Rhizopoda radiaria)*, Berlin: Verlag von Georg Reimer

Haeckel, E. (1876). *The History of Creation: or the Development of the Earth and its Inhabitants by the Action of Natural Causes*, trans. E. R. Lankester, London: Henry S. King Co.

Herrera, A. L. (1942). A new theory of the origins and nature of life. *Science*, **96**, 14.

Huxley, J. (2010). *Evolution: The Modern Synthesis: The Definitive Edition*, Cambridge, MA: MIT Press.

Kamminga, H. (1988). Historical perspective: the problem of the origin of life in the context of developments in biology. *Origins of Life and Evolution of Biospheres*, **18**, 1–10.

Kaufmann, S. A. (1993). *The Origins of Order*, New York: Oxford University Press

Lamarck, J. B. (1914). *Zoological Philosophy*, trans. H. Elliot, London: Macmillan and Co.

Lazcano, A. (2003). Hooke and generation of molds. *Science*, **301**, 1845.

Lazcano, A. (2007). Prebiotic evolution and the origin of life: is a system-level understanding feasible? In: I. Rigoutsos and G. Stephanopoulos, eds., *Systems Biology*, New York, NY: Oxford University Press, pp. 57–78.

Lazcano, A. (2008). What is life? A brief historical overview. *Chemistry and Biodiversity*, **5**, 1–15.

Lazcano, A. (2010). Historical development of origins of life. In D. W. Deamer and J. Szostak, eds., *Cold Spring Harbor Perspectives in Biology: The Origins of Life*, Cold Spring Harbor, NY: Cold Spring Harbor Press, pp. 1–16.

Lazcano, A. (2014). The RNA World: stepping out of the shadows. In G. Trueba, ed., *Why Does Evolution Matter? The Importance of Understanding Evolution*, Newcastle upon Tyne: Cambridge Scholars Publishing, pp. 101–19.

Lehman, I. R., Zimmerman, S.B., Adler, J. *et al.* (1958). Enzymatic synthesis of deoxyribonucleotic acid. V. Chemical composition of enzymatically synthesized deoxyribonucleic acid. *Proceedings of the National Academy of Sciences of the USA*, **44**: 1191–6.

Lehn, J-M. (2002). Toward self-organization and complex matter. *Science*, **295**, 2400–3.

Leicester, H. M. (1974). *Development of Biochemical Concepts from Ancient to Modern Times*, Cambridge, MA: Harvard University Press.

Lewontin, R. C. & Levin, B. (2007). *Biology Under the Influence: Dialectical Essays on Ecology, Agriculture, and Health*, New York, NY: Monthly Review Press.

Mann, T. (1947). *Doktor Faustus*, trans. H. T. Lowe-Porter, New York, NY: Alfred A. Knopf.

Mayr, E. (1997). *This is Biology*, Cambridge, MA: Harvard University Press.

McKay, D. S., Gibson, E. K., Jr., Thomas-Keprta, K. L. *et al.* (1996). Search for past life on Mars: possible relic biogenic activity in martian meteorite ALH84001. *Science*, **273**, 924–30.

Miller, S. L. (1953). A production of amino acids under possible primitive Earth conditions. *Science*, **117**, 528.

Montillo, R. (2013). *The Lady and her Monsters: a Tale of Dissections, Real-Life Dr. Frankensteins, and the Creation of Mary Shelley's Masterpiece*, New York, NY: William Morrow.

Morange, M. (2012). The recent evolution of the question "What is Life?". *History and Philosophy of Life Sciences*, **34**, 425–38.

Muller, H. J. (1926). The gene as the basis of life. *Proceedings of the 1st International Congress of Plant Sciences*, Ithaca, pp. 897–921.
Muller, H. J. (1960). In S. Tax and C. Callender, eds., *Evolution after Darwin: The University of Chicago Centennial Discussions, Panel One. The Origin of Life*, Chicago, IL: The University of Chicago Press, 69–105.
Muller, H. J. (1961). Genetic nucleic acid: Key material in the origin of life. *Perspectives in Biology and Medicine*, **5**, 1–23.
Muller, H. J. (1966). The gene material as the initiator and the organizing basis of life. *American Naturalist*, **100**, 493–502.
Nagy, B. (1975). *Carbonaceous Meteorites*, Amsterdam: Elsevier Scientific Publishing Co.
Oken, L. (1805). *Die Zeugung (The Creation)*, Bamberg-Würzburg: Joseph Anton Goebhardt.
Oparin, A. I. (1924). *Proiskhozhedenie Zhizni* (Mosckovskii Rabochii, Moscow), reprinted and translated in J. D. Bernal (1967) *The Origin of Life*, London: Weidenfeld and Nicolson.
Oparin, A. I. (1938). *The Origin of Life*, New York, NY: McMillan.
Oparin, A. I. (1961). *Life: Its Nature, Origin, and Development*, New York: Academic Press.
Oparin, A. I. (1972). The appearance of life in the Universe, in C. Ponnamperuma, ed., *Exobiology*, Amsterdam: North-Holland, pp. 1–15.
Orgel, L. E. (2008). The implausibility of metabolic cycles in the primitive Earth. *PLoS Biology*, **6**, 18.
Peretó, J. & Catalá, J. (2007). The renaissance of synthetic biology. *Biological Theory*, **2**, 128–30.
Peretó, J., Bada, J. L., & Lazcano, A. (2009). Charles Darwin and the origins of life. *Origins of Life and Evolution of Biospheres*, **39**, 395–406.
Perezgasga, L., Silva, E., Lazcano, A., & Negrón-Mendoza, A. (2003). Herrera's sulfocyanic theory on the origin of life: a critical reappraisal. *International Journal of Astrobiology*, **2**, 1–6.
Saladino, R., Botta, G., Bizzarri, B. M., Di Mauro, E., & García-Ruiz, J. M. (2016). A global scale scenario for prebiotic chemistry: silica based self-assembled mineral structures and formamide. *Biochemistry*, **55**, 2806–11.
Sankaran, N. (2012). How the discovery of ribozymes cast RNA in the roles of both chicken and egg in origin-of-life theories. *Studies in History and Philosophy of Biological and Biomedical Sciences*, **43**, 741–50.
Soyfer, V. N. (2001). The consequences of political dictatorship for Russian science. *Nature Reviews Genetics*, **2**, 723–9.
Steinbock, O., Cartwright, J. H. E., & Barge, L. M. (2016). The fertile physics of chemical gardens. *Physics Today*, **69**(3), 44–51.
Tagliagambe, S. (1978). *Science, Philosophy, and Politics in the Soviet Union 1924–1939*, Milano: Feltrinelli Editore (in Italian).
Urey, H. C. (1952). *The Planets: their Origin and Development*, Chicago, IL: University of Chicago Press.
Watson, J. D. & Crick, F. H. C. (1953). A structure for deoxyribose nucleic acid. *Nature*, **171**, 737–8.
Wisniak, J. (2000). Jöns Jacob Berzelius a guide to the perplexed chemist. *Chemistry Educator*, **5**, 343–50.

Philosophy

5

Science and Philosophy Faced with the Question of Life in the Twentyfirst Century

MICHEL MORANGE

My intention in this contribution is not to provide an overview of the different philosophical conceptions of life that have been proposed since people first started thinking about the issue. My aim is different, and more focused. I will consider the recent transformations of biological knowledge, the changes in theories, models, experimental systems, representations, and show how they restate major philosophical issues on the "nature of life", by which I simply mean a characterization of the differences between organisms and non-living objects (differences shared by all organisms).

The question "What is life?" has not always been present in the writings of biologists. In 1970, François Jacob noted that "Biologists no longer study life today. They no longer attempt to define it" (Jacob, 1973, p. 299). Today, the question is again being asked, more explicitly. Even when this has not been the case, the nature of projects and research, in particular concerning the origin of life and in astrobiology, has yielded implicit answers (Morange, 2008). My objective is to contribute to a dialogue between biologists and philosophers at the beginning of the twentyfirst century.

To reach this objective, I have to overcome two obstacles, one stemming from philosophers and the other from scientists. The dialogue will only be possible if philosophers admit that biology produces significant statements on what life is, and what characterizes living beings. This has not always been the case. In addition to a long tradition in philosophy of reducing the value of scientific knowledge to that of a tool that efficiently predicts and controls natural phenomena, biologists have been accused by phenomenologists of studying organisms, and human beings in particular, from the wrong angle, ignoring the central role of consciousness. Biology and science in general do not say what life is, but no philosophical discourse can ignore the arguments of scientists, and more precisely the description of recent transformations in biological knowledge.

The second obstacle is diametrically opposed. Biologists and scientists have to admit that the question of life is an important one, in general, but also for the development of their own scientific work. A short editorial published in *Nature* in 2007 (Editorial, 2007) will help me to outline the reluctance of many biologists to address this issue and the reasons behind this reluctance, but also the weaknesses of these reasons and the confusion that they reveal. According to the anonymous editorialist, discussions on these issues have always been used by philosophers to oppose scientific progress. This was the case for synthetic biology and research on human embryos. The editorialist considers that life is not a precise scientific concept, and therefore not a useful one. There is no qualitative difference between inanimate objects and organisms, no "threshold" to allow passage between the first and the second. Saying otherwise is tantamount to a resurgence of vitalism. Asking the question of life is in the best case useless, and in the worst case an obstacle to scientific progress.

There is a confusion in the previous statements that originates in the absence of a distinction between life and human life: an embryo is obviously alive; the issue is whether he or she experiences human life. The meanings of the adjective "qualitative" and of the noun "threshold" are also imprecise. No present-day biologist would consider that there is a difference between the nature of organisms and the nature of inert objects, and that the former harbour a "principle" that explains their properties. But this does not mean that there are no important differences that can be precisely described. And this description is particularly important for scientists working on the origin of life and in astrobiology.

First, I will consider how the emergence of molecular biology dramatically transformed the question "What is life?" and yielded answers to questions raised by nineteenth-century biologists. Second, I will describe what has happened in the past 50 years since molecular biology acquired its dominant position among the biological disciplines. Finally, I will address some of the philosophical issues that have been renewed or displaced by these recent transformations.

From Nineteenth-Century Biology to Molecular Biology

Thanks to the work of Georges Canguilhem (Delaporte, 1994) and other historians of biology, the sterile denunciation of vitalism has been replaced by a rich description of the different hypotheses biologists elaborated in the eighteenth and nineteenth centuries to account for the specificity of organisms (Normandin & Wolfe, 2013).

The vision of organisms as "systems" originated in Kant's *The Critique of Judgement* (Kant, 1790) and is found in the writings of Georges Cuvier (Taquet, 2006). Particularly revealing is the caricatural contrast that is often drawn between the attitudes of two eminent French biologists, Louis Pasteur and Claude Bernard. The former is considered a vitalist, having separated the chemical reactions occurring in living cells from those performed by the chemist, whereas Claude Bernard is supposed to be a strong opponent of vitalism and a supporter of a physical-chemical approach to the phenomena of life. The attitudes of these two scientists were far more complex and ambiguous than that. Pasteur considered that some reactions required the presence of living organisms to be accomplished properly, because the internal organization of these organisms oriented the chemical reactions that occurred within them. By using different physical forces (such as magnetism), Pasteur tried to orient *in vitro* chemical reactions in the same direction as those occurring within the organism, by trying to mimic the chemical specificity of life through a well-designed physical-chemical environment.

Claude Bernard rejected vitalism because it was a metaphysical option, believing that metaphysics had no place in science. But he also rejected materialism and was convinced that a chemist cannot reproduce the chemical reactions occurring in organisms because he or she will forever be unable to synthesize the tools used by organisms to perform these reactions (Bernard, 1878).

Both Pasteur and Claude Bernard considered that the specificity of life originated in the organization of the cellular protoplasm, and that this organization had been faithfully transmitted through generations since the origin of life.

Another characteristic of life acquired importance in the second half of the nineteenth century, after the publication of *On the Origin of Species* by Darwin had convinced biologists of the reality of evolution. Whereas the entire universe obeys the Second Law of Thermodynamics and is moving towards increased disorder (and increased entropy), organisms have followed an opposite path during evolution, increasing their complexity and therefore their order, and decreasing their entropy. In summary, for some scientists of the nineteenth century, the question "What is life?" required an explanation of the thermodynamic paradox of organisms, and for all biologists a description of their internal organization and of the way it is faithfully transmitted from one generation to the next.

Molecular biology seemed to provide the explanations and descriptions that were awaited. The explanations of the specific characteristics of organisms were physical-chemical, and for the founders of molecular biology, like Francis Crick and Jacques Monod, this dealt a fatal blow to the vitalistic, metaphysical and teleological concepts that had flourished in biology up to the beginning of the twentieth century (Morange, 2015). Many of the founders of molecular biology turned to biology in the hope of ridding it of metaphysics. But their expectations of this attempt to explain biology naturalistically differed. Some, like Max Delbrück, the leader of the American Phage Group, which was so influential in the rise of molecular biology, foresaw the discovery of new principles and new laws specific to organisms, whereas others only expected metaphysics to disappear in the face of continuous advances in the determination of the molecular and macromolecular structures present in organisms.

The thermodynamic paradox still occupied an important place in Erwin Schrödinger's famous book *What is Life?* (Schrödinger, 1944), but the solution he proposed was not considered satisfactory. The solution finally came when it was realized that the Second Law applies to closed systems like the universe as a whole, but not to open systems such as organisms, which permanently exchange matter and energy with their environment.

The founders of molecular biology considered that the issues raised by nineteenth-century biologists and scientists had been answered by the major results of the new discipline: the structure of macromolecules, their organization and interactions, how the DNA molecule transmits through generations the information necessary to synthesize RNAs and proteins, with a precise description of the mechanisms by which DNA is decoded to generate the synthesis of these macromolecules (in particular proteins) in the different cells of the organism. The founders of molecular biology could reasonably consider that they had unveiled the "secret of life" (Monod, 1971; Judson, 1979).

But the attitudes of molecular biologists were more complex than they may appear. Some of them, such as the French biologist Ernest Kahane, emphasized the significance of the disappearance of the vital principle. He entitled one of his books *Life Does Not Exist*, meaning that the characteristics of organisms and those of inanimate objects could be studied by the same physical-chemical approaches (Kahane, 1962). But other molecular biologists preferred to insist on the specific characteristics of organisms due to the possession of genetic information and of the molecular machinery to decode it. For them, the origin of life coincided with the natural formation of genetic information. The creation in 1960 of the new discipline of exobiology (now known as astrobiology) by Joshua Lederberg, one of the major contributors to the rise of molecular biology in the 1940s and 1950s, was the direct consequence of these discoveries: the search for life on other planets could now be reduced to the search for objects containing informational macromolecules (Lederberg, 1960).

The Previous Answers of Molecular Biology were Provisional

These dominant conceptions of life were provisional. In the past 50 years, a significant transformation has occurred in the answers to the question "What is life?". One of the motors of this transformation was the renewed importance of some disciplines such as biochemistry and ecology, which had been pushed aside by the rapid development of molecular biology. But the most significant impetus originated in the development of molecular biology itself. The informational conception of molecular biology was pushed to its limits, revealing its weaknesses, and generating slightly different answers to the question "What is life?". It is this complex history that I will briefly describe, by gathering these transformations under different headlines.

Fading of the Informational Vision

The emphasis on the existence of genetic information within organisms progressively vanished. The first reason for this is that it became mundane, with school-girls and school-boys being taught about the existence of DNA and of a genetic code very early in their studies. This does not mean that the existence of genetic information in organisms is not important; it simply means that it has lost part of its novelty and attractiveness, opening the door to new concepts.

More significant for the fading of the informational vision was the success of the first genome sequencing programmes, in particular human genome sequencing. So much had been expected from this reading of the "book of life" that the publication of the sequence and the paucity of new answers created immense disappointment. The genetic information of humans had been fully deciphered, and it said nothing! It did not open the new horizons that so many biologists had foretold. This led to the conclusion that knowledge of genetic information was not knowledge of life. Did this mean that life is something other than information?

Before this final blow, the informational vision had already staggered after the discovery at the end of the 1970s that RNAs are able to catalyze chemical reactions as proteins do. This rapidly led to the adoption of a hypothesis that had already been proposed in the 1960s, but now found strong experimental support: the first organisms probably did not possess DNA, RNA and proteins as extant organisms do, but only one macromolecule, RNA, able to catalyze the chemical reactions of metabolism and to self-replicate.

The hypothesis of a primitive RNA world is now widely accepted – although the chemical instability of RNA is a serious issue that has not yet been resolved. This has opened the door to more complex scenarios for the origin of life, in which the RNA world itself would have been preceded by other living worlds based on different types of macromolecules, or even on other types of information-bearing structures. The existence of an RNA world solved a major difficulty in explaining the birth of the present living world: how is it possible to account for the simultaneous appearance of two highly different types of macromolecules, nucleic acids and proteins, and for a precise relation (the genetic code) between them?

But it also undermined the identification of life with genetic information. In an RNA world, RNA molecules are not bearers of information in the same sense as DNA: they do not encode the structure of other macromolecules.

More recently, the increasing number of studies on epigenetic marks (chemical modifications of DNA and of the proteins [histones] surrounding it), on their role in the control of gene expression and on their possible transmission through generations have struck an additional blow against DNA- and information-centred conceptions of organisms. Although epigenetics would not exist without genetics, and its contribution to hereditary phenomena remains to be confirmed, it has helped to balance the picture. If (genetic) information is no longer the main characteristic of organisms, then other characteristics, such as an integrated system of chemical reactions, have to be put forward to explain life.

Interestingly, but not so surprisingly, these difficulties have not prevented evolutionary biologists like Richard Dawkins from comparing the tree of life to a river of genetic information (Dawkins, 1996). Until recent years, most evolutionary biologists used an abstract conception of genes and were uninterested in precise knowledge of their mechanisms of action. This dematerialization of information has been criticized by some philosophers, who consider that the emphasis on information was a return to the conceptions of vitalists, to the belief in the existence of a vital principle (Oyama, 2009).

The Search for Minimal Genomes

For Craig Venter and his collaborators, the search for minimal genomes was the first step to producing a bacterial chassis on which they would add functional modules in order to transform these microorganisms into useful tools for various tasks: depollution, synthesis of molecules for the chemical and pharmaceutical industries. But Venter also considered this work as a way to answer the question of life (Cho *et al.*, 1999). What is the minimum number of genes an organism needs?

This approach yielded two major results. The first was the pinpointing of the importance of metabolism, this ensemble of chemical reactions catalyzed by enzymes that permits the synthesis of the different chemical components present in organisms. In simple organisms, the proportion of genes encoding the enzymes involved in these metabolic reactions is high. This result supported the definition of life formulated by the researchers in artificial life, Humberto Maturana and Francisco Varela: organisms are autopoietic systems able to regenerate their own constituents (Varela *et al.*, 1974). Metabolism regained the place it had had in the first scenarios of the origin of life proposed by Alexander Oparin in the 1920s, which it had lost with the advent of the informational vision of life proposed by molecular biologists. Today, information and metabolism are often considered the two pillars of life.

The study of the minimal genome had a second, unexpected consequence. It was rapidly realized that organisms with minimal genomes were parasites: they had lost some genes involved in the synthesis of the basic components of life and borrowed these basic components from their hosts. Besides pointing out once again the importance of metabolism, this also demonstrated that an organism cannot be defined without describing the environment in which it lives and the other organisms with which it interacts. Life is a property shared by a community of organisms.

This ecological vision of life was supported by new tools that fully characterize in a single step the different organisms (in general microorganisms) living in the same place, in soil or an ocean, or in the gut of a host organism. Metagenomics – the parallel sequencing of all the genomes present in a sample – reveals the existence of these communities of organisms, even though it does not demonstrate the necessity of this association for the survival of the individual organisms. For this reason, as rightly underlined by John Dupré and Maureen O'Malley, this new discipline raises important ontological issues regarding the nature of life (Dupré & O'Malley, 2007).

Paradoxically, the results obtained undermined the initial project. There is not one unique minimal genome, but there are probably many different ones, and their description tells us more about the importance of the environment in which organisms live than it does about the fundamental mechanisms required by these organisms for life.

The Return of the Question: Are Viruses Alive?

The question of the living (or not) nature of viruses was widely discussed after their discovery at the end of the nineteenth century. The evidence progressively accumulated in the 1920s and 1930s that viruses are strict parasites requiring living cells for their replication threw doubt on their living nature. The molecular characterization of viruses between the 1930s and the 1950s, and the subsequent demonstration that they did not contain the molecular machinery necessary for the translation of their genetic information, sounded the death-knell of their living nature.

This issue, which might have been considered as settled, re-emerged some years ago following the convergence of different observations. This renewed interest in viruses is linked to the question of their origin. Are they the remnants of preexisting organisms, the last steps in a regressive process? Or parts of the genetic information of organisms that have escaped from them, and become independent? Whatever the answer, the evidence that most viral gene sequences are specific to the viral world and not present in extant organisms might suggest that viruses are the remnants of a living world that preceded the present one, an RNA world or another type of primitive living world. I do not have enough space here to discuss these issues, but I will mention three other types of observations that nourish the debate about the living nature of viruses (Raoult & Forterre, 2008). The first observation is the abundance and unexpected diversity of viruses (and in particular of bacteriophages, the viruses of bacteria) brought to light by high-throughput sequencing techniques. The second observation is an ensemble of findings suggesting that viruses might have been important for the exchange of genetic information in the living world: circulation between organisms, but also between viruses and organisms. The transition from an RNA world to a DNA world might have resulted from the transfer to primitive cells of genetic information permitting the synthesis of DNA and its replication, a capacity that had previously been limited to viruses. The third observation is that of giant viruses, giant in terms of their size and the amount of information that they contain. They blur the size boundary that from the beginning had pushed scientists to consider viruses as different from other organisms. More importantly, the genes of these giant viruses encode information for the synthesis of enzymes and other cellular constituents that are of no use to a strict parasite. Therefore, the functional border between viruses and bacteria (or Archaea) is also blurred by the discovery and characterization of these giant viruses.

Personally, I am not convinced that the above observations necessitate the reopening of the question of the living nature of viruses. But they clearly show that the border between the living and the non-living is more porous than expected. And, in a way similar to that of the previous observations on minimal life, they emphasize the importance of a community of bacteria and viruses that – alive or not – exchange material and information.

The Projects of Synthetic Biologists

Obtaining a minimal genome is one of the projects of synthetic biologists (Porcar & Pereto, 2014). More ambitious projects are to synthesize an artificial organism by extensively modifying existing ones, or through a bottom-up approach by progressively complexifying simple metabolic systems and/or systems of self-replicating macromolecules (RNAs) to reach the level beyond which these systems might be considered living. Synthetic biology is the legacy of the genetic engineering that developed in the 1970s. What distinguishes its practitioners is the "engineering spirit" that guides their work in synthetic biology and their stated ambitions. These ambitions are the result of the technological progress that has been made during the past 40 years, and even more the result of the accumulation during this period of data suggesting that not only had the principles of life been unveiled (as molecular biologists already thought in the 1960s), but also the details of the mechanisms involved, thus adding flesh to earlier abstract descriptions. Although ambitious, the objectives of synthetic biologists are considered by most biologists to be achievable in the more or less near future. What is interesting for our discussion is that the crossing of the threshold is not seen as the dramatic event that observers outside biology would expect (and fear). Going beyond the threshold separating non-life from life will be a multistep process. It means that synthetic biologists involved in this work will encounter objects of which it will be impossible to say whether they are alive or not. They will be transiently alive at certain phases of their development and not during other phases. The separation between life and non-life will not be a simple border, but a territory the geography of which is poorly defined. It will probably contain not one unique object, but a community of objects balancing between life and non-life, individually and collectively. While the debate on the living nature (or not) of viruses will probably be resolved, the difficulties generated by these future objects produced by synthetic biologists will be much harder to solve. What happened on Earth when life appeared probably created a similar situation, with objects to which it was impossible to attribute a living or non-living nature.

Life as a Contingent Process

Molecular biologists are Darwinians. One of the first and most famous experiments of this discipline was the demonstration by Salvador Luria and Max Delbrück that the emergence of bacteria resistant to bacteriophages was the result of the selection of preexisting random mutants, and not of an active adaptation of bacteria to the presence of bacteriophages (Luria & Delbrück, 1943). Early molecular biologists working on bacteria and bacteriophages widely used the creative power of natural selection to generate variants with new properties.

But most were not interested in the evolutionary history of the macromolecular devices that their studies brought to light. They considered them to be well-designed physical-chemical systems comparable to machines. The search for the way complex structures like flagella formed during evolution was not on their agenda.

This attitude has progressively changed for two different reasons. The first reason is probably the positive consequence of a phenomenon that is far less positive for biology in general: the rise of the hypothesis of so-called intelligent design in the evolutionary process, with the rejection of Darwinism, which is considered to be unable to explain the genesis of these perfectly designed devices. Many examples used by the supporters of the idea of intelligent design were borrowed from biochemistry and molecular biology (Behe, 2007): complex and efficient macromolecular machines for which there is no evolutionary scenario. Many biologists realized that, by neglecting the evolutionary history of these complex devices, they had abandoned ground that was rapidly being occupied by the opponents of Darwinian theory. The recent development of more efficient genome sequencing techniques and comparison of the sequences have provided material for the elaboration of evolutionary scenarios to account for the appearance and progressive complexification of these nanomachines. Such studies have flourished in recent years.

But there was a second, less visible reason. Most molecular biologists initially adopted a progressive vision of Darwinism that was present in the Modern Synthesis. Natural selection has optimized biological systems up to the point where precise knowledge of their formation is useless: they are the best possible systems and their structure is explained by the functions they fulfil and not by the way they progressively formed.

But this Panglossian view of natural selection was criticized by the heirs of the Modern Synthesis, the first being Stephen Jay Gould and Richard Lewontin in an influential article of 1979 (Gould & Lewontin, 1979). The characteristics of an organism are not the simple result of the action of natural selection, but also of the constraints linked to the progressive construction of this organism.

Biochemists and molecular biologists took time to realize that molecular and macromolecular structures are also the result of a mixture of natural selection and structural constraints. These constraints may be physical or historical, the consequence of the contingent history that has produced organisms as they are. One good example is that of the genetic code. There were many efforts after its decipherment to demonstrate that it was "optimal". These efforts were in vain. This does not mean that the genetic code has not been optimized during evolution – to minimize the consequences of mutations for instance – but its main characteristics, including the number of amino acids, are the result of contingent events. Its current characteristics are mainly the result of what evolutionists call "a frozen accident".

Many additional molecular characteristics are the result of a contingent history, such as the relative abundance of some structural motifs found in proteins. Life has explored only one small part of the chemical world compatible with it, which is an argument widely used by synthetic biologists to justify their plans to modify the genetic code or the number of amino acids present in proteins.

The role of contingency in evolution is an important issue actively debated by evolutionists. It is also likely to emerge rapidly as a central issue in work on the origin of life and in astrobiology. It seems evident to many astrobiologists that once favourable conditions were present – temperature, presence of liquid water and of organic molecules – life had to appear. But is this so? A radically different possibility would be that the emergence of organisms was a contingent phenomenon, not the result of a progressive complexification of chemical systems towards life. It is scientifically reasonable to look for a path towards life: if successful this would lead to an explanation of the emergence of life on Earth and justify the research projects of astrobiologists. But success is not guaranteed, and such a research programme might reveal that there is no path to life and generate scenarios including random encounters and interactions. This is a wonderful example of how trying to find natural explanations must not be confused with producing fairy tales. And while failure to find a path to life would not mean that there is no natural explanation for the origin of life, it would mean that this origin will always remain inaccessible to us because it will be impossible to argue in favour of one or other of the numerous contingent events that might have generated life.

Philosophical Consequences of the Recent Transformations of Biological Knowledge

Is the question of "what life is" non-scientific? As I have shown in the introduction, some scientists argue that organisms do not have a special nature but are a category constructed by humans far before any form of science existed, a category that has persisted despite the evidence that there is nothing special about organisms.

As I argued previously, nothing special does not mean an absence of significant differences deserving more thorough study by scientists. It is possible to consider what has happened since the first observations on organisms as the recurring confirmation that this class of objects has a combination of characteristics that are never found together in inanimate objects, a conclusion already reached by Alexander Oparin (Oparin, 1938).

The question of life has been permanently displaced (and not solved!) by the transformations of biological knowledge. Are the twentieth-century transformations such that they represent a new challenge for some philosophers, and a new material to study for others? It is always difficult, even impossible to appreciate the true significance of a recent transformation of knowledge. And it is common today to consider the history of disciplines as a succession of "revolutions", one after the other.

However, I would like to argue that the molecular revolution that occurred in biology in the last century was a truly exceptional event, an unprecedented understanding of the fundamental phenomena associated with life that until then had been lacking. The transformations that I describe in the second part of this chapter were either the consequences of this first wave of discoveries, their replicas, or the results of the difficult dovetailing of new knowledge with old, in particular with the models of evolutionary biology. I have no proof of the value of such a statement. Its only support is the stability of the main components of this "molecular paradigm" since their inception, and the very fast developments in the directed modification of organisms.

I will distinguish three different major philosophical issues raised by the recent transformation of biological knowledge. My objective is not to fully address them, but only to pinpoint their potential interest.

The first becomes visible when we try to appreciate the significance of the work accomplished in synthetic biology. When Craig Venter and his collaborators replaced a bacterial chromosome by a slightly modified copy chemically synthesized *in vitro*, they called it a "creation" (Gibson *et al.*, 2010). Is the term creation appropriate and is the new organism "artificial"? Without discussing the ontological meaning of the terms "creation", "natural" and "artificial", I will simply note that this experiment was done by human beings, organisms, using the knowledge that they had acquired by their study of natural organisms. Are the new organisms produced by synthetic biologists more the extension of the existing living world than the creation of a new one?

Therefore, the importance of this experiment would not reside in the "creation" of new forms of life, but rather in the path that it opens to progressive and indefinite modification of these organisms. The border between present natural and future artificial organisms would be as ill-defined as that between inanimate objects and organisms.

It leads to the second philosophical issue that I wanted to pinpoint. This abandonment of precise borders is probably the most important philosophical consequences of recent work in biology. If the distinction between living and non-living is unambiguous for extant organisms, it is the result of a long historical process, a progressive divergence as described by André Pichot (Pichot, 1991). The same will be true of the artificial organisms created in the future: their "artificiality" will only appear when they have become radically different from extant organisms.

The fact that they result from a long historical process might be also true of thought and consciousness. There is an ancient tradition in philosophy that associates these two characteristics with life: the signs of thought and consciousness are only visible in (not all) organisms. The hypothesis supported by some specialists in computer sciences and Artificial Intelligence that computers will soon become so powerful that they will start to "think" and to be conscious still belongs to the realm of dreams. Computers are everywhere in our daily life and robots are playing an increasing role. It is not obvious that humans place computers and robots above other machines and consider their intelligence to be more than a sign of the intelligence of the humans that have programmed them.

There is a stronger version of the previously described philosophical tradition. Since mind and consciousness belong to the world of organisms, these characteristics are closely associated with the living nature of organisms: even the simplest forms of life must exhibit some signs of intelligence and consciousness.

Biologists were far from being impermeable to these ideas. At the beginning of the twentieth century, some of them looked for signs of intelligence in simple monocellular organisms such as paramecia (Morange, 2006). The language used by biologists underlines this continuity between simple organisms and higher forms with developed cognitive abilities. Bacteria are said to be able to read their environment and to exchange signals with other bacteria. Even macromolecules are able to recognize other macromolecules and to receive signals. Many biologists have not hesitated to see in these molecular mechanisms the first signs of a cognitive process.

Such a comparison raises two difficulties. The first is that molecular recognition is not characteristic of the living world: any reaction between two chemical molecules requires some form of interaction (and recognition). The risk is to attribute affinities and projects to molecules, as Maupertuis did in the eighteenth century (Maupertuis, 1745). The second difficulty will emerge from the efforts made by synthetic biologists to cross the border between non-life and life by a bottom-up approach. If they succeed in the more or less distant future, and we have seen that most biologists think that they will, the preparation of these living entities will demonstrate that scientists, biologists, have acquired a high level of knowledge of the mechanisms at the origin of life, without having progressed in understanding the origin of mind and consciousness. Despite the tremendous recent developments in the cognitive sciences and the use of technologies that provide images of the "thinking brain", the hope of finding a simple explanation of consciousness (and eventually of locating it in the brain) that flourished some decades ago has vanished. The question of life will probably be considered to be solved without any progress being made in our understanding of the mechanisms generating thought and consciousness.

It raises the third philosophical issue that I wanted to emphasize. My hypothesis is that this uncoupling will demonstrate that there is not a strong link between life and consciousness and that it is absurd to look for the first signs of consciousness in simple monocellular organisms. This does not mean that the formation of the first living systems did not create a favourable ground for the development of thought and consciousness. But the latter result from a complex four-billion-year history, a history and not a process that would have automatically started in the first organisms that appeared on Earth.

Consciousness, like life, is the result of history. The possibility of consciousness arose, but was not contained, in the formation of the first organisms. We must allow the creative power of history its full scope.

Acknowledgement

The author is indebted to Dr. David Marsh for critical reading of the manuscript.

References

Behe, M. (2007). *The Edge of Evolution: The Search for the Limits of Darwinism*, New York: Free Press.
Bernard, C. (1878). *Leçons sur les phénomènes de la vie communs aux animaux et aux végétaux*, Paris: Vrin.

Cho, M. K., Magnus, D., Caplan, A. L., McGee, D. & The Ethics of Genomics Group (1999). Ethical considerations in synthesizing a minimal genome. *Science*, **286**, 2087–90.

Dawkins, R. (1996). *River Out of Eden*, New York: Basic Books.

Delaporte, F. (1994). *A Vital Rationalist: Selected Writings from Georges Canguilhem*, New York: Zone.

Dupré, J. & O'Malley, M. A. (2007). Metagenomics and biological ontology. *Stud. Hist. Phil. Biol. & Biomed. Sci.*, **38**, 834–46.

Editorial (2007) Meanings of 'life'. *Nature*, **447**, 1031–2.

Gibson, D. G. et al. (2010). Creation of a bacterial cell controlled by a chemically synthesized genome. *Science*, **329**, 52–6.

Gould, S. J. & Lewontin, R. (1979). The spandrels of San Marco and the Panglossian paradigm: a critique of the adaptationist programme. *Proc. R. Soc. London ser. B*, **205**, 581–98.

Jacob, F. (1973, 1993). *Logic of Life: A Theory of Heredity*, Princeton: Princeton University Press.

Judson, H. F. (1979, 1996). *The Eighth Day of Creation: Makers of the Revolution in Biology*, Cold Spring Harbor: Cold Spring Harbor Laboratory Press.

Kahane, E. (1962). *La vie n'existe pas*, Paris: Editions rationalistes.

Kant, I. (1790, 2002). *The Critique of Judgment*, Indianapolis: Hackett Publishing Company.

Lederberg, J. (1960). Exobiology: approach to life beyond the Earth. *Science*, **132**, 393–400.

Luria, S. E. & Delbrück, M. (1943). Mutations of bacteria from virus sensitivity to virus resistance. *Genetics*, **28**, 491–511.

Maupertuis, P. L. M. de (1745, 1980). *Vénus physique suivie de La lettre sur le progrès des sciences*, Paris: Aubier Montaigne.

Monod, J. (1971). *Chance and Necessity: An Essay on the Natural Philosophy of Modern Biology*, New York: Knopf.

Morange, M. (2006). Ciliates as models... of what? *J. Biosci.*, **31**, 27–30.

Morange, M. (2008). *Life Explained*, New Haven: Yale University Press.

Morange, M. (2015). Monod and the spirit of molecular biology. *C. R. Acad. Sci. Paris Biol.*, **338**, 380–4.

Normandin, S. & Wolfe, C. T. (2013). *Vitalism and the Scientific Image in Post-Enlightment Life Science, 1800–2010*, New York: Springer.

Oparin, A. I. (1938). *The Origin of Life*, New York: MacMillan.

Oyama, S. (2009). Compromising Positions: The Minding of Matter. In A. Barberousse, M. Morange & T. Pradeu, eds., *Mapping the Future of Biology: Evolving Concepts and Theories*, Dordrecht: Springer, pp. 27–45.

Pichot, A. (1991). *Histoire de la notion de vie*, Paris: Gallimard, coll. Tel.

Porcar, M. & Pereto, J. (2014). *Synthetic Biology: From iGEM to the Artificial Cell*, Dordrecht: Springer.

Raoult, D. & Forterre, P. (2008). Redefining viruses: lessons from mimivirus. *Nature Res. Microbiol.*, **6**, 315–9.

Schrödinger, E. (1944). *What is Life?*, Cambridge: Cambridge University Press.

Taquet, P. (2006). *Georges Cuvier: naissance d'un génie*, Paris: Odile Jacob.

Varela, F., Maturana, H. & Uribe, R. (1974). Autopoiesis: the organization of living systems, its characterization and a model. *Biosystems*, **5**, 187–96.

6

What is Life? And Why is the Question Still Open?

CLAUS BEISBART

Introduction

Is there life on some of the exoplanets that have recently been detected? And what is the origin of life? These questions are at the center of much research in biology and adjacent fields. But, obviously, the answers to these questions turn on how we reply to a more fundamental question: "What is life?". We can only detect life, and we can only gain insights about its origin, if we know to some extent what it is.

What then is life? This question has occupied mankind for ages. Many answers have been given, and philosophers and scientists, ranging from Aristotle to Ernst Schrödinger and Ernst Mayr, have devoted their attention to the question (see Bedau & Cleland, 2010, for an anthology of important works about the nature of life). As it happens, their answers differ quite a lot, and we face an embarrassment of riches. While some take it that life is characterized by phenomena such as nutrition, growth or reproduction (cf. Aristotle's account of life in *De anima*, 412a), others submit that life is subject to Darwinian evolution (see e.g. Joyce, 1994) or require that it have metabolism (e.g. Gánti, 2003, p. 103; see Sagan, 1970, for a taxonomy of accounts of life).

Why then is there so much disagreement about the question of what life is? Are we concerned with one of the age-old metaphysical questions that are unlikely ever to be answered by human beings? Or is the problem simply that we still lack empirical knowledge to answer the question definitely? Or is the question itself ambiguous such that the various answers reflect different understandings of the question?

This paper aims to contribute to our understanding of why the question is so difficult to answer and how we can make progress. The key assumption of this paper is that attempts to answer the question try to provide a definition of life. My strategy thus is to bring insights about definitions to bear upon the problem of telling what life is.

To this purpose, I first look more closely at the question of what life is. I formulate conditions that every answer to the question should obey. I also take the opportunity to introduce some useful terminology. My conditions on a definition of life can be fulfilled by a descriptive definition of "life", a type of definition that is often contrasted to stipulative definitions. Then comes a section that is devoted to descriptive definitions of "life". I argue that descriptive definition has recently been too much underrated. A very promising route to a descriptive definition is analysis, but, as I show, an analysis in terms of necessary and jointly sufficient conditions is unlikely to succeed in the case of life. This does not imply that the notion of life cannot be taught or understood. But we do have a problem if we want a principled, general account of what life is. This problem is presented as a theoretical dilemma at the end of the section. The following section is devoted to attempts to escape the dilemma. I discuss Cleland's recent quest for a theoretical identity statement concerning life and come to reject it. Instead, I propose Carnapian explication as a sensible midway between a descriptive definition of "life" and a more stipulative, but principled account of life. My proposal can also make sense of a plurality of definitions of life.

The relation of this paper to existing research literature is as follows: I can draw on some philosophical work about definitions. Unfortunately, the most recent important monograph about this topic that I have found – Robinson (1950) – is already quite old, but I find it still useful.[1] Additionally, I can draw on the overview by Gupta (2015). As far as life is concerned, there are obviously a lot of works about the question of what life is, but I need not review this literature here, since my main aim is not to give a new answer to the question. The past few years have seen some systematic reflections about the possibility of defining life (see e.g. the contributions to the special issue in *Synthese* **185**, edited by Mark A. Bedau). In particular, Cleland and Chyba (2002), Cleland and Chyba (2007) and Cleland (2012) have offered very interesting thoughts about the definability of life. I nevertheless disagree with them on a number of counts, and I will critically engage with their work and other relevant research in due course.

What is the Question?

Consider the question "What is life?". What exactly does the question ask for and what sort of answer can we expect? Naively and pre-theoretically, we may demand that the answer be a statement that fulfills at least the following conditions.[2]

– The statement is about life.
– It is essentially or necessarily true of life.
– It is complete, i.e., it implies everything that is essentially or necessarily true of life.
– The statement does not have redundant parts.
– The statement is illuminating.

The point of the *first* condition is obvious. As far as truth is concerned, the *second* condition is obvious too. The qualification "essentially or necessarily" is motivated by the fact that not every true statement about life really tells us what life is. For instance, the statement that life on our planet is more than one billion years old is true, but does not get us closer to what life really is. It is not part of the essence of life and only contingently true of it. To express the quest for knowing what life really is, we use the traditional notions of essence and necessity. These notions and their relationship are controversial (for a famous proposal about the relationship see Fine, 1994), but for a first shot they are fine, and we will presently turn back to the problem of what they really require. The *third* condition is meant to guarantee that we capture completely what life is. The *fourth* condition is a matter of efficiency: No part of the answer may be dropped. Thus, if some part of the answer follows from other parts of it, the fourth condition is violated. Finally, the *fifth* condition excludes definitions that are circular or otherwise not informative, e.g. if life is defined in terms of living beings or by using an expression that is synonymous to "life" (cf. Robinson, 1950, pp. 94–6).[3]

As noted, the second condition requires that the answer to our question hold essentially or necessarily true. But it is often denied that statements can hold essentially true (e.g. Robinson, 1950, pp. 154–5). Necessity is a controversial idea, too; Hume, for instance, suggested that necessity is in the eye of the beholder (Hume, 1739, I.3.xiv, p. 166). Fortunately, there is at least one variety of necessity that is fairly uncontroversial, which is conceptual necessity. A statement is necessarily true in this sense if it holds in virtue of the concepts that are used in it.

Let us thus talk about concepts. For our purposes, we may say that a concept is what a predicate, e.g. the expression "unicorn" or "is a unicorn", designates in a specific type of use.[4] The idea is, for example, that "unicorn" or "is a unicorn" stands for a certain concept, at least if it is used in a specific way. The qualification "used in a specific way" is needed since some expressions are ambiguous and stand for several concepts, depending on the context and, maybe, the intention of the speaker. Thus, a unique concept is only designated in a specific type of use of a predicate, where types of use are just individuated in such a way as to establish unique designation. Let us further say that each type of use gives rise to a meaning. If a predicate designates a concept in a specific type of use, then a so-called *nominal definition* is supposed to tell this very concept (Robinson, 1950, p. 16; Gupta, 2015, Sec. 1.1). A nominal definition thus specifies the meaning of a designator of a concept.[5] In this way, a nominal definition expresses what some sort of thing designated by a certain predicate is necessarily because the predicate is simply meant to designate this sort of thing. Nominal definitions are often contrasted with so-called real definitions (see Robinson, 1950, p. 16 and Ch. VI).

Now turn back to our question of what life is. If "life" designates a concept, then we can provide a nominal definition to answer the question with a statement that holds necessarily true of life. "Life" does indeed always designate a concept, but it is not always the same because the expression has several meanings. For instance, there is life as property that all living beings share, but also life that cannot be shared because it is the life of an individual living being and, thus, may be better thought of as a process. "Life" also has other meanings; it may e.g. stand for an aggregate of living beings, etc. (see e.g. the entry for "life", n., in the OED). It is clear though that the above-mentioned discussion about life is focused on the first meaning of life. In this paper, I thus restrict myself to the corresponding concept. I assume that this concept is also designated with the predicate "is a living being". There are good reasons to focus on this predicate because the "grammar" of a noun is often notoriously unclear, unless it has been specified what sort of concept it designates. Nevertheless, to stick with the conventions of the debate, I shall continue to speak of life too.

Philosophers distinguish between several sorts of nominal definitions. Following Gupta (2015, Sec. 1), we may in particular discriminate between stipulative and descriptive nominal definitions. While the former stipulate the meaning of a word, the latter report how it is used (in a particular way) as a matter of fact.[6] Now, we are here not interested in pure stipulation. Let us thus leave the stipulative definition aside.[7] Since there are no other types of nominal definitions that are attractive for our purposes, we end up with the proposal that the question of what life is asks for a descriptive definition of the word "life".

This proposal accords well with a certain conception of philosophy according to which philosophers have the task to obtain a priori knowledge, i.e., knowledge that is independent of experience. This knowledge then is claimed to coincide with analytic knowledge, i.e., knowledge of statements that hold true in terms of the words or concepts involved. The consequence is that philosophers investigate what our basic concepts, e.g. "life", mean, while science is aimed at obtaining synthetic, empirical knowledge about, for example, life. Such a division of labour was roughly intended by logical positivism (e.g. Neurath *et al.*, 1929).

At this point, an objection looms large. When we ask what life is, we want to get at the real thing and not just learn the meaning of the word "life" in a particular type of use. Concepts, it may be said, are invented by humans as representational devices. But we are not interested in representational devices when we ask the question of what life is. What we want to know is what life really is.

This type of critique has recently been voiced by Cleland and Chyba (2002), Cleland and Chyba (2007) and Cleland (2012). According to Cleland and Chyba (2007, p. 326), "the idea that one can answer the question "What is life?" by defining 'life' is mistaken, resting upon confusions about the nature of definition and its capacity to answer fundamental questions about natural categories [...]."

As becomes clear in their paper (Cleland & Chyba, 2007, pp. 328–9), the authors are talking here about nominal, in particular descriptive, definitions. To justify their claim, Cleland and Chyba (2002, pp. 389–90) and Cleland and Chyba (2007, p. 330) argue as follows: "life" most plausibly refers to a natural kind, as does "water", for example. Here, natural kinds are not arbitrary concepts that humans have cooked up, but rather "categories carved out by nature", as Cleland and Chyba (2007, p. 330) put it. We do not generally specify what the natural kind in question is by providing a descriptive definition. In the example of water, a descriptive definition will characterize water in terms of properties that are accessible to human observers: it is fluid, transparent, etc. But, really, water is H_2O, i.e., a chemical substance of a specific molecular structure. We will never get to this by pointing out what "water" means, or so the argument goes.[8]

But is the project of a descriptive definition of "life" in fact misconceived? As to the argument presented by Cleland and Chyba, we may first note that it is built upon two strong assumptions, namely that there are natural kinds and that "life" refers to one of them. I will critically assess the plausibility of the second assumption below. Furthermore, the argument presumes that a descriptive definition is not related to natural kinds. To object against this assumption, we may first claim that, as a matter of fact, some descriptive definitions pick a natural kind. Nothing that has been said about descriptive definitions prevents them from doing so. However, this objection does not get us very far because Cleland and Chyba can rightly reply that descriptive definitions will not in general pick natural kinds. There is, however, a more subtle objection to Cleland and Chyba. Suppose, for the sake of argument, that there is in fact a natural kind that corresponds to our concept of life. So we can characterize life in terms of this natural kind. But recall that we are after a statement about life that holds necessarily. So is it necessary that life is this natural kind? Might it not be a fluke that life is this natural kind? It is only not, if "life" was meant to refer to a natural kind from the very outset. But this is something that a descriptive definition of "life" should bring out.[9]

As to the general worry that a descriptive definition of "life" does not get us to the real thing, we may first note that this presupposes a pessimistic view of concepts. Even if all concepts are created by humans, it does not follow that they are arbitrary. On the contrary, many concepts have been developed on the basis of experience and understanding. It is a commonplace in post-Kuhnian philosophy of science that concepts shift during the history of science. Often, these shifts reflect crucial insights about a topic. It thus seems that well-entrenched concepts need not be arbitrary, but may rather incorporate substantial insights.

There is, further, the question of whether the quest for the real thing can be made sense of without any reference to concepts. Life is not a real thing in the sense of one material being located in space and time. Rather, life is instantiated many times, and thus something general. Human beings use the word "life" to group certain things together. Concepts can do something like this, so it is natural to say that we need to specify a concept, if we wish to answer the question of what life is.

There is also the following challenge: We are after a statement about life that holds necessarily. Descriptive definitions hold true necessarily in a rather uncontroversial manner. If somebody claims to answer the question of what life is without providing a descriptive definition, she has to explain how her answer can hold necessarily or essentially true about life.[10] I am not saying that this challenge cannot be met, and Cleland and Chyba have an interesting proposal in this regard (see below). But we should at least admit that they face a challenge.

Work for a Descriptive Definition of "Life": Analytic Definition

Analytic Descriptive Definitions

Let us thus try to provide a descriptive definition of "life". There are several ways to do so. Quite generally, we need to distinguish between the purpose and the method of a definition because the same purpose (e.g. to give a stipulative definition) may be realized in various ways (Robinson, 1950, p. 15). As regards descriptive definition, the most promising method is analysis (Robinson, 1950, pp. 96–8): The meaning of an expression is characterized by analyzing the thing that is designated by the expression. In our case, the thing is a concept, so we need to analyze this very concept. This is only possible if the concept is complex, i.e., if it consists of parts. Of course, we are not here talking of the parts of a material object; rather, the parts are frequently concepts themselves, which may be combined in one way or the other.

Descriptive definitions that proceed by the method of analysis are very useful because they give insight about the thing designated by a word (Robinson, 1950, p. 97). According to Robinson (1950, pp. 189–90), analytic descriptive definitions have often been taken as real definitions, i.e., as telling what something really is (see also Robinson, 1950, pp. 171–8).

Very often, the concept to be analyzed is specified in terms of necessary and jointly sufficient conditions, where the conditions are spelled out by using other concepts. A good example is the analytic definition of "vixen" as a female fox. Here, the designatum of "vixen" is analyzed using the concepts of being a fox and being female. They are glued together by using the idea of co-instantiation or the logical "and". Each condition is necessary because every vixen needs to be a fox and female. They are jointly sufficient, because every female fox is necessarily a vixen.[11]

In this paper, I will concentrate on analysis in terms of necessary and jointly sufficient conditions. It is arguable that every analysis can be cast in this form: Each analysis identifies the components that jointly constitute a thing. Now, each component can be stated in a necessary condition, and we have jointly sufficient conditions if the components together really constitute the thing. Conversely, if we cannot state necessary and jointly sufficient conditions on something, why should we claim to have analyzed it? If we cannot provide necessary conditions, why claim that we have identified components? If we cannot provide jointly sufficient conditions, why claim that we have identified components that really constitute the thing?

But for the purposes of this paper, I need not assume that every analysis can be given in terms of necessary and jointly sufficient conditions. My argument goes also through for possible other types of analyses if they fulfill the following conditions. First, they are relatively simple and thus provide a principled account of life. Second, there is strong evidence that such an analysis fails for life.[12]

Analytic Definitions of Life

In the history of science and philosophy, a lot of analytic definitions of life have been proposed.[13] The recent working definition of life given by the Exobiology Program of the NASA may be taken as an example: "*life* is a self-sustained chemical system capable of undergoing Darwinian evolution" (see e.g. Joyce, 1994, p. xi). Here, life is characterized by the two conditions of being a self-sustained chemical system and the capability of undergoing a certain type of evolution. The so-called metabolic definition characterizes life by the condition of metabolism, which, in turn, may be explained a bit further (see e.g. Sagan, 1970; Sagan, 2016). More generally, most, if not all conditions that have been proposed in definitions of life are about the following sorts of things.

– Functionings and abilities, i.e., things that living beings do or can do: e.g. reproduction or the ability to reproduce; growth or the ability to grow, etc.
– Material composition, i.e., the material parts of which living beings consist. Typically, the idea here is that life is based upon certain molecules that contain carbon.
– Structure, i.e., the way in which the material parts that constitute living beings (at some level, e.g. at the level of cells) are structured and interact with each other. A famous proposal in this respect advanced by e.g. Kant (1790, § 65) is that living beings are integrated wholes.[14]
– The laws to which life is subject. In this respect, Darwinian evolution is often mentioned.[15]

That many analytic definitions of life have been proposed is not per se a problem. There is no presumption that there is only one correct analytic definition of something. There may be several ways to analyze something into components. But then the definitions should be compatible in the following sense: Each definition of "life" determines a list of all and only those (actual or possible) things that instantiate life. This is the extension proposed for the concept of life. Two definitions of life are compatible if they pick the same extensions. But this condition is not fulfilled because the proposed definitions disagree on their extensions. There are two types of disagreement: (i) two definitions of life disagree with respect to a real item known to exist, i.e., a thing that is known to exist or to have existed in the actual world is classified as living by one definition, while not by the other; and (ii) two definitions of life disagree with respect to a thing whose existence is at least logically possible, but which is not known to exist or to have existed. Disagreements of the latter sort are thought to be problematic too because a definition of "life" is not only supposed to cover real living beings and because it seems appropriate to be open to the possibility that there are other living beings on other planets that markedly differ from those that live on Earth.

Since most proposed definitions of life are incompatible with each other in one or the other way, they cannot be accepted at the same time, and so we find objections against each definition. Typical criticism has it that a proposed definition is too inclusive or too exclusive. In the former case, it has things instantiate life that are not really living beings. For instance, according to Cleland and Chyba (2007, p. 327), at least some proposals that define life in terms of metabolism are too inclusive because they allow that fire or cars are living too. In the latter case, real or possible living beings are not covered by the definition, which often gives rise to the objection of chauvinism. For instance, a definition according to which life is carbon-based falls under C. Sagan's verdict of "carbon chauvinism" (Sagan, 1973, p. 46).

Such objections against proposed definitions of life presume that certain things are indeed uncontroversial examples of (non-)living beings. This condition is not always fulfilled. For instance, it is not clear whether so-called artificial life really instantiates life (see e.g. Lange, 1996; Sober, 2003; Pennock, 2012). Let us say that a concept is vague if, and only if, there are particulars called "borderline cases" for which it is not settled whether they instantiate the concept or not (Sorenson, 2016). Our notion of life, or what is meant by "life", is certainly vague in this sense, but this does not necessarily prevent descriptive definition. For we may be able to analyze our concept of life into other concepts that, due to their own vagueness, properly represent the vagueness of the former notion. Suppose, for instance, that we define life in terms of metabolism. If the notion of metabolism is vague too (see Boden, 1999, for a related discussion) and if the definition can explain the vagueness of the concept of life by using the vagueness of the notion of metabolism, then the definition is fine despite of the vagueness.[16] Alternatively, we may allow that the analysis given by a putative definition of "life" only holds up to the vagueness inherent in the notion, which is to say that we neglect how the definition deals with borderline cases (Cleland & Chyba, 2007, p. 330).

The problems with the definitions of "life" do not (only) turn on borderline cases though (Cleland & Chyba, 2007). This is clear from the examples of definitions given above and the respective criticisms. For instance, the problem with a metabolic definition of "life" is simply its implication that some entities that are clearly not living beings are living creatures. In the following, I will assume that all proposed analytic definitions of "life" fail to account for some clear examples of living or non-living beings. This assumption explains the ongoing controversy about the question of what life is.

Now this does not imply that no principled descriptive definition of "life" can be provided. There are other methods to obtain (descriptive) definitions; in particular, a synthetic definition of "life" may be given. Such a definition specifies the thing a word designates by relating the thing to other things (see Robinson, 1950, pp. 98–106). But synthetic definitions do not really help here. First, they need not specify the meaning of a linguistic expression (Robinson, 1950, p. 102). They may thus be subject to counterexamples mentioning possible objects that are misclassified according to the definition. Second, as far as I can see, no uncontroversial proposal of a synthetic definition of "life" has been made thus far. More generally, my argument in this section does not depend on the assumption that a descriptive definition of "life" needs to provide an analysis. The plurality of proposed definitions and the related objections against the proposals cast doubt on the idea that we can obtain any descriptive definition of "life" that is given in terms of necessary and jointly sufficient conditions or principled in some other way.

Beyond Analytic Definition? A Dilemma

Now, if we cannot provide an analytic descriptive definition of "life" in terms of necessary and jointly sufficient conditions or some other sort of principled definition of "life" – and I will assume this in what follows – this is not too much of a scandal. It does show that we cannot analyze our concept of life into other concepts. But there are other predicates too that do not have an analytic definition. For instance, some words, e.g. "ultramarine", stand for some basic sort of experience. In the terms of John Locke, the concept of ultramarine counts as a simple idea (Locke, 1690, II.2.i), and David Hume (1739, I.1.i) would call it a simple impression. The best, maybe the only, way to introduce a word that designates this particular shade of blue is to point to objects that have this color – you had rather see it. This is an example of the method of ostensive definition (see Robinson, 1950, pp. 117–26). Now, "life" certainly does not stand for a basic experience because the perception of living beings is complex and can be analyzed in terms of more basic experiences. More generally, the concept of life, and thus, in a sense, life, is not simple in the sense of having no components of any sort. In some sense, metabolism may be counted as a component of life, because many, if not all, living beings have a metabolism. What the failure of analysis shows is only that there is no way of breaking down life in its components such that we can say: This combination of components *is* life.

But there are other words that do not have an analytic or another sort of principled definition and that function in a more similar way as "life" does. For instance, Wittgenstein famously suggested that the things called "games" do not share a common core of features (Wittgenstein, 1953, § 66). Rather, there are family resemblances, i.e., each two games share some relevant similarities, but these need not be the same for all pairs of games. Now as we have seen, so far, there is no unproblematic proposal for necessary and sufficient conditions on life. So, maybe, the expression "life" works as does "game": It picks items that share family resemblances. This suggestion is made by Pennock (2012, p. 17) who proposes to say that there are several life forms that are connected to each other in the tree of life (the suggestion is also discussed by Bedau, 2010, p. 1012; Shields, 2012, pp. 109–10). According to a similar proposal, "life" designates a homeostatic property cluster, i.e., a bundle of properties that are frequently co-instantiated due to certain causal processes (see e.g. Diéguez, 2013, especially p. 184; who follows Boyd, 1999, see Boyd's pp. 143–4 for his definition of homeostatic property clusters).

This is not the place to discuss such proposals in detail. What is important for our investigation is only this: It is plausible that the word "life" does not allow for an analytic definition because it refers to individuals that are only related by family resemblances or because it denotes a homeostatic property cluster. It may then still be characterized by using some sort of descriptive definition. This is as it should be, because we can tell other people what we mean by life, and we do understand each other when we talk about life. But if the concept of life does not allow for analysis for the reason given, then there is no common essence or nature to all living beings, and a definition of "life" will not really answer our question in a principled way. The definition does not give any general account of life.

One reaction to this problem is to give up the project of a descriptive definition of "life" altogether. The idea is to provide a unifying account of life, but one that is not supposed to reflect the meaning of "life" any more. In this regard, it is particularly attractive to base the account on scientific knowledge, in particular from biology. The last couple of years have seen a number of proposals in this respect. Bedau (2010, p. 1013) for instance pleads for what he calls an Aristotelian account of life, which is supposed to explain various "phenomena of life", in particular several hallmarks of life, borderline cases and puzzles about life (Bedau, 2010, pp. 1013–15). In a similar vein, Ruiz-Mirazo et al. (2004, p. 325) contrast descriptive definitions with "essentialist definitions" that "characterize a given phenomenon in terms of its most basic dynamic mechanisms and organization". They require such a definition to have explanatory power. Their own attempt to give such a definition tries to account for life in terms of open-ended evolution and autonomy.

But giving up the project of descriptive definition of "life" faces the following challenge. If somebody proposes an account of life by providing, for example, what Ruiz-Mirazo et al. (2004) call "essentialist definition", she somehow specifies an essence and thus something general that all living beings and only them are supposed to share. But this essence can be stated in necessary and sufficient conditions or in a similar principled way. For instance, the proposal by Ruiz-Mirazo et al. (2004) consists of two necessary and jointly sufficient conditions on life, namely open-ended evolution and autonomy. It then is very likely that the account does not agree with the concept of life, simply because we have seen that attempts to define "life" in terms of necessary and sufficient conditions or in a similar principled way have had a hard time. There are likely objects that count as living beings according to our concept of life that are not included in the account, or the other way round. If there are such counterexamples, then our second condition is violated because the account does not turn out to be true, not to mention that it cannot be essentially or necessarily true.

It may be objected that a scientifically informed account of life can provide necessary and jointly sufficient conditions for being alive, simply because science can introduce new concepts that may ultimately allow for stating necessary and sufficient conditions of life, as it is now understood. Suppose, for instance, that, on the basis of empirical research, scientists find out that all living beings, and only them, have a certain property F-ness, which was not known before (we do not need to bother with how the account deals with examples for which it is unsettled whether they instantiate life or not). We can then turn this statement into an account of life by saying that the essence of life is F-ness. This objection is not very convincing though. What empirical findings can tell us is that all *observed* living beings, and only them, share F-ness. But this does not yet tell us that all living beings must share F-ness. The account is thus subject to counterexamples: We may well be able to imagine living beings that do not have F-ness. Of course, the counterexamples may not refer to systems that exist somewhere in our Universe. Nevertheless, they are counterexamples, if we are looking for features that living beings necessarily have.

Altogether, we seem to be caught in the following theoretical dilemma. We could try to answer the question of what life is in terms of a descriptive definition, but then we are unlikely to obtain an analysis in terms of necessary and sufficient conditions or any principled account of life. Alternatively, we may try to characterize life in a more principled way, e.g. in terms of necessary and sufficient conditions, but then we have to admit that the account is subject to counterexamples and that our account does not hold true due to conceptual necessity. A further question then is whether we are still talking about life? The threat is that we are changing the topic and introducing a stipulative definition of "life" (this objection is discussed by Machery, 2012, p. 157).[17]

Can We Escape the Dilemma?
Life as a Natural Kind

In recent work, Cleland and Chyba have made a proposal that may be seen as an attempt to escape the dilemma (Cleland & Chyba, 2007; Cleland, 2012). To motivate the proposal, Cleland (2012, pp. 132–3) raises the following question: Does not the history of science present us with discoveries in which people have obtained new insights that were incompatible with what they used to think before about the very same topic? For instance, it may be said that Einstein's views about mass are not at all compatible with those by Newton; nevertheless, both physicists are talking about the same thing.

To make conceptual space for this possibility, Cleland and Chyba draw on ideas by Putnam and Kripke. The key idea is that there are natural kind terms and that related predicates function as pointers to natural kinds (Cleland & Chyba, 2007, pp. 335–6; Cleland, 2012, pp. 135–8). The reference of such an expression (and the extension of the corresponding concept) need not be determined by a description, but rather by certain events in which the reference is fixed by relating the expression to appropriate samples of the kind. This allows for the possibility that we do not initially know the characteristics of the natural kind and only come to learn about the kind as we gain more empirical knowledge about it. Also, we may for some time mis-classify some things since we do not yet know what exactly the natural kind is like. The theoretical dilemma stated above may thus be avoided in the following way: If "life" designates a natural kind, then we can find a principled account of life. Even if this account is incompatible with quite a lot we have thought about life, we cannot say that we have changed the topic, because "life" was supposed to stand for the natural kind from the outset.

This proposal may also be understood as an answer to our challenge to explain how life can necessarily have certain traits: The idea is that theoretical identity statements hold necessarily according to Kripke's theory of rigid designation (Kripke, 1980).[18]

As an example of a natural kind, Cleland and Chyba (2007) and Cleland (2012) use water. A plausible answer to the question of what water is has it to be H_2O. According to Cleland and Chyba, this is a theoretical identity statement (Cleland & Chyba, 2007, p. 330). It does not specify the meaning of "water" or enfold the concept of water, but rather picks the natural kind that water is. Likewise, they suggest, we should look for a theoretical identity statement for life. The authors suggest that such an identity may follow from a "general theory of life" (Cleland & Chyba, 2007, p. 332; see Cleland, 2012, pp. 138–9 for an argument why a theory is needed). They admit that we do not yet have this theory and point to problems in searching for it (Cleland & Chyba, 2007, pp. 332–3).

To examine Cleland and Chyba's proposal, we may first distinguish between two claims, which can aptly be named by drawing on the terminology that Smith (1994, pp. 63–4) uses in relation to J. L. Mackie's work. The first claim is conceptual, the idea being that "life" *is meant* to stand for a natural kind. The other is substantive in that it asserts the existence of such a natural kind. The first claim has some plausibility; for instance, according to Kitcher (1982, pp. 342–3), scientists, in their talk, intend to designate natural kinds quite generally, and such an intention may also guide a lot of talk in ordinary language. Let me thus grant the first, conceptual claim and focus on the substantive claim and on the question of whether there is a natural kind life (to be fair, we should note that Cleland and Chyba only very tentatively suggest that there is such a kind; for Cleland 2012, this is in fact an assumption that she regards plausible, but that she does not argue for, see e.g. her p. 127).

For a first comment, recall that Cleland and Chyba hope for a theory of life that answers the question of what life is. But there is a disanalogy to the example of water. The theory that entails the theoretical identity statement about water is not a theory of water only, but a theory with a much broader scope, which covers various sorts of chemical bondings between elements from the periodic system.[19] More generally, a theory that allows for the theoretical identity statement "X = Y" need not be a theory of X. For instance, Newtonian physics allows for an identification of energy with a certain well-defined physical characteristic, but is not for this reason a theory of energy. There is thus no reason to think that it is a theory *of life* that will provide a suitable theoretical identity statement about life.

Second and relatedly, the identity "water = H_2O" (which I grant for the sake of argument) instantiates a general scheme of identities in which a substance is characterized in terms of its chemical composition. There are many other instantiations of this scheme, e.g. "iron = Fe". A plausible explanation is that natural kinds come in groups. This is in fact true of many natural kinds (chemical substances, elementary particles from particle physics, biological species – if the latter are indeed natural kinds). Thus, as a rule, natural kinds seem to come in groups that define larger classification schemes in which we can sort things. If this is true, then which classification scheme is life part of? What are the other natural kinds that operate on the same level? I do not see any attractive answer to this question.

Finally, what sort of natural kind may life be? With what sort of thing may it be identified (cf. Cleland, 2012, p. 140)? With a certain sort of chemical composition? But clearly, life does not have a unique molecular structure, as water arguably has. Or with certain functionings, e.g. motion and reproduction? But can functionings characterize natural kinds? More generally, the suggestions that come to mind if we wonder what is characteristic of a putative natural kind of life lead us back to conditions that have been mentioned in well-known definitions of life. Now there is evidence that no simple combination of these conditions gives as an adequate analytic descriptive definition of "life" (see the discussion above). The question then is whether the conditions can be used to characterize a putative natural kind of life. One problem with well-known definitions of life is that they are subject to counterexamples. Now these counterexamples need not be a decisive objection against an account of a putative natural kind because it is possible that we have thus far been mistaken about whether certain things belong to the natural kind (think of the example of whales, which are not fish). But there should not be too many counterexamples because, otherwise, the question is whether the expression "life" was really meant to refer to the proposed natural kind. Another problem was that there is a plurality of incompatible definitions of life. If life is a natural kind, then one of the definitions or a simple combination of conditions mentioned thus far should characterize life. But how can we find out which combination of conditions is characteristic of the natural kind? The most plausible answer is that a certain combination of conditions has most explanatory power regarding typical examples of life. But so far, it is not clear what types of conditions have most explanatory power. A lot of functionings can be in some sense explained in terms of molecular biology, but our molecular biology is irrelevant if the material constitution of living beings may be based on different molecules. Darwinian evolution does not explain the material constitution of living beings and their reproduction. So I do not quite see how a certain combination of conditions should have so much explanatory power that it characterizes a natural kind. This is not a decisive argument against Cleland and Chyba's suggestion but, taken together, the considerations listed in the last few paragraphs cast doubt on the idea that life is a natural kind.

An Explication of Life

As an alternative way to escape the dilemma, I would like to suggest the method of explication, as proposed by Carnap (1962, §§ 2–3). An explication replaces a concept that is unclear, often called "explicandum", by a new, more exact one, called "explicatum" (Carnap, 1962, p. 3). Explications are subject to the following desiderata (see Carnap, 1962, p. 7, for a concise statement).[20] First, the explicatum should be similar to the explicandum. Second, the explicatum must be such that it can be characterized in an exact way, e.g. by an explicit definition. One, if not *the*, crucial point of exactness is to decide borderline cases of the explicandum.[21] Third, the explication is supposed to be fruitful. For Carnap, it is so if the explicatum figures in statements of laws and of other generalizations, but there are other ways to spell out fruitfulness. Fourth, the explicatum should be simple in the following way: Its characterization should be simple, and it should allow for simple generalizations.

Obviously, the desiderata on explications can pull in different directions. Typically, we will gain on the last three criteria if we steer away from the original explicandum and pay regarding similarity.[22] Thus, trade-offs are needed between the criteria to decide whether an explication is overall better than another. There are at least two ways to think about the trade-offs. One way is to say that the trade-offs depend on weights put on the desiderata and that it is only relative to such weights that an explication can be superior to another.[23] We can then fix the weights by relying on a prior purpose, in which case it turns on the purpose of whether one explication is better than another one. As an alternative (which is inspired by what Lewis, 1973, pp. 73–4, proposes for best systems), we may claim that there is a range of objectively reasonable methods to set the weights and that all of them lead to the same results.[24] Accordingly, one explication may be better than another one, full stop. Either way, we may hope to obtain robust results, i.e., results that are stable under slight variations of the weights (see Lewis, 1973, p. 74).

My claim now is that explication can help us to escape the dilemma specified above. To see this, note first that the dilemma can be understood in terms of Carnap's desiderata. The first horn of the dilemma, namely a descriptive definition of "life", is completely driven by the first desideratum, namely similarity because a descriptive definition should fully reflect the ways in which we talk of life in the relevant sense. The second horn of the dilemma, namely a stipulative account of life in terms of, for example, necessary and jointly sufficient conditions, is driven by the desiderata of exactness and simplicity. As mentioned above, explicit definition is one way to grant exactness, and a stipulative definition in terms of, for example, necessary and jointly sufficient conditions is explicit. It is arguably likely to allow for less borderline cases than, for example, an ostensive definition. Further, other things being equal, simplicity favors a definition in terms of such conditions or in a similar, principled form because we thus obtain a very simple form. Considerations of fruitfulness may also pull us in the direction of a stipulative definition in terms of such conditions, but this turns on what fruitfulness is.

But since explication is subject to the desiderata of similarity *as well as* exactness and simplicity, it offers a third option. Using explication, we can try to give an account of life that steers a middle course between a faithful representation of our concept of life and a fruitful, exact and simple, but stipulative account in terms of conditions of life. The proposal then is to replace our notion of life for the purposes of scientific inquiry with a notion that we may call "s-life" that strikes a fair compromise between Carnap's desiderata. Such a compromise would be reached by a concept of s-life that may be defined in terms of necessary and jointly sufficient conditions, that is as close as possible to our existent concept of life and that is otherwise maximally fruitful, simple and exact. But, in what follows, I will not stick to the demand that we need necessary and jointly sufficient conditions for s-life; other ways to fulfill exactness and simplicity will be accepted too.[25] In any case, the proposal urges a third way between a faithful account of the existing concept of life and a fruitful stipulation. The idea is to find an optimal way to do justice to both demands that are implicit in the horns of the dilemma.[26]

Quite generally, there are various ways to deal with a dilemma. One is to argue that one horn of the dilemma is not really problematic. Another tries to show that the options constituting the dilemma do not exclude each other. My proposal pursues neither of these strategies. In my view, the dilemma persists as a conflict between several desiderata on explication that pull in different directions. What my proposal achieves is to embed the dilemma in a well-known theoretical framework to give us a better understanding of it. The proposal also points out that we can strike a balance between the desiderata and take a middle course between the horns of the dilemma.

My proposal has a number of virtues. First, it is plausible that a lot of recent proposals to define life (e.g. Ruiz-Mirazo *et al.*, 2004; Bedau, 2010) are indeed attempts at providing an explication. They are certainly guided by the standards of exactness, fruitfulness and simplicity. For instance, Ruiz-Mirazo *et al.* (2004, pp. 325–6) specify desiderata on their account of life. When they require explanatory power, but also when they ask for a general notion of life, this may be understood as contributing to fruitfulness. Likewise, when they ask for "conceptual elegance" (Ruiz-Mirazo *et al.*, 2004, p. 326) and the avoidance of redundancy, this is a quest for simplicity. And their requirement of clear criteria with which we may judge life, may be understood in the sense of exactness.[27] As far as I can see, the authors do not call for closeness to our ordinary notion of life, but I think, they would ultimately have to agree with this requirement too. One indication for this is as follows: As Machery (2012, pp. 147–8) observes, in the debate about what life is, authors often appeal to intuitions about what counts as life. Machery argues that they must thus appeal to our ordinary concept of life. If they do so, they show that closeness to our ordinary conception of life matters to them. This should not come as a surprise given that we should not move too far beyond our conception of life if we do not want to change the very topic we are interested in.

Cleland's quest for a theoretical identity statement about life can also be seen as urging an explication. This is because natural kinds allow for generalizations, and if life is a natural kind, then a related explicatum is particularly fruitful. Thus, if life is a natural kind, then a Carnapian explication that puts enough weight on fruitfulness should hit the natural kind, and quite robustly so, i.e., quite independently of how much weight exactly is put on it.[28]

But second, my proposal can also account for the plurality of proposed definitions. There will be a reasonable plurality of definitions of life if the criteria may be reasonably balanced in various ways and if what the best result is depends on the weights of the criteria, at least given our present empirical knowledge.

My proposal also raises some questions. To finish this section, I would like to address two of them. First, how exactly can we steer a middle course between staying faithful to our entrenched concept of life and between giving a simple, exact and fruitful account of life? The answer depends on the weights that we place on the four desiderata. If there are few reasonable methods to set the weights, which lead to the same results, we can obtain an explication that is overall most appropriate. By contrast, if we are not much constrained in the ways in which we may reasonably put weights on the desiderata, we are free to choose them according to our purposes. For instance, we may give priority to similarity and only accept an explication that is very similar to our concept of life in that it reproduces what we think about a range of examples specified in advance. Alternatively, we may be more relaxed about similarity and care more about the possibility to obtain laws about life. Note though that an explication of life has to put some weight on similarity, because, otherwise, there would be no reason to speak of an explication of *life*. Note also that it seems sensible to weigh the desiderata in such a way as to make sure that the result is robust, i.e., that it does not change if the weights of the desiderata are slightly modified.

Second, where does an explication of life lead us? Will the explicatum roughly have the same extension as our present concept of life? Or will it be very different (cf. Gayon, 2010, p. 243)? To some extent, the answer turns on the weights of the desiderata in the explication again. But it also depends on our notion of life and on the possibilities to obtain laws for something like what we call life. For it is these factors that determine what sorts of compromises between the desiderata are possible. The question is, for example, whether a fairly minor revision of our notion of life would allow for a powerful theory with many laws. Our second question thus needs to be answered in close collaboration between philosophers and scientists.

Conclusions

What is life, and why is the question still open? The problem is not that there are no proposals on the table; rather, we face an embarrassment of riches. Many diverging accounts of life have been suggested. According to the diagnosis developed in this paper, this is so because our ordinary notion of life does not allow for an analysis in terms of necessary and sufficient conditions (or for a similarly principled descriptive definition of life). A useful and unifying account of life may nevertheless be given as a Carnapian explication. The advantage of this theoretical framework is that it allows for a stipulative element without giving up the connection to our original notion of life. It also helps to make manifest methodological requirements that have been assumed for a definition of life. My proposal is not thought to replace our concept of life in everyday parlance too. It is not clear to me whether there is a need for reform in our everyday thinking about life. The concept of life may fulfill roles that go beyond what can be investigated by the sciences (see e.g. Toepfer, 2005).

Acknowledgements

I am very grateful to an anonymous referee and to Georg Brun for extremely helpful and constructive comments. Thanks also to the editor of this volume for his patience.

Notes

1 As of 1950, Robinson did not know about Quine's skepticism concerning the distinction between synthetic and analytic knowledge (Quine, 1951) and Putnam's (1975) and Kripke's (1980) works about meaning and reference. I will take into account crucial ideas by Kripke and Putnam in due course. For a useful collection of papers about definition see Fetzer *et al.* (1991).
2 For the sake of simplicity, I assume that the whole answer to the question is absorbed into one statement, which is expressed in one sentence.
3 This shows that the question of what life is does not ask for an arbitrary definition (which may be given by a synonym; Robinson 1950, pp. 94–6), but rather for a certain type of definition.
4 Predicates are here taken to be linguistic expressions (e.g. words or combinations of words). As usual in philosophy, I use quotation marks, if a linguistic expression is mentioned. Predicates are meant here in the logical sense; such predicates can, but need not take the position of the grammatical predicate of a sentence. In what follows, I use the expression "designation" to specify the relationship between a predicate and a concept. I will further say that a predicate *refers* to the instances or the set of instances of the concept that is designated by the expression. To say that a predicate designates a concept is not meant to exclude that it picks a natural kind too.
5 This characterization of a nominal definition is not general, since it only refers to predicates, but suffices for our purposes. Since nominal definitions can be said to define linguistic expressions, I will speak of the definition of e.g. "life", if I refer to a nominal definition.
6 Gupta (2015, Sec. 1.2) mentions dictionary definitions as a further type of nominal definition, but thinks that they are not the sorts of definitions that philosophers are interested in. See Robinson (1950, p. 19) for a similar classification.
7 Stipulative definitions cannot be true properly speaking: They do not describe existing use, but rather prescribe the use of a word in the future (Robinson, 1950, Ch. IV, particularly p. 62).
8 Cleland and Chyba may thus be said to urge for a real definition of life.
9 It may be objected that the identity of life with a specific natural kind is a matter of nomic necessity. This is not a plausible objection though, because natural laws do not connect words and respective concepts with natural kinds out there in nature.

10 Of course, stipulative definitions hold true by necessity too. But stipulation does not help here because we are here not interested in a stipulative definition (see above).
11 Note though that not every definition in terms of necessary and jointly sufficient conditions need provide an analysis since there is no guarantee that the conditions analyze the thing designated by a word.
12 Note that something that cannot be analyzed may still have components in some sense, viz. components that are not strictly necessary (see below). Cleland and Chyba (2002, p. 389) call a descriptive definition in terms of necessary and jointly sufficient conditions "ideal definition"; Shields (2012, p. 108) dubs an account of life in terms of such conditions "univocal", while Bedau (2010, p. 1011) calls it "Cartesian". Machery (2012, p. 147) speaks of a "classificatory definition" in this respect.
13 We are here only interested in analytic *descriptive* definitions. Thus, even if an analytic definition is not meant to be descriptive, I will here assume it to be descriptive. I turn to non-descriptive definitions in due course.
14 The focus of Kant's *Critique of the Teleological Power of Judgement* (1790) is on what he calls natural ends, which he takes to be organized beings (§65; Kantian terms are here translated following the 2000 edition by P. Guyer), but he does not offer his account of natural ends as an analysis of life. It is nevertheless appropriate to make his ideas fruitful for the understanding of life, since he clearly has living beings in mind when he talks about natural ends and organized beings (see in particular his examples in §64; cf. also §73).
15 It is arguable that some of the items from the list, in particular Darwinian evolution, do not lead to an analytic definition because they do not name constituents of the concept of life. For the remainder of this paper, this does not really matter.
16 It is true though that such a definition will not be of any help to settle controversial examples that turn on the vagueness of metabolism. But a descriptive definition is not necessarily assumed to be so.
17 A similar dilemma is claimed by Machery (2012). For him, life is either a folk concept or a scientific one. Machery argues that most folk concepts escape an illuminating definition quite generally, while he takes it to be unlikely that a unique scientific concept of life is forthcoming since various disciplines are working on the notion of life. Thus, in either of both cases, a definition of life is unlikely. The dilemma here is a slightly different one. What ultimately pulls in different directions in my view is the quest to be faithful to the well-entrenched conception of life and the demand that life be explained in a principled way.
18 Kripke's theory underwrites talk about essences too, but we need not discuss this here.
19 It is not clear what exactly this theory is. For instance, it is controversial whether the periodic system qualifies as a theory at all (see Weisberg, 2007, pp. 213–15 for a recent discussion). But my argument goes through for all theories that cover the molecular structure of water and of other substances.
20 Carnap (1962, p. 5) himself speaks of requirements instead of desiderata, but I prefer "desiderata" because a desideratum may be fulfilled to higher or lesser degrees.
21 See Carnap (1963, p. 936) for additional remarks about exactness and Brun (2015, Sec. 2.3) for a proposal of how to understand Carnap's desideratum exactness.
22 Carnap (1962, p. 7) assumes that fruitfulness has lexical priority over simplicity, but in what follows, I will not adopt this assumption.
23 Here and in the following, the terms "weight" and "to weigh" are used in a broad sense for every method to decide in cases in which the desiderata conflict.
24 Compare Shields (2012, p. 107) who rejects the idea that accounts of life must be purpose-relative.
25 Carnap himself does not even require a definition to satisfy exactness (see Carnap, 1962, p. 3, and Carnap, 1963, p. 935).
26 My proposal is in accordance with Shields (2012, p. 107, fn. 5) who argues against the alleged dichotomy that life is either a natural kind or that its definition is only a matter of how we want to talk.
27 They also ask for minimality, but this may be seen as related to fruitfulness or simplicity.
28 Carnap (1962, pp. 4–5) suggests sodium chloride (NaCl) as an explication of salt.

References

Aristotle (2016). *De anima*, trans. C. Shields, Oxford: Clarendon Press.
Bedau, M. A. (2010). An Aristotelian account of minimal chemical life. *Astrobiology*, **10**, 1011–21.
Bedau, M. A. & Cleland, C. E., eds. (2010). *The Nature of Life. Classical and Contemporary Perspectives from Philosophy and Science*, Cambridge: Cambridge University Press.
Boden, M. A (1999). Is metabolism necessary? *The British Journal for the Philosophy of Science*, **50**(2), 231–48.
Boyd, R. (1999). Homeostasis, species, and higher taxa. In R. A. Wilson, ed., *Species: New Interdisciplinary Essays*, Cambridge, MA: MIT Press, pp. 141–85.
Brun, G. (2015). Explication as a method of conceptual re-engineering. *Erkenntnis*. DOI: 10.1007/s10670–015–9791–5.
Carnap, R. (1962). *Logical Foundations of Probability*, 2nd edn, Chicago: University of Chicago Press.
Carnap, R. (1963). Replies and systematic expositions. In P. A. Schilpp, ed., *The Philosophy of Rudolf Carnap*, Chicago, Il: Open Court, pp. 859–1013.
Cleland, C. & Chyba, C. (2002). Defining 'life'. *Origins of Life and Evolution of the Biosphere*, **32**(4), 387–93.
Cleland, C. & Chyba, C. (2007). Does 'life' have a definition?. In T. Woodruff, I. Sullivan and J. Baross, eds., *Planets and life: The emerging science of astrobiology*, reprint in: M. A. Bedau and C. E. Cleland, eds., *The Nature of Life*, Cambridge: Cambridge University Press, 2010, pp. 326–39.
Cleland, C. (2012). Life without definition. *Synthese*, **185**, 125–44.
Diéguez, A. (2013). Life as a homeostatic property cluster. *Biological Theory*, 7, 180–6.
Fetzer, J. H., Shatz, D. and Schlesinger G. N., eds. (1991). *Definitions and Definability: Philosophical Perspectives*, Dordrecht: Kluwer Academic Publishers.
Fine, K. (1994). Essence and modality: The second Philosophical Perspectives lecture. *Philosophical Perspectives*, **8**, 1–16.
Gánti, T. (2003). The Principles of Life, ed. by J. Griesemer and E. Szathmary, reprint in: M. A. Bedau and C. E. Cleland, eds., *The Nature of Life*, Cambridge: Cambridge University Press, 2010, pp. 102–12.
Gayon, J. (2010). Defining life: Synthesis and conclusions. *Origins of Life and Evolution of Biospheres*, **40**, 231–44.
Gupta, A. (2015). Definitions, *The Stanford Encyclopedia of Philosophy* (Summer 2015 Edition), Edward N. Zalta, ed., http://plato.stanford.edu/archives/sum2015/entries/definitions/.
Hume, D. (1739, 1978). *Treatise of Human Nature*, L. A. Selby-Bigge, ed., 2nd edn, Oxford: Clarendon Press.
Joyce, G. F. (1994). Foreword. In D. W. Deamer and G. R. Fleischaker, eds., *Origins of Life: The Central Concepts*, Boston, MA: Jones & Bartlett, xi–xii.
Kant, I. (1790, 2000). *Critique of the Power of Judgment*, trans. P. Guyer and E. Matthews, Cambridge : Cambridge University Press.
Kitcher, P. (1982). Genes. *British Journal for the Philosophy of Science*, **33**, 337–59.
Kripke, S. A. (1980). *Naming and Necessity*, Oxford: Blackwell.
Lange, M. (1996). Life, "artificial life," and scientific explanation, *Philosophy of Science*, **63**(2), 225–44, reprint in M. A. Bedau and C. E. Cleland, eds., *The Nature of Life*. Cambridge: Cambridge University Press, 2010, pp. 236–48.
Lewis, D. K. (1973). *Counterfactuals*, Oxford : Basil Blackwell.

Locke, J. (1690, 1894). *An Essay Concerning Humane Understanding*, Oxford: Clarendon Press.
Machery, E. (2012), Why I stopped worrying about the definition of life...and why you should do as well. *Synthese*, **185**(1), 145–64.
Neurath, O. A. *et al.* (1929). *Wissenschaftliche Weltauffassung, Sozialismus und Logischer Empirismus*, reprint in R. Hegselmann, ed., Frankfurt am Main: Suhrkamp, 1979, 81–101.
Quine, W. V. O. (1951). Two dogmas of empiricism. *The Philosophical Review*, **60**, 20–43, reprint in: *From a Logical Point of View*, Cambridge, MA: Harvard University Press, 1953, 20–46.
Pennock, R. T. (2012). Negotiating boundaries in the definition of life: Wittgensteinian and Darwinian insights on resolving conceptual border conflicts. *Synthese*, **185**(1), 5–20.
Putnam, H. (1975). The meaning of meaning. In H. Putnam, *Mind, Language and Reality: Philosophical Papers*, Cambridge: Cambridge University Press, 215–71.
Robinson, R. (1950). *Definition*, Oxford: Clarendon Press.
Ruiz-Mirazo, K., Peretó, J. & Moreno, A. (2004). A universal definition of life: autonomy and open-ended evolution. *Origins of Life and Evolution of the Biosphere*, **34**, 323–46.
Sagan, C. (1970). Life. In *Encyclopædia Britannica*, reprint in: M. A. Bedau and C. E. Cleland, eds., *The Nature of Life*, Cambridge: Cambridge University Press, 2010, pp. 303–6.
Sagan, C. (1973). *The Cosmic Connection: An Extraterrestrial Perspective*, reprint in: Sagan, C., *Carl Sagan's Cosmic Connection: An Extraterrestrial Perspective*, produced by J. Agel, Cambridge: Cambridge University Press, 2000.
Sagan, D. (2016). Life, In *Encyclopædia Britannica. Encyclopædia Britannica Online*. Encyclopædia Britannica Inc., 2016. Web. 14 May. 2016, http://www.britannica.com/topic/life.
Shields, C. (2012). The dialectic of life. *Synthese*, **185**, 103–24.
Smith, M. (1994). *The Moral Problem*, Oxford: Blackwell.
Sober, E. (2003). Learning from Functionalism, prospects for strong artificial life. In: C. G. Langton, C. Taylor, J. Doyne Farmer, and S. Rasmussen, eds., Artificial Life II, pp. 749–765, reprint in: M. A. Bedau and C. E. Cleland, eds., *The Nature of Life*, Cambridge: Cambridge University Press 2010, pp. 225–35.
Sorensen, R. (2016). Vagueness, *The Stanford Encyclopedia of Philosophy* (Spring 2016 Edition), Edward N. Zalta, ed. http://plato.stanford.edu/archives/spr2016/entries/vagueness/.
Toepfer, G. (2005). Der Begriff des Lebens. In U. Krohs and G. Toepfer, eds., *Philosophie der Biologie. Eine Einführung*, Frankfurt am Main: Suhrkamp, pp. 157–74.
Weisberg, M. (2007). Who is a modeler? *British Journal for Philosophy of Science*, **58**, 207–33.
Wittgenstein, L. (1953). *Philosophical Investigations*, trans. G. E. M. Anscombe, Oxford: Basil Blackwell.

7

Is the Origin of Life a Fluke? Why the Chance Hypothesis Should Not be Dismissed Too Quickly

CHRISTIAN WEIDEMANN

Origin of Life Research and the Chance Hypothesis

'Life belongs to the very fabric of the world. Were it not an obligatory manifestation of the combinatorial properties of matter, it could not possibly have arisen naturally. By ascribing to chance an event of such unimaginable complexity and improbability [...] [one] does, in fact, invoke a miracle [and] sides with the creationists'. *(de Duve, 1991, p. 217)*[1]

Despite their widely divergent suggestions as to how life[2] might first have arisen, most contemporary researchers on the origin of life share the following three convictions (cf. White, 2007, pp. 453–4)[3] with Nobel laureate de Duve.

(1) The origin of life on Earth cannot be explained scientifically with reference to purposeful agency.

It is true that some researchers privately believe in a supernatural designer. This belief, however, does not play any explicit role in their scientific hypotheses. Methodological naturalism is largely taken for granted and intentional or teleological explanations are almost universally frowned upon.[4]

(2) Life on Earth did not emerge by mere chance.

This second claim is admittedly vague and needs to be interpreted with care. Its negation should be seen as compatible with both indeterminism *and* determinism. Our universe is deterministic if, and only if, its natural laws conjoined with the state of the universe at any one time entail the state of the universe at all other times. Determinism, thus construed, is neutral with respect to how much of the future is, in principle, knowable by us (e.g., Earman, 1986, ch. 2). This leaves room for talk of chance, *in some sense*. Such a variety of chance would suffice for the purposes of this paper: Obviously, even in a deterministic universe there can be games of chance, like roulette. The outcome of the spinning of a roulette wheel is due to chance, because even a very slight and, for all practical purposes, indistinguishable variation in the initial conditions (a small deviation in the movement of the croupier's hand or in the speed of the wheel, a little breeze, an almost indiscernible bump...) would have resulted in a completely different result. In a similar vein, scientists who deny that life on Earth arose by chance are best understood as claiming that the emergence of life on Earth was *not* highly sensitive to precise and (epistemically) improbable initial conditions (White, 2007, pp. 456–7).[5] Likewise, for the case in which there are indeterministic processes in our world, they deny that life on our planet arose by genuine random events that are extremely improbable. According to the contemporary standard view, it is out of the question that the first formation of the fundamental building blocks of life (such as genes, proteins, let alone cells...) was a mere fluke brought about by the random shuffling of molecules in a primordial soup.[6] Manfred Eigen (1992, p. 10; cf. Salisbury, 1969, p. 342) calculated that for a single gene containing a thousand nucleotides the number of possible variants is roughly 10^{602}, almost all of them non-functional. Only a minuscule percentage of the possible combinations could be tried out during the history of the (observable) universe,[7] let alone Earth's history. Consequently, Iris Fry (2000, p. 194) proclaimed, 'The improbabilities involved in a chance scenario are just too high for serious consideration'.

By denying that the emergence of life on Earth was due to a very happy accident, one is, of course, not committed to the view that the formation of life on Earth was *inevitable* (de Duve, 1991), nor forced to negate that chance processes were crucial for life's origin. Rather, (2) should be understood as saying that, given the *general* physical and chemical conditions on Earth, the emergence of life was not *too* unlikely. Opinions among origin of life researchers may differ on how much improbability can be tolerated in a satisfactory account of life's origin on Earth. However, there is a widely shared agreement[8] that no arbitrarily low value of probability will do.

(3) Even if all hitherto proposed theories on the origin of life were shown to be false by new evidence, the justification for believing that the emergence of life on Earth was not due to a mere fluke would remain largely unaffected.

If all the dozens of proposed theories on who killed J. F. Kennedy or on the true function of the Stonehenge monument were strongly discredited by evidence, we would still *know* that Kennedy was murdered and Stonehenge was built with an intention. This knowledge is *epistemically prior* to the relative success or failure of particular theories on Kennedy's death or the meaning of Stonehenge. Similarly, 'the conviction that life did not arise largely by chance is treated [by most people in the field] as *epistemically prior* to the development of alternative theories' (White, 2007, p. 453; cf. p. 458).

Although I will comment in passing on the plausibility of the first two convictions, my discussion will focus on the last presupposition. I shall argue that it is ill founded for at least two reasons.

Firstly, given a sufficiently big and variegated ensemble of universes (i.e. a 'multiverse'),[9] defended by a growing number of cosmologists, *any* finite chance of life emerging on Earth-like planets will suffice to guarantee the formation of life somewhere in the cosmos. An observational selection effect could then explain why the cosmic neighbourhood we observe is, in fact, a very (very!) special place. Even if only one planet in the whole cosmos harboured (intelligent) life, we would find ourselves on just that planet. Call this the *Vast Cosmos Argument*.

Secondly, even if the cosmos were not big enough to render the random formation of life (and the subsequent evolution of intelligent observers) in the universe more probable than not, there still seems to be no conceivable a priori reason to suppose that processes by which complex molecules arise are more likely to be (unintentionally) biased towards self-replicating, life-producing molecules than to other types of molecules. Call this the *Indifferent Nature Argument*.

A philosopher by training, I am of course poorly equipped to evaluate particular scientific hypotheses on life's origin. My goal is more modest. I will try to show that the chance hypothesis should not be dismissed *prior* to the evaluation of alternative explanations. *Unless* a plausible alternative explanation is available (or life that has emerged independently on another planet is discovered), the chance hypothesis remains a serious contender. My amateurish impression is that all accounts of the origin of life proposed so far clearly fall short of providing a satisfactory *and* complete story. Therefore, it seems to me, the chance hypothesis would be on the table even after a careful examination of alternative explanations.

Let me emphasize that little of what I have to say will be new. The Vast Cosmos Argument, which, in a certain sense, can be traced back to ancient Epicureanism, has been put forward in modern times by astrophysicist Michael H. Hart (1982) and others (e.g., Barrow & Tipler, 1986, ch. 8; Koonin, 2007). The Indifferent Nature Argument, too, has an eminent history. It has been given a precise form by MIT philosopher Roger White (2007).

The Vast Cosmos Argument

Weak Anthropic Principle and a Vast Cosmos

'When I consider the short duration of my life, swallowed up in the eternity before and after, the little space which I fill, and even can see, engulfed in the infinite immensity of spaces of which I am ignorant, and which know me not, I am frightened, and am astonished at being here rather than there; for there is no reason why here rather than there, why now rather than then'.

(Pascal, 1670, frg. 205/68; see also frg. 693/198; numbering: Brunschvicg / Lafuma)

The scientific revolution that started in the sixteenth century is often credited with the establishment of a 'Copernican Principle'.[10] The principle is usually presented as saying that 'we do not occupy a privileged position in the universe' (e.g., Jones & Lambourne, 2003, p. 214). Although very convenient and in perfect agreement with the spirit of Blaise Pascal's gloomy remarks, such a formulation is highly misleading.

It takes sunlight roughly four hours to reach Neptune, the farthest acknowledged planet of the solar system. Our distance to the next star, Proxima Centauri, however, is already more than four light-years; while the diameter of our galaxy is 'only' about 100 000 light-years, the next major galaxy, the Great Andromeda Nebula, is 2.5 million light-years away, and so on. Given that the observable universe is dominated by gigantic, almost empty voids, our position seems to be very privileged indeed! In the light of this objection, the Copernican Principle needs to be interpreted as merely implying that, *viewed on a large enough scale*, we do not occupy a privileged position (Hawking & Ellis, 1973, pp. 134–5; p. 350). However, assuming that a certain vast region of the universe – at least as big as a supercluster of galaxies – in which we find ourselves, is not special, because the universe is supposed to be approximately spatially homogeneous and isotropic when viewed on a large enough scale,[11] is obviously very different from claiming that we do not occupy a special place *at all*. The Copernican Principle, at least in its above formulation, becomes even more problematic when we consider our position in time. A few seconds after the Big Bang the universe looked very different from the way it looks now. Even worse, during most of its future existence the universe will, while slowly approaching heat death, contain nothing but photons and leptons. The era of stars and planets comprises only a tiny part of the overall history of our universe.[12]

So, why do we not find ourselves in an otherwise empty void between galaxies or at a much earlier or much later time in the history of the universe? The answer to this is the same as the answer to why we do not dwell on the surface of stars or in the direct vicinity of massive black holes. *Nobody does*. Nobody can.[13]

Intelligent beings can make only observations that are compatible with their existence as observers. This is the *weak anthropic principle*.[14] If we find ourselves in a place (or at a time) very atypical for the universe as a whole, this fact is in no need of a special explanation, given that the prerequisites for the emergence of life are only provided by such a place (or time).

Now suppose that the universe is very, *very* big, though – to keep things simple for the moment – finite. Imagine that it contains 10^{100000} Earth-like planets. (A planet is Earth-like if and only if it is located within the habitable zone of a Sun-sized star, if its chemical and geological characteristics are very similar to those on Earth four billion years ago, etc.) In such a universe, functional genes, proteins and even cells, very probably, would form spontaneously on some planets, even if the chances of such random events on any individual planet were close to zero.[15] Now, according to most modern cosmologists, the visible universe *is* indeed only a tiny part of reality. A growing number of physicists even endorse an infinite-multiverse theory.[16] In such a multiverse, the most improbable scenarios (almost) necessarily take place somewhere.[17]

Given that our universe is so big, it is obviously wrong that 'the improbabilities involved in a chance scenario [of life's emergence] are just too high for serious consideration', as Fry (2000, p. 194) wants us to believe. Quite the contrary, the tiny probability of a spontaneous formation of living cells on a single occasion (let's say, in a primordial soup) *plus* the huge number of occasions nicely explain why there is life in the cosmos. Even if this were correct, you might object, it would remain an unbelievably lucky accident that we find ourselves at one of the rare cosmic spots where the conditions are exactly right for life. Applying the weak anthropic principle, however, shows that such an objection is deeply confused. We could not be anywhere else! What we observe is restricted by the conditions necessary for our presence as observers. Owing to this *selection effect*, the observation that our planet is teeming with organisms tells us very little about the frequency of life in the cosmos. If the preconditions for intelligent life were such that only one planet meets them, we would find ourselves on just this planet.

Now, of course, the vast cosmos hypothesis is just a hypothesis. If there is independent evidence for the hypothesis, we can use it as a premise and argue that the probability for life is not small, given the vastness of the cosmos. If so, no further explanations of life's emergence would be needed. Some will question, however, whether we have enough independent evidence for adopting the hypothesis that the cosmos is so vast that life is very probable even if the initial planetary conditions and/or indeterministic processes required for life are extremely unlikely. Then, the vast cosmos hypothesis and the chance hypothesis may be considered as a package that needs to be judged in light of the evidence. Still, I see no reason why the package should not provide an at least equally good explanation of life's existence than alternative theories.

Objections

(a) Explaining Life on Earth vs Explaining Life in the Cosmos

The vastness of the cosmos may help us to understand why life exists at all. It does not contribute, however, to an explanation of the emergence of life *on Earth*. How could it? Suppose you are throwing five dice. On the first roll, you get five sixes. Clearly, *this* remarkable outcome cannot be explained by the fact that all over the world billions of dice have been rolled in the past and will be rolled in the future.[18] The outcome of *this* roll is not causally related to the outcome of former or later rolls. Neither can the fact that life evolved on Earth be explained by invoking the vast number of Earth-like planets in the universe, even if there really were such a vast number. Earth-like conditions that obtain elsewhere in the universe have no significant causal impact on terrestrial events.[19]

Response: The question 'Why did life originate on *Earth*?' is indeed the wrong kind of question. According to the chance hypothesis, life is the result of a fluke. It is due to a very special set of initial conditions highly sensitive to even the slightest variations or due to a series of genuine random events that is extremely unlikely. The vastness of the cosmos makes the emergence of life *somewhere* highly probable, even given its dependence on highly sensitive conditions. The weak anthropic principle tells us that we should not be surprised about finding ourselves at *some* place where the right conditions obtain. Unlike the dice roller, we could not have observed a different outcome.[20] Admittedly, all this fails to explain why Earth, of all planets, is fit for life. Is this a serious shortcoming of the chance plus vast cosmos hypothesis? I do not think so. If we abstract from all features of Earth that are essential to its ability to produce living creatures, there is otherwise nothing special about it. It is hard to see how the demand to explain why life emerged *on this particular planet*, Earth, is supposed to be motivated (Manson & Thrush, 2003, p. 73). We do have an explanation for the fact that there are life-containing planets. We know that, if there are observers at all, they have to be located on one of these planets (interplanetary spacecraft ignored). We happen to find ourselves on Earth. There is nothing left to be explained.

It might be objected that this response violates the 'Requirement of Total Evidence' (Carnap, 1962, p. 211), namely that no relevant information should be ignored in determining the degree of confirmation of a hypothesis. Suppose I am sitting in a taxi-cab. I feel nauseous. This entails that *someone* in the car feels nauseous. If I *only* knew the latter proposition this would confirm the hypothesis that the driver is intoxicated. However, given my total evidence, I have no reason whatsoever to think that the driver is intoxicated. Now, as I do not only know that *someone* in the car feels nauseous, so I do not only know that life exists *somewhere*: I know that life exists *on Earth*. And as it is misguided to look for an explanation of the fact that someone in the car feels nauseous, if I know of one particular passenger that he suffers from this condition, so it is a mistake to ask for an explanation that life exists somewhere, if I know that it exists here.

To see that this objection is not compelling either, consider the following example:

'Suppose that we are given a one-kilogram sample of matter and are told to determine how many uranium atoms it contains, if any. Unfortunately, our Geiger counter is broken. Luckily for us, however, we have at our disposal an amazing resource: a uranium oracle. This gifted individual knows the state of each and every uranium atom, and even has names for them. We leave the oracle in a room with the sample and come back an hour later. The oracle tells us that just one uranium atom decayed: Fred. From the fact that Fred decayed we deduce that one uranium atom decayed. Can we proceed to use half-life calculations to estimate the number of uranium atoms in the sample? Not if we are required to reason from the fact that Fred decayed rather than the fact that some uranium atom or other decayed. Since the presence of other uranium atoms makes it no more likely that Fred should have decayed, the fact that Fred decayed doesn't confirm the hypothesis that the sample contains the calculated number of uranium atoms. Indeed, we are not even entitled to conclude that there is more than one uranium atom in the sample'.   (Manson & Thrush, 2003, p. 74)

This would be an unacceptable result. The moral to be drawn is that there are cases in which we are allowed to ignore parts of our total evidence. The extra information that it was Fred who decayed does not disconfirm the hypothesis that the sample contains the expected number of uranium atoms (confirmed by the fact that some uranium atom decayed). Therefore, it can be ignored. In a similar vein, the extra information that life exists on Earth does not disconfirm the hypothesis that the cosmos contains a vast number of Earth-like planets (confirmed by the fact that life exists at all), and, therefore, can be ignored.

Or so it seems. However, this still is not the end of the story. Let us modify the uranium oracle thought experiment.

'Suppose from the beginning of the experiment the oracle tracked only Fred's progress. In that case, we would err in inferring that there are other uranium atoms nearby, because the presence of other uranium atoms would not make Fred's decay any more likely. The proposition that Fred decayed would not disconfirm the hypothesis that there are many uranium atoms, but we would be wrong in taking Fred's decay as confirming the hypothesis that there are many uranium atoms. The problem is that in our modified story the oracle preselects Fred, whereas in our original story the oracle is open to any decay event'.
(Manson & Thrush, 2003, p. 75)

Do we not have to say that Earth was preselected, too, namely by the fact that we could not possibly have observed life on a different planet (interplanetary spacecraft ignored)? According to the necessity of origins hypothesis (cf. Kripke, 1980, fn. 56–7), we could not possibly have had different parents. It seems plausible to infer from this that we neither could have originated from a different planet. We do not 'happen to find ourselves on Earth'. The only alternative for our being born on Earth (or in its cosmic neighbourhood) is not being born at all. If so, the existence of other planets indeed seems irrelevant for an explanation of *our* observation that life exists.[21]

There are two problems with this argument.

Firstly, and most importantly, the necessity of origins hypothesis is very implausible with respect to home planets. Suppose that, due to a very slight variation in the initial conditions of the universe, no life emerged on Earth. There still could have been other planets, with exactly the same initial conditions and exactly (or almost exactly) the same prehistory as actual Earth. Why should we think that it would have been impossible for members of *homo sapiens*, including you and me, to evolve on one of these planets?[22]

Secondly, even if we concede the necessity of origins hypothesis with respect to home planets, the Vast Cosmos Hypothesis can still help to make the existence of life on Earth less surprising (Leslie, 1989, pp. 143–4). A lottery has a thousand million tickets. Louise wins. Although she was very lucky, she was not unbelievably lucky. But now suppose she learns that, in fact, she was the *only* participant. Nobody else bought a ticket! It seems wrong to argue that such a discovery would be irrelevant, because the number of tickets sold had no causal impact on Louise's winning prospects. Although Louise could have won only with her ticket, the bigger the number of losing tickets sold, the less in need of explanation is her win.

In a similar vein, although (according to the necessity of origins hypothesis) we could not find ourselves on a planet different from Earth, the bigger the number of other planets, the less in need of explanation is the existence of life on Earth.

(b) Explanatory Impotence of Multiverse Hypotheses

Could we not explain absolutely everything by invoking a sufficiently big multiverse? In such a multiverse, even under conditions extremely hostile to life, very (very!) rarely observers or even complete biospheres would pop up as random quantum fluctuations. There is, for example, a finite (though, of course, astronomically small) probability that a black hole will 'emit a television set or [a doppelganger of] Charles Darwin' (Hawking & Israel, 1979, p. 19) as Hawking radiation.

The complication of these 'freak observers' (Bostrom, 2002, p. 51) or 'Boltzmann brains'[23] (Albrecht & Sorbo, 2004) shows that multiverse theories could be used to explain everything, and therefore, in fact, explain nothing. They are untestable, since no experiment can rule out a theory that allows for all possible outcomes. Multiverse hypotheses are not a proper part of science (e.g., Steinhardt, 2014).

Response: Although, according to the chance hypothesis, the emergence of life on Earth-like planets and the subsequent evolution of intelligent beings are very rare events, they are, I suppose, nowhere near as improbable as the emission of conscious observers[24] by black holes. By assuming a vast, but not too vast, cosmos, we may assign to the spontaneous emergence of life in a primordial soup on Earth-like planets and the evolution of intelligent life a probability very close to 1, while at the same time assigning a significantly lower probability to the existence of freak observers.

However, it may seem ad hoc to stipulate that we live in a vast, but finite, cosmos, if the data also allow for an infinite world. Let us therefore consider the possibility that the cosmos is infinite in size, as many leading cosmologists indeed tend to think (cf. Greene, 2011; Tegmark, 2014; see also Leslie, 2001). In an infinite cosmos with an infinite number of random quantum fluctuations and an infinite number of black holes, the probability of any possible observation being made would indeed be 1. So how can theories that entail an infinite cosmos have *any* observational consequences at all? How can we non-arbitrarily choose between very different kinds of multiverse theories, if they all give the same conditional probability to every observation statement, namely 1 [P (Observation | Multiverse Hypothesis$_i$) = 1]?

The Weak Anthropic Principle is obviously of no help here. We need a more sophisticated principle, the *Principle of Mediocrity*.

(Principle of Mediocrity) One should reason as if one were a random sample from the set of all intelligent observers in one's reference class.[25]

Why is this sensible advice? Imagine the following two thought experiments.[26]

(i) An opaque urn contains 1000 balls. A tombola is arranged and 1000 tickets are sold. Peter buys a ticket. He knows that there are only black and white balls in the urn and that the winning ball is the only one of its kind. He does not know, however, which colour the winning ball is. He draws a black ball. He concludes that, very probably, the winning ball is white.

(ii) Ms X awakens in an empty and windowless room suffering from retrograde amnesia. She has lost all her personal memories; she does not even remember what the language she speaks is called. Her demographical knowledge, however, has remained intact. To start somewhere, she begins wondering what her nationality might be. She concludes that there is a great chance of her being Chinese or Indian.

These inferences are perfectly reasonable, notwithstanding that someone might insist, 'But couldn't it be that Peter actually drew the winning ball or that Ms X was born in Liechtenstein?'. It could be, of course, but it is highly unlikely.

In a similar vein, we might say that, though (an infinite number of) freak observers do exist in an infinite cosmos, it is still true that the overwhelming majority of all observers are regular observers who have evolved by Darwinian processes. We need to interpret the term 'majority' with care, however, since, according to Cantorian set theory, the infinite classes of freak and regular observers have the same *cardinality* (like, say, the class of natural numbers and the class of natural numbers divisible by $10^{1000000}$). Nonetheless, there is surely an intuitive sense in which 'almost all' observers in our (infinite) universe have emerged via regular evolutionary processes (or in which 'almost no' natural number is divisible by $10^{10000000}$). Spelling out this intuitive sense is no easy matter, but I see no reason, why we should not be confident that it can be done, in principle. It must be conceded, however, that, so far, there is no widely agreed on solution to this 'measure problem' and it is unclear how such a solution could be achieved.[27] According to Max Tegmark (2014, p. 314), the 'measure problem' even constitutes 'the greatest crisis in physics today'.

Given a solution to the problem, theories that entail an infinite universe can be tested (though perhaps not strictly falsified) by human observations via the *Principle of Mediocrity*. For example, a theory that leads us to expect that there are 'more' freak observers than regular observers should be rejected.[28] Otherwise we would have to accept either that we are untypical observers or, even worse, that we are freak observers (without noticing it).

(c) Accounting for the <u>Early</u> Emergence of Life on Earth

If life on Earth is due to a fluke, why did it emerge so early (in astronomical terms), almost immediately after its crust had cooled enough for the formation of rocks and plate tectonics? For many authors this fact virtually excludes the possibility of chance. For instance, palaeontologist David Raup (1985, p. 32) asserts, '[b]ecause life originated very early in the history of the earth, its beginning may not have been difficult'. And Carl Sagan (1995, p. 2) crisply states, 'The origin of life must be a highly probable circumstance; as soon as conditions permit, up it pops!'.

Response: Raup and Sagan overlook a possible observation selection effect. The evolution of *intelligent* life depends on a number of crucial steps, the first emergence of microbial life being only one of them (e.g., Schopf, 1992). However, the time window for the evolution of intelligent life on a planet is strictly limited by the average lifetime of the supporting star. If the necessary later transitions are very unlikely to happen during the lifetime of the Sun,[29] we should expect life to begin on Earth rather early, regardless of how easy or difficult its beginning was. This may sound paradoxical, but without an *early* emergence of life the chance that intelligent observers will evolve decreases in an exponential way. Thus, given the improbability of later steps towards intelligent life, we should expect that on our home planet microbial life emerged early, no matter whether the chance hypothesis is correct and whether such an early occurrence was highly improbable in itself or not.[30]

To illustrate the point, let us imagine a special kind of lottery. Ten million dollars are locked in a vault. To get to the money, players have to pick six combination locks through random trial and error. Though the expected time needed to pick each lock is about ten hours, the player only has one hour to pick all six locks, without getting a second chance.[31] When a player finally succeeds, we certainly should not be surprised to find her picking the first lock pretty early, say, within the first ten minutes of the allotted time.

In the light of this example, one could try to raise a different objection: We would expect the successful lottery player to consume almost her entire allotted time. Analogously, if the expected duration of the evolution of intelligent life on a planet were much longer than the lifetime of the supporting star, we would expect it to take almost the entire habitable time of the planet for this improbable event to occur. However, intelligent life evolved on Earth roughly one billion years before the habitable window will eventually close. Moreover, if the evolution of intelligent life is such an improbable event, why do we find ourselves in a solar system with a G-type main-sequence star – fusing hydrogen for approximately 10 billion years – although smaller K-type main sequence stars are not only three times more abundant in the galaxy, but also provide a much longer time window for evolution, fusing hydrogen for up to 30 billion years, thereby exponentially increasing the likelihood of the evolution of intelligent life?[32]

Though these objections are worth exploring, I do not think that they are devastating. Firstly, note that intelligent life evolved only in the last fifth of Earth's habitable era, which is still rather late. Secondly, there is some evidence that, even before dramatic anthropogenic effects set in, the biomass as well as biodiversity on Earth had begun to decline. Human beings may manage to survive on Earth for another 500 million years or even longer, but the conditions will become *very* uncomfortable much earlier (see, e.g., Ward & Brownlee, 2003). It appears unlikely that, during a time of shrinking diversity, a new intelligent species would evolve. Thirdly, at least as far as I can say, we know too little about the factors that make planets (and moons) habitable, in principle, in order to reliably infer that intelligent life is more likely to be found in the orbit of K-type main sequence stars than in the orbit of G-type stars. For instance, the 'habitable zone' of K-type stars is smaller and the risk of a tidal lock (one hemisphere of a planet constantly facing the star) is bigger.

(d) Putting an End to Scientific Research on the Origin of Life

Is the endorsement of the chance hypothesis not detrimental to scientific progress? According to Iris Fry (2000, p. 216; my emphasis), for example, only 'a better understanding of both the primordial geochemical setting and the molecular basis of life, *while abolishing the option of a chance-like emergence*, is giving rise to new creative approaches to the origin-of-life problem'.

Response: As an objection to the chance hypothesis, Fry's statement obviously begs the question. *If* the chance hypothesis is correct, there is simply no point in looking for 'new creative approaches to the origin-of-life problem'. And *if*, because of the evidence for a huge cosmos (and the lack of evidence for extraterrestrial life), we have very little reason to expect a non-chance explanation of life's origin, abandoning origin of life research could save a lot of time, money and intellectual energy that might be better invested elsewhere.

Let us consider a possible rejoinder. Remember the intellectual climate that enabled Darwin's and Wallace's discovery of natural selection. At the end of the eighteenth century and the beginning of the nineteenth century, the opposition to Epicureanism united design theorists like Paley and early evolutionists like Erasmus Darwin and Lamarck, who all had a great impact on Charles Darwin. According to Epicurus and his followers, the emergence of complexity in nature can roughly be explained as follows: atoms whirling at random in a void eventually combined to produce stable and functional arrangements.[33] Unlike Epicurus, however, most natural philosophers of the Enlightenment believed the universe to be (fairly!) small and young. Accordingly, the chance that a complex configuration, such as the human eye, had occurred randomly was deemed tiny. Now imagine that, for independent reasons, at the beginning of the nineteenth century, an infinite multiverse theory had been firmly established. Would not Darwin and Wallace have lost any incentive to come up with the idea of natural selection?

No. Let us apply the Principle of Mediocrity. If we assume that in our world the first exemplars of *all* species were produced by the random movements of molecules in a primordial soup or void, what should we expect to see? Well, we certainly should *not* expect to observe an enormously diverse biosphere with millions of different species. If all species in our cosmos were produced by random molecule movements, diverse biospheres would be very, *very* sparsely distributed. Neither should we expect to observe the astonishing degree of adaptation that surrounds us. A much less diverse and much less adapted biosphere would have sufficed to permit intelligent life to exist.[34]

By contrast, whether we assume that the *first* emergence of life on Earth was inevitable, reasonably probable, or the result of a fluke, makes little or no difference to our rational expectations with respect to biological diversity or adaptation.

Though the Vast Cosmos Argument may account for the first emergence of life, it provides no suitable explanation for the marvellous abundance of biological forms and behavioural patterns we observe. Therefore, we have no reason to suspect that Darwin and Wallace would not have come up with the idea of natural selection, if they had believed in an infinite universe.

Conclusion

Although the mentioned objections highlight important intricacies, none of them succeeds. As far as I can see, only the truth of one of the following claims would render the Vast Cosmos Argument obsolete.

(i) We have strong evidence that extraterrestrial life that originated independently from life on Earth exists in our *cosmic neighbourhood*.[35]
(ii) A well-confirmed and largely complete non-chance explanation of life's origin is available.
(iii) The reasons that led most contemporary cosmologists to believe that the observable universe is only a tiny (perhaps infinitesimally tiny) part of reality are erroneous.

Prior to an evaluation of these claims, the rejection of the chance hypothesis is unjustified. However, only a minority of scientists subscribe to any of the three claims. In other words: the chance hypothesis is alive and kicking.

Indifferent Nature Argument

Let us now suppose for the sake of argument that the cosmos does *not* extend far beyond the visible horizon of our universe. If so, the Vast Cosmos Argument will cease to work. Given the chance hypothesis, there would not be enough suitable planets to make the existence of intelligent life at least reasonably probable. Would this not eliminate the chance hypothesis as a serious contender in the origin of life debate? Surprisingly, the answer seems to be negative, as I will try to show relying on an argument given by Roger White (2007).

Accepting Improbable Chance Events

To grasp the intuitive idea of his argument, let us start with an uncontroversial fallacy.

Peter has watched the roulette table for a while. 'Isn't it amazing? The last numbers have been 12, 34, 5, 18 and 27. You know what the probability for this sequence S was? 1:69 343 957. A conspiracy must be going on. The improbabilities involved in a chance scenario are just too high for serious consideration'.

Peter is wrong. Not everything that is highly improbable, given a chance scenario, stands in need of special explanation. *Some* numbers had to fall. Nothing in the outcome suggests that it was not the result of a chance process. Two claims must not be conflated

a) It is improbable that chance would generate sequence S. [$P(S/\text{Chance}) \approx 0$.]
b) It is improbable that S is the result of chance. [$P(\text{Chance}/S) \approx 0$.]

While (a) is certainly true, (b) does not follow from it. Here is a similar example:

It was astronomically improbable that the genes of my parents would combine in a way that yields me. *It is probable*, however, that my existence is due to chance (at least given a finite universe).

Compare this case with the origin of life problem. According to the chance hypothesis, initial planetary conditions suitable for life were extremely unlikely to occur on Earth or elsewhere. However, as we have learned from the roulette and genome examples, from this alone it does not follow that it is improbable that the emergence of life was due to chance. *Some* initial conditions had to obtain on Earth! Moreover, every other set of initial conditions would have been just as unlikely, given the chance hypothesis. No less than Stephen Jay Gould (1985, p. 183) maintains, 'Any complex historical outcome – intelligent life on earth, for example – represents a summation of improbabilities and becomes thereby absurdly unlikely. But something has to happen, even if any particular "something" must stun us by its improbability'.

Doubting Improbable Chance Events

Gould's reaction is premature, however. Although there are clear cases in which we have no reason to doubt the chance character of highly improbable outcomes, there are equally clear cases of outcomes that cry out for an explanation, *although every other outcome would have been just as improbable*, given a chance scenario.

To see this, let us somewhat modify the genome example. Suppose that I have made an astonishing discovery while searching a genetic sequence database: a certain Mike seems to have a genome almost identical to mine. I ask the database officials to explain this coincidence. My suspicion is that either Mike is my forgotten twin, or he is a clone, or, most probably, something got mixed up. The officials' answer, however, denies all these possibilities, 'The sheer fact that you have the genome you have and Mike has the genome he has, is nothing to be surprised about. Every other combination of your two genomes would have been just as improbable. Please don't waste our time'. Of course, nobody would (and should) accept such an answer. The actual combination of Mike's and my genome is not just improbable; it *stands out* against almost all other possible combinations and therefore cries out for an explanation.

Here are three similar examples.

(1) Peter has watched the roulette table for a while: 'Isn't it amazing? The last numbers have been 1, 2, 3, 4 and 5'.
(2) Jenny hits the lottery jackpot. Two weeks ago her husband had already won a million. She is the best friend of lottery announcer Miranda.
(3) Sarah drops her pencil, it falls to the ground (rather than flying away, disappearing, etc.). She repeats the procedure a dozen times with the same result.

In what respect do these cases differ from the ones before? According to John Leslie (1989, p. 10), 'A chief (or the only?) reason for thinking that something stands in [special] need [of explanation], i.e. for justifiable reluctance to dismiss it as how things just happen to be, is that one in fact glimpses some tidy way in which it might be explained'.[36] In our examples, an alternative explanation (forgotten twin, manipulation, physical regularity) indeed springs to mind almost immediately. This does not mean that further analysis could not show that the chance hypothesis was correct after all. The fact that we can 'glimpse some tidy way of explaining' the outcomes at stake, however, seems to shift the burden of proof onto those who believe in a chance scenario.

Comparing Three Hypotheses

The crucial question then is whether the first emergence of life on Earth should be lumped together with cases where we have reason to accept improbable chance events, or whether we should treat it like cases where we justifiably doubt that an outcome was due to chance.

Following Roger White (2007) again, it will be useful for our discussion to discriminate between three mutually exclusive and exhaustive hypotheses.

Chance Hypothesis (CH): The process that resulted in the emergence of life was just a matter of chance, i.e. it was *not biased* toward some elementary possibilities (e.g., toward the combination of certain kinds of molecules in a primordial soup) over others.

Design Hypothesis (DH): The process that resulted in the emergence of life was *intentionally biased* toward some elementary possibilities over others.

Unintentional Bias Hypothesis (UH): The process that resulted in the emergence of life was *unintentionally biased* toward some elementary possibilities over others.

Let us consider two questions in turn.

First, is the fact that life exists on Earth more probable given the Design Hypothesis than given the Chance Hypothesis? [P (Life on Earth / DH) > P (Life on Earth / CH).] If so, the existence of life favours DH over CH.

Second, is the fact that life exists on Earth more probable given the Unintentional Bias Hypothesis than Given the Chance Hypothesis? [P (Life on Earth / UH) > P (Life on Earth / CH).] If so, the existence of life favours UH over CH.

It may be objected that the answers to these questions are not decisive. Even if the existence of life strongly favoured DH (or UH) over CH, it would not necessarily follow that we should *prefer* DH (or UH) to CH, of course. We may have good reasons to assign to DH (or UH) a much lower prior probability.[37] The fact that my socks used to get lost in mysterious ways is made very likely by the hypothesis that there exists a nefarious species of invisible socks-stealing goblins. However, this does not give me a sufficient reason to believe in goblins.

But I do not think that this caveat is of great importance here. As long as there is no well-confirmed and largely complete account of life's origin available and as long as we do not discover extraterrestrial life, there is little reason to assign significantly different prior probabilities to UH and CH (for DH see next paragraph). Let us thus try to answer both questions.

Design (Intentional Bias) vs Chance

Whether life on Earth supports the Design Hypothesis crucially depends on which, if any, motives we can justifiably attribute to an enormously skilful designer. In order to see why this is important, suppose a poker player accuses a fellow player of having repeatedly fixed the deck when dealing. The accused player is known as an accomplished magician. Consider the following two cases.

> *Case 1*: The accused player dealt himself lots of valuable hands (such as royal flush, full house, etc.)
>
> *Case 2:* The accused player dealt himself only worthless hands.

While the Design (= Cheating) Hypothesis is confirmed by the data in the former case, it is not in the latter, because the dealer lacks any plausible motive to deal himself worthless hands (Manson, 2003, pp. 16–17; Weidemann, 2010, pp. 189–90).

Similarly, whether the fact that life exists on Earth favours DH over CH depends on whether a designer with the power to determine or influence the early history of the Earth would have had a motive to bias Earth history *towards life*. It is controversial whether we are justified in attributing such a motive to a superhuman or divine designer or group of designers whose psychology might have very little in common with ours.[38] But at least it is certainly not beyond the pale to think that a powerful designer is presumably more interested in bringing about macromolecules that would eventually give rise to an abundant biosphere and to intelligent beings than macromolecules that would not.

There are, of course, other problems with the design hypothesis (its purported ad hoc character, the problem of evil, etc.; for a short overview see Weidemann, 2014a). Moreover, as already mentioned at the beginning of the paper, the overwhelming majority of origin of life researchers do not consider the design hypothesis a permissible option within science. Alternatively, we may say that they assign to the design hypothesis a prior probability equal or very close to zero. For the purposes of this paper, I will thus not consider it to be a serious rival to the chance hypothesis.[39]

Unintentional Bias vs Chance

Imagine a real-life poker game (cf. again White 2007, pp. 468–9). Linda has been dealt four aces two times in a row. She reasons as follows, 'I doubt that those hands were dealt by an ordinary random shuffling. I know the dealer. He lacks the skills to cheat and does not even like me. Therefore, I think that there is probably an explanation involving impersonal physical mechanisms'.

This would be an absurd reaction, of course. It is perfectly possible, and perhaps even plausible, that, for physical reasons, certain configurations of cards are more likely to occur than others. But if so, why are these configurations supposed to benefit Linda? Why should anyone expect that an impersonal mechanism is biased towards dealing aces to brunettes? We can have a reasonable debate about the probable preferences of powerful designers, but there is simply no point in arguing about 'probable preferences' of nature. At least if we assume that the cosmos just 'is', not itself being the product of design, nature is indifferent to the emergence of life or to the well-being of living things or persons (though, once there are any, it is not indifferent with respect to their inclusive fitness).[40]

Now consider UH. Is the existence of life more probable given an unintentional bias toward some kinds of macromolecules in a primordial soup than given to chance?

This would only be true if there were some a priori reason to suppose that processes by which complex molecules arise are more likely to be unintentionally biased towards self-replicating, life-producing molecules than to other types of molecules. But why should we think so? When origin of life researchers reject the chance *and* design hypothesis from the outset, they seem to commit the same blunder as Linda in the poker example, albeit in a more subtle form.

An origin of life researcher may react by reconsidering the design hypothesis. If, however, on reflection, she still finds it unacceptable, she lacks any justification to dismiss the chance hypothesis prior to the development of largely complete and well confirmed alternative scientific explanations, as long as no extraterrestrial life has been discovered. Given that neither such explanations nor strong evidence of extraterrestrial beings are available, she lacks any reason to dismiss the chance hypothesis as a serious contender in the origin of life debate.

Concluding Remarks

I have presented two arguments why the chance hypothesis should not be dismissed too quickly, i.e. *prior* to the development of plausible and complete alternative theories or to the discovery of extraterrestrial life. To my mind, each of these arguments is successful individually. In combination, they are devastating. There is even more to be said in favour of the chance hypothesis. Most importantly, it dissolves the dreaded Fermi paradox: If our galaxy is teeming with (intelligent) species, where are they?[41] The obstacles to interstellar and even intergalactic spacecraft do not seem to be unsurmountable, cosmic habitats tend to be rather short-lived etc., and yet we have seen no sign of alien life so far.[42]

7 Is the Origin of Life a Fluke?

Presumably, most origin of life researchers and astrobiologists will stick to their conviction that life is a common phenomenon in the observable universe, undeterred by the Fermi paradox, the Vast Cosmos and the Indifferent Nature Argument. Unless they have a plausible and largely complete account of the origin of life in their hands, however, they should at least acknowledge this conviction for what it is: an act of faith.[43,44]

Notes

1 Here is a small list of other prominent quotes. 'Such a question [...] "as could life have originated by a chance occurrence of atoms" clearly leads as our knowledge, and also the limitations of time and space available, increase, to a negative answer. This answer would seem to me, combined with the knowledge that life is actually there, to lead to the conclusion that some sequences other than chance occurrences must have led to the appearance of life as we know it' (Bernal, 1965, p. 53). 'In the evolution of life on Earth we are dealing with millions of different life forms, each based on many genes. Yet the mutational mechanism as presently imagined could fall short by hundreds of orders of magnitude of producing, in a mere four billion years, even a single required gene. [...] Surely the biological community should apply all its resources for producing creative ideas to some of these problems if our teachings are to remain internally consistent' (Salisbury, 1969, pp. 342–3). 'The genes found today, cannot have arisen randomly, as it were by the throw of a dice. There must exist a process of optimization that works toward functional efficiency. Even if there are several routes to optimal efficiency, mere trial and error cannot be one of them' (Eigen, 1992, p. 12).
2 I have no definition of 'life' to offer, but am confident that an intuitive understanding will suffice for my purposes. I will use the term 'intelligent life' to refer to living beings who can reflect on their own existence and origin.
3 White draws in particular on Fry (2000, ch. 13).
4 There are only two noteworthy (and notorious) examples of a rejection of (1) by respectable contemporary scientists. (a) Fred Hoyle and Chandra Wickramasinghe's version of the *panspermia theory* according to which life originated in outer space, and comets brought it to Earth. To answer the objection that such a scenario would simply shift the origin-of-life problem to another planet, the authors assume that life, as we know it, was first brought about by a superintellect (e.g., Hoyle & Wickramasinghe, 1981). (b) Michael Behe's *intelligent design theory*, according to which 'irreducible complexity', as exemplified by even the most basic biological systems, could not have emerged via Darwinian or other blind natural mechanisms and, therefore, is best explained by intelligent design (Behe, 1996, esp. ch. 8; see also Bradley, 2004).
5 Alternatively, we can say that, according to the chance hypothesis, the probability for the emergence of life on Earth was extremely tiny, given complete knowledge of the laws of nature and detailed (but incomplete) knowledge of the early (pre-life) conditions on Earth.
6 During the first half of the twentieth century, the idea that life started with a highly improbable fortuitous 'first event' was popular. However, the biochemical basis of life was only poorly understood and the complexity of 'primitive life' grossly underestimated. This led researchers like Leonard Troland to think that the improbable event needed to get the evolution of life started would, *very probably*, nevertheless take place somewhere, at some time, given the hundreds of millions of years available during Earth's history (cf. Fry, 2000, pp. 74–6; pp. 193–4).
7 Given the number of protons, neutrons and electrons in the observable universe ($\approx 10^{80}$), its age ($\approx 4 \times 10^{16}$ s) and the maximum number of transitions from one physical state to another per second ($\approx 10^{45}$), the maximum number of events in the observable universe so far is $\approx 10^{141}$. This mind-bogglingly big number still has 461 zeros less than the number of possible gene variants calculated by Salisbury and Eigen.
8 At least among biologists whose work did not focus on the origin of life there have been prominent distractors, most notably Monod (1971, p. 144), 'Life appeared on earth: what,

before the event, were the chances that this would occur? The present structure of the biosphere far from excludes the possibility that the decisive event occurred only once. Which would mean that its a priori probability was virtually zero'. Similar views can be found in Mayr (1982, pp. 583–4) and Conway Morris (2003, ch. 4).

9   What physicists count as a *universe* within the multiverse is a vague affair. For our purposes it may suffice to define universes as 'large space-time regions that are causally (fairly) disconnected from each other'. For a typology of multiverse theories, see Greene (2011) and Tegmark (2014).

10   Copernicus himself almost certainly would not have subscribed to the 'Copernican Principle'. He emphasized, for instance, that according to his model, Earth *almost* remained at the center of the universe; 'the distance is as nothing, particularly in comparison with the sphere of the fixed stars' (Copernicus, 1543, p. 13).

11   This assumption is usually called the 'Cosmological Principle'.

12   The steady state theory, advocated by Hermann Bondi, Thomas Gold and Fred Hoyle, was indeed characterized by a *perfect cosmological principle* that states that the universe is homogeneous and isotropic in space *and* time. The theory was discredited by the discovery of cosmic microwave background radiation in 1965. Its temporary popularity can perhaps partly be explained by an exaggerated and even dogmatic adherence to strong versions of the Copernican Principle (see, e.g., Carter, 1974, p. 125).

13   At least I assume so, for the moment. But see my later remarks on 'freak observers' and 'Boltzmann brains'. See also Feinberg and Shapiro (1980), in which the renowned authors speculate about bizarre forms of life, including 'plasma life within stars'.

14   Dubbed and explicitly stated for the first time by Brandon Carter (1974). A very influential, though highly misleading discussion of anthropic principles can be found in Barrow and Tipler (1986, pp. 15–26). Whereas Carter's weak anthropic principle was meant only as a useful (and often overlooked) tautology, Barrow and Tipler, in addition, introduced a group of much stronger anthropic principles according to which life *necessarily* exists in the cosmos. These principles, however, are nothing but speculative metaphysics for which the authors give little, if any, arguments (see, e.g., Gardner, 1986). For intriguing philosophical discussion see Leslie (1989) and Bostrom (2002).

15   Richard Dawkins (1986, pp. 144–5) is admirably clear on this point. 'Yet if we assume, as we are perfectly entitled to do for the sake of argument, that life has originated only once in the universe, it follows that we are *allowed* to postulate a very large amount of luck in a theory, because there are so many planets in the universe where life *could* have originated. [. . .] To be more precise: the maximum odds against the origin of life on any one planet that our theories are allowed to postulate, is the number of available planets in the universe divided by the odds that life, once started, will evolve sufficient intelligence to speculate about its own origins.' Unfortunately, Dawkins underestimates *how* big the cosmos probably is (see next note). This leads him to insist that any plausible account of the origin of life needs to make the emergence of life on Earth *much* more probable than a fluke produced by the random shuffling of molecules (but see Dawkins, 2006, pp. 137–8).

16   Here is just a small sample of popular books on the topic: (Rees, 1999; Vilenkin, 2006; Carr, 2007; Greene, 2011; Tegmark, 2014). It is worth emphasizing that, even according to George Ellis (2011, p. 38), one of the most prominent critics of more ambitious multiverse theories, 'nearly all cosmologists' accept at least a 'Level 1 Multiverse', i.e. the idea that beyond the cosmic visible horizon, there are 'many – perhaps infinitely many – domains, much like the one we see.'

17   Postulating that our cosmos is infinite in space (and mass) alone does not suffice to account for the emergence of life, of course. Additional (in my eyes, rather innocuous) assumptions have to be made, for example that the number of *kinds* of fundamental constituents is finite, that laws of nature hold uniformly, that there is a proper mix of homogeneity and heterogeneity, i.e. an infinite number of planets with sufficiently varying initial conditions etc.; for some useful discussion see Mash (1993, esp. pp. 208–10). (Thanks to Claus Beisbart for the pointer.) Mash also thinks that proponents of the Vast Cosmos Explanation of Life need to assume that

physicalism is true and that determinism is false. This strikes me as wrong, but I will not elaborate on that here.
18 If you think otherwise, you are committing an Inverse Gambler's Fallacy (see Hacking, 1987).
19 At least if we ignore the possibility of panspermia, see note 4.
   A similar kind of objection is usually discussed in the context of the 'fine tuning' for life of natural constants and cosmological (rather than planetary) initial conditions. I think that in this context objections against the Vast Cosmos (Multiverse) Argument are somewhat more plausible (see especially White, 2003; and most recently Landsman, 2016), though, in the final analysis, equally unjustified.
20 This is a *very* important disanalogy (see, e.g., Leslie, 1988).
21 I am grateful to an anonymous reviewer for raising this objection.
22 It must be conceded, however, that intuitions may strongly vary here. We wade into deep metaphysical waters, regarding the identity conditions of persons and planets. For a recent discussion of origin essentialism, see Mackie (2006, especially ch. 6).
23 Named after Austrian physicist Ludwig Boltzmann (see, e.g., Boltzmann, 1897, §4).
24 By the way, such an emission is only physically possible if we assume that mental states (nomologically) supervene on physical states – a position widely, though not universally, shared in the philosophy of mind.
25 This formulation is almost identical with Nick Bostrom's 'self-sampling assumption' (Bostrom, 2002, p. 57). Bostrom thinks the formulation can be further improved, but this complication need not concern us here. To my knowledge, the term 'Principle of Mediocrity' was coined by astrophysicist Alexander Vilenkin (see, e.g., Vilenkin, 1996).
   A characteristic misrepresentation of the principle of mediocrity can be found in Mash (1993, p. 213). Mash quotes a couple of alleged instances of the mediocrity assumption: e.g. 'our surroundings are more or less typical of any other region of the universe' (Shklovskii & Sagan, 1966, p. 357), and 'our own evolution is typical' (Tipler 1985, p. 134). But only the second statement is covered by the principle of mediocrity. A typical *observer* may find herself at a place in the cosmos that is very atypical for the cosmos as a whole. In fact, Shklovskii and Sagan refer to a very problematic version of the Copernican Principle (see my discussion above).
26 For similar examples, see Leslie (1996, ch. 5). I have already used the text examples in Weidemann (2016, p. 125).
27 See, e.g., Aguirre (2007) and Smeenk (2014).
28 Here are just two more examples of papers that discuss related issues: Page (2006) and de Simone *et al.* (2010).
29 How likely the evolution of intelligent beings on Earth-like planets is, given that primitive life has already emerged, is a controversial issue. As far as I can say, the great majority of evolutionary biologists consider the evolution of intelligence highly improbable. See only the statements of two giants in the field (Dobzhansky, 1972; Mayr, 1994). There are diverging voices, however (see, e.g., Conway Morris, 2003). Two things should give pause to 'optimists': (i) the transition from prokaryotic to multicellular life on Earth took two billion years, (ii) of all the hundreds of millions of species during natural history, only one achieved self-conscious intelligence. This indicates either that this kind of intelligence is very hard to acquire, or that its acquisition is almost always maladaptive (or both).
30 For technical details of this intuitive idea, see Hanson (1998).
31 See Bostrom (2002, pp. 79–80) for a similar, though more fanciful, example (with princess and beheaded suitors).
32 Main sequence stars, i.e. stars that are currently fusing hydrogen into helium, are classified under the Morgan–Keenan (MK) system using the letters O, B, A, F, G, K and M, ranging from the hottest (and rarest) O-type to the coolest (and most frequent) M-type. The Sun belongs to the G-type. Most astrobiologists believe that only G- and K-type main sequence stars provide a realistic habitat for life to evolve.
33 David Hume (1779) mentions the 'Epicurean Hypothesis' in his *Dialogues Concerning Natural Religion*, part 8. Sober (2008, pp. 116–25) provides some useful discussion on Epicureanism and its differences to intelligent design theories and Darwinism.

34 If we assume not only that the cosmos is infinitely big, but also that there was an infinite amount of time for species to emerge, we face a different problem. Why do we not see unicorns and mermaids? Why do we observe a taxonomic hierarchy rather than a continuum of forms? Ironically, nobody has worded these and other objections to a chance explanation of biological 'contrivances' more masterfully than the often-loathed William Paley (1802, especially pp. 37–40); cf. Sober (2008, pp. 124–5).

35 If we have strong evidence that the cosmos is very big or infinite, we also have strong evidence for the existence of extraterrestrial life *somewhere*. This does not disconfirm the chance hypothesis, of course.

36 For a more technical account of Leslie's idea, see van Inwagen (2015, p. 204). For some critical discussion, see Bostrom (2002, pp. 23–32).

37 $P(DH / CH / \text{Life on Earth}) > P(CH / \text{Life on Earth})$ if and only if $P(DH) \times P(\text{Life on Earth} / DH) > P(CH) \times P(\text{Life on Earth} / CH)$, where '$P(DH)$' and '$P(CH)$' refer to the probability of DH resp. CH prior to the life on Earth evidence.

38 Richard Swinburne, in particular, has argued extensively that an omnipotent, omniscient and morally perfect being would have good reasons to create corporeal persons (see, e.g., Swinburne, 2003, pp. 107–14); for skeptical views see, e.g., Narveson (2003) and Sober (2008, pp. 141–8); see also Venn (1888), pp. 258–64.

39 For the record, I have strong doubts whether the reasons usually given either for simply ignoring the design hypothesis in the text or for assigning an astronomically low prior probability to it, are tenable. On the other hand, I think that cosmic fine-tuning (Weidemann, 2010; 2014a) is a much better candidate for suspecting design than 'biochemical' or 'planetary' fine-tuning. Unfortunately, I cannot discuss these issues here.

40 An anonymous referee has correctly pointed out to me that Linda's reaction would be perfectly rational, if she were dealt four aces, say, 30 or 300 times in a row, assuming she can be rationally certain that the distribution is not intentionally biased. However, I do not see how this detail could possibly undermine my (and White's) point. I am *not* trying to show that there *could not* be an unintentional bias towards life, but only that we have no reason to assume so a priori. If Linda observed four aces 30 times in a row she would obtain a very strong a posteriori reason for suspecting unintentional bias. *If* biochemists, while simulating the conditions on Earth four billion years ago, succeeded in producing life '30 times in a row' or if astrobiologists discovered extraterrestrial life that emerged independently on other planets, the unintentional bias towards life hypothesis would indeed be triumphant (given design could justifiably be ruled out).

41 There are, of course, other solutions to the Fermi paradox. Most of these solutions, however, suffer from the defect that a plausible explanation why *some* civilizations never developed interstellar spacecraft or succeeded in communicating (they went extinct before, lost interest etc.), is implausibly applied to *all*. More convincing than these sociological solutions is the hypothesis that, though life may be ubiquitous in the cosmos, intelligent life is extremely rare. A good overview of the problem is provided by Webb (2002).

42 The chance hypothesis also nicely explains why there is relatively little consensus among origin-of-life researchers (that is, beyond their dislike of the chance hypothesis), and even benefits Christian apologetics – if intelligent life were widespread in the universe, why of all planets did God choose to become incarnate on Earth? (See Weidemann, 2014b.)

43 Consider, for instance, the following statement by astrobiologists Louis Neal Irwin and Dirk Schulze-Makuch (2011, p. 302), which reminds one more of an Article of Faith than scientific prose: '[...] it is near inconceivable to us that the evolution of conspicuously intelligent, technologically advanced forms of life have never evolved anywhere else in the [observable] universe. On the contrary, we believe that this must have happened many times in many places'. What the authors have to offer as an argument for this belief amounts to little more than an *argument from analogy* (or generalisation from a single case), 'Even if Earth-like planets are rare, given enough solar systems [in the observable universe], planets as inviting for life as Earth must exist elsewhere. Since we know that life exists on Earth, we're most likely to find it and recognize it on other worlds that resemble Earth' (Irwin & Schulze-Makuch, 2011, p. 2).

44 I am indebted to Claus Beisbart and an anonymous reviewer for many helpful suggestions. Work on this paper was generously supported by the John Templeton Foundation ('Randomness and Divine Providence'-project).

References

Aguirre, A. (2007). Making predictions in a multiverse: conundrums, dangers, coincidences. In B. Carr, ed., *Universe or Multiverse?* Cambridge: Cambridge University Press, pp. 367–86.

Albrecht, A. & Sorbo, L. (2004). Can the universe afford inflation? *Physical Review D*, **70**(6), 063528.

Barrow, J. & Tipler, F. (1986). *The Anthropic Cosmological Principle*, Oxford: Oxford University Press.

Behe, M. (1996). *Darwin's Black Box. The Biochemical Challenge to Evolution*, New York, NY: Free Press.

Bernal, J. D. (1965). Discussion of Peter T. Mora. The Folly of Probability. In S. W. Fox, ed., *The Origins of Prebiological Systems and of Their Molecular Matrices*, New York, London: Academic Press, pp. 52–6.

Boltzmann, L. (1897). Zu Hrn. Zermelos Abhandlung „Über die mechanische Erklärung irreversibler Vorgänge". *Annalen der Physik*, **60**, 392–8.

Bostrom, N. (2002). *Anthropic Bias. Observation Selection Effects in Science and Philosophy*, London: Routledge.

Bradley, W. L. (2004). Information, entropy, and the origin of life. In W. M. Dembski and M. Ruse, eds., *Debating Design*, Cambridge: Cambridge University Press, pp. 331–51.

Carnap, R. (1962). *Logical Foundations of Probability*, 2nd edn, Chicago, Il: University of Chicago Press.

Carr, B., ed. (2007). *Universe or Multiverse?*, Cambridge: Cambridge University Press.

Carter, B. (1974). Large Number Coincidences and the Anthropic Principle in Cosmology, reprint in J. Leslie, ed., *Physical Cosmology and Philosophy*, New York, London: Prentice Hall, 1990, pp. 125–33.

Conway Morris, S. (2003). *Life's Solution. Inevitable Humans in a Lonely Universe*, Cambridge: Cambridge University Press.

Copernicus, N. (1543). *On the Revolutions of Heavenly Spheres*, trans. C. G. Wallis, Amherst: Prometheus, 1995.

Dawkins, R. (1986). *The Blind Watchmaker*, London: Longman.

Dawkins, R. (2006). *The God Delusion*, London: Bantam.

de Duve, C. (1991). *Blueprint for a Cell*, Burlington (N.C.): Neil Patterson Publishers.

de Simone, A., Guth, A. H., Linde, A. *et al.* (2010). Boltzmann brains and the scale-factor cutoff measure of the multiverse. *Physical Review D*, **82**(6), 063520.

Dobzhansky, T. (1972). Darwinian evolution and the problem of extraterrestrial life. *Perspectives in Biology and Medicine*, **15**(2), 157–76.

Earman, J. (1986). *A Primer on Determinism*, Dordrecht: D. Reidel.

Eigen, M. (1992). *Steps towards Life. A Perspective on Evolution*, Oxford: Oxford University Press.

Ellis, G. F. R. (2011). Does the multiverse really exist? *Scientific American*, **305**, 38–43.

Feinberg, G. & Shapiro, R. (1980). *Life Beyond Earth. The Intelligent Earthling's Guide to Life in the Universe*, New York, NY: William Morrow.

Fry, I. (2000). *The Emergence of Life. A Historical and Scientific Overview*, New Brunswick, NJ: Rutgers University Press.

Gardner, M. (1986). WAP, SAP, PAP & FAP. *New York Review of Books*, **33** (8th May), 22–5.

Gould, S. J. (1985). Mind and Supermind, reprint in J. Leslie, ed., *Physical Cosmology and Philosophy*, New York, London: Prentice Hall, 1990, pp. 181–8.

Greene, B. (2011). *The Hidden Reality. Parallel Universes and the Deep Laws of the Cosmos*, New York, NY: Alfred A. Knopf.

Hacking, I. (1987). The Inverse Gambler's Fallacy: The argument from design. The anthropic principle applied to Wheeler universes. *Mind*, **96**(383), 331–40.

Hanson, R. (1998). *Must Early Life Be Easy? The Rhythm of Major Evolutionary Transitions*, (unpublished manuscript), http://mason.gmu.edu/~rhanson/hardstep.pdf.

Hart, M. H. (1982). Atmospheric Evolution, the Drake Equation, and DNA: Sparse Life in an Infinite Universe, reprint in J. Leslie, ed., *Physical Cosmology and Philosophy*, New York, London: Prentice Hall, 1990, pp. 256–66.

Hawking, S. W. & Ellis, G. F. R. (1973). *The Large Scale Structure of Space-Time*, Cambridge: Cambridge University Press.

Hawking, S. W. & Israel, W. (1979). An introductory survey. In S. W. Hawking and W. Israel, eds., *General Relativity*, Cambridge: Cambridge University Press, pp. 1–23.

Hoyle, F. & Wickramasinghe, C. (1981). *Evolution from Space*, London: Touchstone.

Hume, D. (1779). *Dialogues Concerning Natural Religion*, London: Penguin.

Irwin, L. N. & Schulze-Makuch, D. (2011). *Cosmic Biology*, New York, NY: Springer.

Jones, M. H. & Lambourne, R. J., eds. (2003). *An Introduction to Galaxies and Cosmology*, Cambridge: Cambridge University Press.

Koonin, E. V. (2007). The cosmological model of eternal inflation and the transition from chance to biological evolution in the history of life. *Biology Direct*, **2**(15).

Kripke, Saul (1980). *Naming and Necessity*, Cambridge, MA: Harvard University Press.

Landsman, K. (2016). The fine tuning argument. Exploring the improbability of our existence. In K. Landsman and E. van Wolde, eds., *The Challenge of Chance. A Multidisciplinary Approach from Science and the Humanities*, Heidelberg: Springer, 111–29.

Leslie, J. (1988). No Inverse Gambler's Fallacy in Cosmology. *Mind*, **97** (386), 269–72.

Leslie, J. (1989). *Universes*, London, New York, NY: Routledge.

Leslie, J. (1996). *The End of the World*, London, New York, NY: Routledge.

Leslie, J. (2001). *Infinite Minds. A Philosophical Cosmology*, Oxford: Oxford University Press.

Mackie, Penelope (2006). *How Things Might Have Been. Individuals, Kinds, and Essential Properties*, Oxford: Oxford University Press.

Manson, N. A. (2003). Introduction. In N. Manson, ed., *God and Design*, London: Routledge, pp. 1–23.

Manson, N. A. & Thrush, M. (2003). Fine-tuning, multiple universes, and the "This Universe" objection. *Pacific Philosophical Quarterly*, **84**, 67–83.

Mash, R. (1993). Big numbers and induction in the case for extraterrestrial intelligence. *Philosophy of Science*, **60**(2), 204–22.

Mayr, E. (1982). *The Growth of Biological Thought*, Cambridge, MA: Belknap Press.

Mayr, E. (1994). Does it pay to acquire high intelligence? *Perspectives in Biology and Medicine* **37**(3), 337–8.

Monod, J. (1971). *Chance and Necessity. An Essay on the Natural Philosophy of Modern Biology*, New York, NY: Vintage.

Narveson, J. (2003). God by design? In: N. A. Manson, ed., *God and Design*, London: Routledge, 88–104.

Page, D. (2006). *Is our Universe Likely to Decay in 20 Billion Years?* Web: http://arxiv.org/pdf/hep-th/0610079v2.pdf.

Paley, W. (1802). *Natural Theology or Evidence of the Existence and Attributes of the Deity, Collected from the Appearances of Nature*, reprint M. D. Eddy and D. Knight eds., Oxford: Oxford University Press, 2006.

Pascal, B. (1670). *Pensées*, trans. W. F. Trotter, New York, NY: E. P. Dutton, 1958.

Raup, D. M. (1985). ETI without intelligence. In E. Regis Jr, ed., *Extraterrestrials. Science and Alien Intelligence*, Cambridge: Cambridge University Press, pp. 31–42.

Rees, M. (1999). *Just Six Numbers*, London: Weidenfeld & Nicholson.
Sagan, C. (1995). The abundance of life bearing planets. *Bioastronomy News*, 7(4), 1–4.
Salisbury, F. B. (1969). Natural selection and the complexity of the gene. *Nature*, **224**, 342–3.
Schopf, W., ed. (1992). *Major Events in the History of Life*, Boston, MA: Jones & Bartlett.
Shklovskii, I. S. & Sagan, C. (1966). *Intelligent Life in the Universe*, San Francisco, CA: Holden Day.
Smeenk, C. (2014). Predictability crisis in early universe cosmology. *Studies in History and Philosophy of Modern Physics*, **46**, 122–33.
Sober, E. (2008). *Evidence and Evolution. The Logic Behind the Science*, Cambridge: Cambridge University Press.
Steinhardt, P. (2014). Big Bang blunder bursts the multiverse bubble. *Nature*, **510**, 9.
Swinburne, R. (2003). The argument to God from fine-tuning reassessed. In N. A. Manson, ed., *God and Design*, London: Routledge, pp. 105–23.
Tegmark, M. (2014). *Our Mathematical Universe*, New York, NY: Alfred A. Knopf.
Tipler, F. J. (1985). Extraterrestrial intelligent beings do not exist. In E. Regis, Jr., ed., *Extraterrestrials, Science and Alien Intelligence*, Cambridge: Cambridge University Press, pp. 133–50.
van Inwagen, P. (2015). *Metaphysics*, 4th edn, Boulder, CO: Westview Press.
Venn, J. (1888). *The Logic of Chance*, 3rd edn, London: Macmillan.
Vilenkin, A. (1996). Quantum cosmology and the constants of nature. In K. Sato, T. Suginohara & N. Sugiyama, eds., *The Cosmological Constant and the Evolution of the Universe*, Tokyo: Universal Academy Press, pp. 161–8.
Vilenkin, A. (2006). *Many Worlds in One. The Search for Other Universes*, New York, NY: Hill and Wang.
Ward, P. & Brownlee, D. (2003). *The Life and Death of Planet Earth*, New York, NY: Henry Holt.
Webb, S. (2002). *Where is Everybody? 50 Solutions to the Fermi Paradox and the Problem of Extraterrestrial Life*, New York, NY: Copernicus.
Weidemann, C. (2010). Zufall, Gott oder Multiversum? Die Feinabstimmung der Naturkonstanten und die Erklärungsbedürftigkeit des Lebens in der modernen Kosmologie. In H.-G. Nissing, ed., *Natur. Ein philosophischer Grundbegriff*. Darmstadt: WBG, 181–96.
Weidemann, C. (2014a). Paralleluniversen, Gott und unser Platz im Kosmos. Das Feinabstimmungsproblem. *Zur Debatte* 2/2014, 35–6.
Weidemann, C. (2014b). Christian soteriology and extraterrestrial intelligence. *Journal of the British Interplanetary Society*, **67**, 418–25.
Weidemann, C. (2016). Did Jesus die for Klingons too? In P. Levinson & M. Waltemathe, eds., *Touching the Face of the Cosmos: On the Intersection of Space Travel and Religion*, New York, NY: Fordham University Press, pp. 119–29.
White, R. (2003). Fine-tuning and multiple universes. *Nous*, **34** (2000), 260–76, reprint (with a new postscript) in N. A. Manson, ed., *God and Design*, London: Routledge, pp. 229–50.
White, R. (2007). Does the origins of life research rest on a mistake? *Nous*, **41**, 453–77.

8

Some Contemporary – and Persistent – Fallacies and Confusions about Astrobiology

MILAN M. ĆIRKOVIĆ

> There is another world, but it is in this one.
> – W. B. Yeats

Introduction: The Astrobiological Revolution (in the Making)

Since 1995, we have witnessed an explosive growth of the nascent multidisciplinary field of astrobiology, which deals with three canonical questions: *How does life begin and develop? Does life exist elsewhere in the universe? What is the future of life on Earth and in space?* (see the "Astrobiology Roadmap" of NASA, Des Marais *et al.*, 2008). A host of important discoveries and advances has been made during the past decade or so, the most important certainly being a discovery of a large number of extrasolar planets; the existence of many extremophile organisms possibly comprising the "deep hot biosphere" of Thomas Gold; the discovery of subsurface water on Mars and the huge ocean on Europa, and possibly also Ganymede and Callisto; the unequivocal discovery of amino acids and a large number of other complex organic compounds inside meteorites; modeling organic chemistry in Titan's atmosphere; the quantitative treatment of the Galactic Habitable Zone; the development of a new generation of panspermia theories, spurred by experimental verification that even terrestrial microorganisms easily survive conditions of an asteroidal/cometary impact; and progress in methodology of SETI studies, etc. (for reviews and textbooks, see Des Marais & Walter, 1999; Darling, 2001; Grinspoon, 2003; Chyba & Hand, 2005; Horneck & Rettberg, 2007; Benner, 2009; Bennett & Shostak, 2011). However, the epistemological and methodological basis of astrobiological and SETI studies presents us with a hornet's nest of issues that, with a few exceptions, have not been tackled in the literature so far. In addition, there are older prejudices and fallacies that somewhat paradoxically have survived and sometimes have mutated into new and internet-friendly strains, which still put obstacles in the path of improving the image of the study of cosmic life.

8 Some Contemporary – and Persistent – Fallacies and Confusions

Every revolution is accompanied by its counter-revolution. So is astrobiology – although due to its wide, multidisciplinary nature this is not always obvious (e.g., Rinaldi, 2007). On the account of its very width, the opposition often comes "from within", i.e., from a field or discipline or a school of thought which is part of the astrobiological synthesis and harbors conservative researchers suspicious of crossing disciplinary fences and barbed wire. A striking example was the distinguished microbiologist and historian of science Howard Gest (1921–2012) who, in several articles posted on web archives, charged astrobiology as being pseudoscience and propaganda ploy (Gest, 2006, 2011). While his attack was perhaps too vitriolic to be taken entirely seriously, it is significant that it attracted much attention on the web and in the social media, confirming that Gest just openly said what many thought.

I shall not tackle here a number of different fallacies often spread by the media that do not belong to the realm of proper science or philosophy. Such is, for instance, the case with the widely encountered prejudice that astrobiology and SETI are expensive scientific endeavours.[1] Even a cursory acquaintance with the real data shows that it is simply untrue and has never been true, even in the time of the greatest debates surrounding ill-fated NASA's *High Resolution Microwave Survey* (Garber, 1999). Furthermore, the alleged high cost of the "search for little green men" is hardly a plausible conjecture even without any empirical data. However, the public image is often that of an extremely expensive and exotic enterprise that is a luxury for even the richest societies. This fallacy has already inflicted significant damage on actual science, but requires a separate and more detailed analysis and refutation. For the moment, we can conclude that the public reception of the search for life and intelligence elsewhere is in sharp opposition to reality – the astrobiological enterprise has been among the cheapest of scientific endeavors of our time. This is valid even within the context of funding of science, not to mention wider societal context, where squandering sums many orders of magnitude greater on senseless military and political adventures is still considered "reasonable" and "common sense".

Another popular hostile mispresentation is that astrobiology is somehow a specific NASA ploy, aimed at upholding concrete bureaucracy, funding and the general standing of that particular organization. The previously mentioned criticism from Howard Gest belongs to this category, but it is one of the most-often repeated conspiratorial claims in the "dark corners" of forums, blogs and social media. Nobody even cursorily informed will doubt that NASA has much bureaucratic baggage and that it is far from always being the efficient and future-oriented organization it claims to be. To assert, however, that a whole new multidisciplinary field expanding throughout the world is a consequence of particular business policy of a single US agency approaches the levels of the most egregious conspiracy theories and general pseudoscience.[2]

Why is it so important to point out confusions and fallacies surrounding the still young astrobiological enterprise? Would it not be enough to leave the passage of time itself to sift and filter truly important and true ideas? This old view is a consequence of deterministic views of history of science – which are a consequence of deterministic views of history as such – and has been more openly abandoned in recent years. In contrast, viewing history in general as "chaostory", in which the trajectory is extremely sensitive to small perturbations, has become important, if not the dominant mode of thinking (Ferguson, 1999; Bunzl, 2004). Virtual history gives us new tools to assess truly important moments and ground-breaking decisions or discoveries. What I wish to propose here, in passing, is that we apply the standards of this counterfactual way of thinking to the astrobiological revolution in evaluating its fruits so far and its prospects for the future. While immensely fascinating in its own right, astrobiology has important lessons for the future of humanity, on Earth and elsewhere; this practical aspect should not be forgotten (Baum, 2010).

So, each research program needs to be evaluated on its current merit in the actual societal and cultural climate. In order to achieve this, persistent analysis of errors and fallacies, and their debunking and elimination, are necessary rather than optional and luxurious activities. We need *proactive* relation, including discussing and debunking of fallacies circulating in scientific media, in popular science, and in wider social media. The world of astrobiology is rapidly becoming so complex that it is no surprise that people occasionally find themselves confused and bewildered; but we need to continue forward – and damn the torpedoes!

Geocentrism – Old Variety

"There is no astrobiology because there is no instance of confirmed extraterrestrial life". That's an adage often heard among scientists and laypeople alike, first presented in a forceful manner in an influential paper by George Gaylord Simpson in 1964. Simpson's stature as probably the most distinguished paleontologist of the twentieth century and one of the leaders of the "second wave" of neo-Darwinian evolutionary synthesis has certainly contributed to this claim being taken more seriously. Of course, there was no astrobiology as such in Simpson's time; he wrote the following.[3]

There is even increasing recognition of a new science of extraterrestrial life, sometimes called exobiology – a curious development in view of the fact that this "science" has yet to demonstrate that its subject matter exists!... Yet there is not a scrap of evidence that "life as we do not know it" actually exists or even that it could exist – evidence, for example, in the form of detailed specifications for a natural system that might exhibit attributes of life without the basis of life as we do know it. (Computers and other artifacts that mimic some features of the life of their makers are not really pertinent to this question.)

On this extremely constraining epistemological criterion, much of what has happened in the past half a century, even in biology, was not science! This has been aptly commented upon by Chyba and Hand (2005).

If exobiology (or astrobiology) were understood to mean solely the study of extraterrestrial life – which it is not – Simpson's criticism would remain strictly true but might nevertheless seem bizarre to many astronomers or physicists. Astrophysicists, after all, spent decades studying and searching for black holes before accumulating today's compelling evidence that they exist... The same can be said for the search for room-temperature superconductors, proton decay, violations of special relativity, or for that matter the Higgs boson.

Note that the Higgs boson was empirically detected at CERN's LHC only in 2013, at the aggregated cost of the order of $\$10^{10}$! Interestingly enough, all the examples listed by Chyba and Hand were either defined or at best became prominent after Simpson's essay. One can add items to the list, more or less specialized: gravitational lenses (between 1924 and 1979), gravity waves, cosmological inflation, molecular assemblers, extremophiles, dark matter particles, global climate change (until recently) and, the most pertinent of all from the point of view of Simpson's topic, *extrasolar planets themselves* (until 1995). Since the existence of extrasolar planets was purely a matter of theoretical faith and not empirical evidence in Simpson's time, would it be justified for him to claim that "statements about extrasolar planets are not really about anything – or, at the very least, they are not science"?

Of course, as argued by Chyba and Hand, the real subject of astrobiology is *cosmic life*, not just extraterrestrial life (even disregarding that the notion of "extraterrestrial" is today hard to define clearly, since the Earth is not a closed-box system and the exchange of matter with its cosmic environment is evident). To the extent that such research deals with testable hypotheses and comparing their predictions with existing evidence, there is no epistemological anomaly in astrobiology compared to other scientific fields. The lack of anomalies is visible, among other things, from the timely appearance of books such as *Astrobiology of Earth* (Gale, 2009), or the many chapters, reviews and research articles on apparently "terrestrial" astrobiological topics.

There was, however, a powerful intellectual movement that postulated sharp division between the terrestrial and cosmic realms of study. It was (of course) Aristotelianism. Over the course of many centuries it has been the ruling dogma that "translunar" and "cislunar" space are subject to different natural laws and even different kinds of material constituents (aether being predominant in the translunar sphere).[4] While sophisticated Aristotelians always maintained that it is the translunar realm that is in fact superior in axiological sense, this is of secondary importance in comparison to the central thesis of *discontinuity*. In addition, a vulgarized version of Aristotelianism subsequently emerged and became powerful in part because it catered to anthropocentrism and human chauvinism. The demise of Aristotelianism in the course of the Scientific Revolution inaugurated by Copernicus should have taught us that the difference between "terrestrial" and "extraterrestrial" is rather arbitrary, the boundary between the two domains being at best diffuse and uncertain, at worst the product of human bureaucracies. In contrast, the fact that the Earth is *an open system* in both the technical physical meaning[5] and in the vernacular has been gradually acknowledged in the course of the past century. Long *before* the advent of the astrobiological revolution we realized that even such a key component of the Earth system as water came to our planet from outside, notably in icy asteroids and/or comets (Abe & Matsui, 1985; for modern, astrobiology-illuminated view see, e.g., Sleep, Zahnle & Neuhoff, 2001; Drake, 2005). The same acknowledgement of the "cosmic connection" applies to most of the other components of the Earth system. The most important of these in the present context is a possibility of interplanetary panspermia, especially the exchange of biotic material between Earth and Mars, which has become an entirely legitimate hypothesis with several potential empirical tests (Davies, 2003).

So, strictly speaking, there is no boundary whatsoever between astrobiology and terrestrial biology. While humans and their institutions may continue to hold to the disciplinary borders (and even enforce it occasionally by building walls and employing border patrols!), nature itself revels in continuity and openness. And this is exactly the pivotal lesson of the astrobiological revolution – not just that it affirms the openness of the Earth system, which we have already realized in the geosciences and space sciences, but that it gives us a strong basis for the continuation of the Copernican revolution and for a decisive rejection of anthropocentrism in the very heart of human institutions, academic, publishing, and all. What Simpson regarded as a fatal weakness has indeed become a major source of strength in the new century.

Isn't It Just Science Fiction?

A particular criticism, somewhat connected with the issue of geocentrism, that is often (over)heard is that the ideas and concepts of astrobiology are either entirely science fictional, or at least very close to science fiction or sound like science fiction or some similar locution, which makes them allegedly less respectful if not openly pseudoscientific. This issue has two main aspects. The first is related to the issue of "reality" of phenomena under consideration. This need not be some high-brow ontological issue; it is enough to accept rather mundane criteria of usual scientific realism. It is claimed that entities assumed or considered within astrobiological discourse are not real in the scientific sense and are products of fertile imagination and wishful thinking.

The second aspect of the science fiction "charge" can be construed as an epistemological problem (where do some of the concepts/entities/ideas discussed by astrobiologists of today come from?), but it is in fact more a question of *legitimacy*. Is it legitimate to use creative art not only as a source of somewhat vague "inspiration" in science, but as a source of useful conjectures and working hypotheses, playing a specific, well-defined role in specific research programs? In a trivialized form, is it allowed to quote literary sources in your "hard" scientific papers and reports? (This scientometric issue gains relevance in many practical terms related to the increased administrative needs of realistic scientific institutions.) Traditionally, the answer has been not only negative, but disparaging as well.

I posit that it is high time to bite the bullet and assert that literary discourse presents a treasure-trove of potentially useful scientific hypotheses, or at least proto-hypotheses. Depending on the specifics of the discipline, the lack of quantitative precision may or may not be detrimental for the epistemological status of the ideas presented, but even in the case of highly quantitative fields, like nuclear physics, literary pre-figuration of important hypotheses is part of the real historical legacy. The example at hand is the famous prediction of a chain-reaction within fissile material as a way of releasing nuclear power by Herbert George Wells in his *The World Set Free* (1914). It stopped Leo Szilard on his walk in the midst of a London drive and almost cost him his life (Rhodes, 1986). A far less known, but probably even more spectacular – and astrobiology-related – example is ascribing the concept of "smart dust" to a 1964 novel by Stanislaw Lem in a premier research journal in the field of textile technology (Farrer, 2010).[6]

In fact, astrobiology is an excellent ground for evaluating this approach of science fiction as a source for science, although it occasionally surfaced in other contexts as well (for particularly illuminating examples dealing with cognitive sciences, see Swirski, 2000, 2006). We can intuitively understand why that is so. Milan Kundera argued: "In art, the classic metaphysical questions – Where do we come from? Where are we going? – have a clear, concrete meaning, and are not at all unanswerable".[7] This could be understood and interpreted on several levels but, for our present purpose, an even very superficial understanding seems to be enough. An artist creates her own universe, which can serve as a model even for the toughest philosophical questions; those scientific fields that are young and in which philosophical preferences still play an important role are likely to find such models useful as thought experiments and generators of hypotheses. (There is another take-home lesson for astrobiology from Kundera's brilliant essays. He writes: "How to define 'provincialism'? As the inability (or the refusal) to see one's own culture in the *large context*".[8] How large context exactly do we need?

Contemporary research suggests: at least the Galactic one. Therefore, we must grow out of our *Galactic provincialism*!)

Consider, for example, an ingenious solution to Fermi's paradox offered by the Canadian science-fiction author Karl Schroeder in his novel *Permanence* (Schroeder, 2002). We ask ourselves "*where are other intelligent beings?*" without really understanding our expectations of the phenomenon of intelligence – which is, above all, a biological phenomenon. Our expectations, Schroeder suggests, are wrong. Intelligence is significant only insofar as it offers an evolutionary advantage, a meaningful response to the selective pressure of the fluctuating environment. It is an adaptive trait, like any other. Adaptive traits are bound to disappear once the environment changes sufficiently for any selective advantage which existed previously to disappear. Ancestors of whales and dolphins had legs once – and used them for walking. In the long run, intelligence is bound to disappear, as its selective advantage is temporally limited by ever-changing physical and ecological conditions. According to Schroeder's idea, intelligent species emerge, expand, conquer many worlds and many ecosystems, but eventually succumb to dysgenic pressures, degenerate and revert to the pre-intelligent stage. This is a *novel* solution in both senses of the word – it blends discursive philosophical and scientific thinking with a poetic expression appropriate for its format; it has not been seriously analysed in the astrobiological research literature – and it works as a possible solution to Fermi's puzzle (Ćirković, 2005). Its qualitative nature devoid of precision and numerical apparatus should not occlude its cognitive value.

Note the double standard often employed by a more conservative scientific audience toward philosophical criteria for scientific methodology: on one hand, attempts to define "proper science" a priori are pretentious and naive, science is "what scientists do"; one the other hand, when scientists do take important inspiration from pop-cultural discourse, it is not "really proper science". In contrast, approaching astrobiology through pop-cultural context (spearheaded obviously by science-fictional discourse) has an additional advantage in making the work in public outreach, promotion and popularization of science significantly easier. On one hand, everybody nowadays complains that the public is not sufficiently acquainted with important scientific and technological issues and is, therefore, incapable of making informed, rational decisions about both financing science and the issues where scientific knowledge is essential (climate change, nuclear waste storage, biosafety, etc.). On the other hand, many still consider public outreach work to be somehow less important than "proper science" – this is again operating a double standard. Is it not much better to use an important resource – pop-cultural references – whenever thematically appropriate and available? And it is clearly arguable that these two criteria, *thematic appropriateness* and *availability*, are very much present exactly in astrobiology.

Evolution?

There has been much confusion about the issue of whether the theory of biological evolution conflicts with the existence of extraterrestrial life and intelligence. It is still often the case that the old adage of George Gaylord Simpson and Ernst Mayr that astrobiologists "may think of themselves as biologists, but they tend to know more about physics, chemistry, and biochemistry than they do about evolutionary biology" (Simpson, 1964; Mayr, 1993) is unthinkingly repeated, especially in the popular press.[9] Alfred Russel Wallace was the first to argue in print that the theory of biological evolution suggests that life on Earth, and especially humans as intelligent beings, are unique in the universe (Wallace, 1903; see also Heffernan, 1978). The fact that he concluded that on the basis of a totally wrong cosmological model should have given all astrobiological sceptics a pause.

In a nutshell, not only is Darwinian evolution perfectly compatible with the Galaxy (and indeed the visible universe) being full of various living and even intelligent beings, but even examples occasionally brought forward against (neo)darwinism are erroneous and misguided when observed in light of astrobiology. For instance, Popper (1974) suggested that finding a *simple* ecosystem on Mars, if it originated and evolved independently of the terrestrial biosphere, would constitute a counterexample to Darwinian evolution as a universal biological phenomenon, since it obviously created a very complex biosphere on Earth. A timely response to this charge was given by Ruse (1977), but the wider, more general question still looms: do we expect the evolutionary mechanisms that have shaped the terrestrial biosphere in the course of the past 3.8 billion years to be truly universal – arguably constituting natural law(s) – or does their relative importance vary in time and space, while we observe only an instance, perhaps a happenstance, of their action? In other words, do we expect life on other planets to evolve by natural selection and genetic drift as on Earth, or may we expect worlds where Lamarckism or orthogenesis or saltations or some entirely unknown and unthought-of mechanism take precedence? This is obviously an important epistemological problem for the life sciences (Bromham, 2016) – and one which cannot be empirically resolved *without* help from astrobiology. (One might even speculate – somewhat tongue-in-cheek – that resistance of at least some evolutionists to astrobiology indicates a fear that the ultimate testing ground might show that their favorite evolutionary mechanism is not *truly* universal.)

Of course, the claim about evolutionary biology allegedly not supporting astrobiological and SETI research could also be understood as a sociological claim *about biologists*. Thus, Barrow and Tipler (1986, p. 133) wrote: "there has developed a general consensus among evolutionists that the evolution of intelligent life... is unlikely to have occurred on any other planet in the entire visible universe. The consensus view has been defended by... Dobzhansky, Simpson, Francois, Ayala *et al.* and Mayr". However, this is all at the anecdotal level; while there was no scientific poll among all biologists, or even just evolutionary biologists, there are many counterexamples than can easily be found. Vocal public proponents of astrobiological *research* – though not necessarily the assumptions or the philosophy usually employed by astrobiologists – included distinguished paleontologists Steven Jay Gould (e.g., Gould, 1987) and David Raup (e.g., Raup, 1992). Even those openly skeptical toward astrobiology and SETI, such as Simon Conway Morris, do engage in the debate, publish in astrobiological journals, and participate in conferences and meetings having astrobiology as their primary topic (e.g., Conway Morris, 2003a, 2003b, 2011). This testifies, time and again, on the great unifying power of astrobiology.

Recently, there appears to have been a shift in views toward the opposite position: that it is exactly astrobiology which could, at least in principle, provide the ultimate testing grounds for Darwinism (Chela-Flores, 2003; Rospars, 2011; see the elaboration by the present author in Ćirković, 2012). While remaining limited to the terrestrial biosphere and the single history of life, we cannot truly assess the extent to which Darwinian mechanisms are indeed universal. This is more general still than just the question of whether the *observed* macroevolutionary patterns in our biosphere's history are typical (being products of general, law-like regularities) or an outlier (being just happenstance products of contingency or stochasticity). Even the world with the same set of evolutionary mechanisms could look drastically different – this is the famous "rewinding the tape" thought experiment of Gould (1989).[10] But the conundrum cannot be definitely resolved without astrobiological input.

There are other classic puzzles in the philosophy of biology where ongoing astrobiological research could be of help. The problem of the level of selection assumes the existence of a biological hierarchy, but this working assumption is misleading, since it is exactly the origin of this hierarchy – on Earth and, presumably, other inhabited places in the universe – which needs to be explained through evolutionary theory. Samir Okasha, a distinguished philosopher of biology, recently wrote (Okasha, 2006, p. 76):

Ideally, we would like an evolutionary theory which explains how lower-level entities became aggregated into higher-level entities, e.g. how independent genes joined up to form chromosomes, how organelles came to be incorporated into prokaryotic cells, how single celled organisms gave rise to multi-cellular ones... we want to know how the biological hierarchy got there in the first place, rather than just treating it as a given.

In this particular problem, the help of astrobiology seems to be invaluable again – as with Popper's problem of the origin of diversity and the contingency vs convergence problem. Okasha proceeds to contrast "synchronic" and "diachronic" views of the levels of selection, where the former takes levels appropriate for the existing taxons for granted while the latter attempts to show their interdependence and emergence in time from well-motivated evolutionary processes. Since the majority of the grand evolutionary transitions (with the exception, perhaps, of those involving intelligent observers) occurred in the distant geological past, the "diachronic" view is at an obvious disadvantage, barring either the invention of time travel or an exploration of other biospheres. I submit that the latter is both conceptually and practically much likelier.

It is safe to conjecture that, at this stage of development, almost any breakthrough in astrobiology would have significant consequences for evolutionary theory. This could be the discovery of living or fossil life forms on Mars or in subglacial oceans, the discovery of a "shadow biosphere" (remnant of an independent abiogenesis) on Earth, discovery of an oxygen-rich atmosphere on an extrasolar planet, deciphering a radio message of an extraterrestrial civilization, detecting an extraterrestrial artefact like a Dyson shell, or any other similar occurrence. Besides making headlines, any such discovery would become an important argument in any number of the debates about evolutionary theory. While most evolutionists are ready to admit that much when asked in principle, the ramifications of the discovery of extraterrestrial life still have not generated remotely as much theoretical and speculative activity as one might have expected. But the confluence of numerical models, evolutionary theory and a rapid stream of astrobiological discoveries will likely bring about some exciting sailing.

This particular channel of relevance of astrobiology has only recently surfaced. For instance, Jean-Pierre Rospars (2011) argues in a recent paper that the existence of universal biological laws would have a crucial impact on astrobiology in general and on SETI in particular. After reviewing some of the arguments in favor of laws – for example, the efficiency of the universal genetic code in comparison to a large number of random or simulated variations – and some in favor of convergence, Rospars makes an interesting proposal that this immediately implies the necessity of *adding* another metric to the usual spatio-temporal distance between hypothetical intelligent species in the Galaxy, namely the cognitive distance. While this is not new in itself – indeed, Lem suggested something similar in *Summa Technologiae* more than 40 years ago (Lem, 1974) – the novelty lies in the very chain of reasoning, which starts entirely locally with the philosophical issue of biological laws and a new perspective on convergence. As distinguished paleontologist Douglas Erwin recently noted:[11]

[T]he nascent field of astrobiology is predicated on sufficient regularities in the nature of life that we can employ our understanding of life on Earth to make predictions about both the probability of life and its nature elsewhere in the universe.

Such expansion of the explanatory mandate of evolution is a – perhaps, triumphant – consequence of the Darwinian universalism of those dignified concluding words of *On the Origin of Species* (Darwin, 1859, pp. 489–90):

> It is interesting to contemplate an entangled bank, clothed with many plants of many kinds, with birds singing in the bushes, with various insects flitting about, and with worms crawling through the damp earth, and to reflect that these elaborately constructed forms, so different from each other, and dependent on each other in so complex a manner, have all been produced by laws acting around us.... There is grandeur in this view of life, with its several powers, having been originally breathed into a few forms or into one; and that, whilst this planet has gone cycling on according to the fixed law of gravity, from so simple a beginning endless forms most beautiful and most wonderful have been, and are being, evolved.

After all, the basic impetus for Darwin's great theory came from traveling on board HMS *Beagle* (and Wallace's analogous stays in South America and the Mollucas) – which are travels *in space*, and from the point of view of a Victorian gentleman are analogous to space travel today.

Geocentrism – New Variety

The history of science has its own wakes and tides in which individual concepts and memes often get caught like seashore detritus. A new age of geocentrism was inaugurated at the very end of the twentieth century in the celebrated book *Rare Earth* by Peter Ward and Donald Brownlee, whose appearance heralded the birth of a new astrobiological paradigm (Ward & Brownlee, 2000). They have expounded a view that, while simple microbial life is probably ubiquitous throughout the Galaxy, complex biospheres, like the terrestrial one, are very rare due to the exceptional combination of many distinct requirements. These ingredients of the *rare-Earth hypothesis* are well known to even a casual student of astrobiology.

- **Circumstellar habitable zone**: a habitable planet needs to be in the very narrow interval of distances from the parent star.
- **"Rare Moon"**: having a large moon to stabilize the planetary axis is crucial for the long-term climate stability.
- **"Rare Jupiter"**: having a giant planet ("Jupiter") at the right distance to deflect much of the incoming cometary and asteroidal material enables sufficiently low level of impact catastrophes.
- **"Rare decaying isotopes"**: Radioactive *r*-elements (especially U and Th) need to be present in the planetary interior in sufficient amount to enable plate tectonics and the functioning of the carbon–silicate cycle.
- **"Rare Cambrian-explosion analogs"**: the evolution of complex metazoans requires exceptional physical, chemical and geological conditions for episodes of sudden diversification and expansion of life.

Each of these requirements is *prima facie* unlikely, so that their combination is bound to be incredibly rare and probably unique in the Milky Way. Thus, we are forced to return to a pre-Copernican stance, at least regarding habitability. In addition, Ward and Brownlee broke new ground by pointing out the importance of hitherto downplayed factors, like the importance of plate tectonics, inertial interchange events, or "Snowball Earth" episodes of global glaciation for the development of complex life. In many ways, the rare-Earth hypothesis has since become somewhat of a default position in many astrobiological circles, and – since it predicts the absence of rationale for SETI – a mainstay of SETI scepticism. Its challenge to Copernicanism has been largely accepted (although there seem to be lower prices to pay on the market of ideas) as sound in mainstream astrobiology. Particular rare-Earth hypotheses (insofar as we may treat them as separate) are difficult to assess, lacking first-hand knowledge of other Earthlike planets, but some of the difficulties have been exposed in the literature thus far.

While the rare-Earth hypothesis is going to generate controversy and research interest for quite some time to come, it is important to emphasize that the last word will be given by observational work. There is a whole array of potential near-future instruments and tests which could assess the predictions of the rare-Earth paradigm (e.g., Edwards, 2004; Lin, Gonzalez Abad & Loeb, 2014). This is not only because astrobiology, in spite of its revolutionary nature, conforms to the basic tenets of scientific methodology, although that point has sometimes been lost on its critics, since Simpson's time; even more, as time passes, there are more sophisticated models and predictions (e.g., temporal windows in the work of Chopra & Lineweaver, 2016).

There are many criticisms rightly raised against the rare-Earth hypothesis, but here I shall mention just one, which holds significant philosophical interest. It is a particular case of the wider fallacy of neglecting the context. If we are certain that A & B entail X (while X does not entail A & B), but for some reason we can observe or take into account only A – because of observational limitations, theoretical incompleteness or computational intractability – then it is often the case that observation of some A' ≠ A would lead us to believe that ¬X. But the reasoning is incorrect, since it might be the case that some other (unobservable) B' holds, and A' & B' also entail X. In this simplified model, B is the *ceteris paribus*, which we wish to retain in order for our counterfactual requirements ("without Jupiter, Earth would have been less habitable", etc.) to be meaningful. However, it might also be that the state of affairs (A', B) is incoherent.

In the context of rare-Earth theorizing, the lack of *ceteris paribus* is visible in many places, since many of the individual rare-Earth requirements are formulated in terms of Earth-related counterfactuals: *if such and such were different, the Earth would not be hospitable to complex life forms*. Thus, since such and such is improbable, this decreases the overall probability of complex life forms evolving anywhere. I submit that most of these arguments are unsound, since the implicit *ceteris paribus* cannot be maintained, which is visible in the well-known instance of "rare Jupiter". Is Jupiter the optimal "shield" of Earth from cometary/asteroidal bombardment? A smaller part of the problem is that the rare-Earth claim might be empirically wrong. The common-sense conclusion about the role of Jupiter – employed by rare-Earth theorists – has been brought into question by the recent work of Horner and Jones (2008, 2009), who use massive numerical simulations to show that the conjecture that Jupiter acts as a shield against bombardment of the inner Solar System is untrue in a large part of parameter space. Moreover, they conclude "that such planets often actually increase the impact flux greatly over that which would be expected were a giant planet not present". If the results of Horner and Jones withstand the test of time and further research, it is a serious blow to the rare-Earth hypothesis and the related anti-Copernican way of thinking. However, it would still be a major score for astrobiology as a field, since it will demonstrate maturity of the discipline in which intuitively solid prejudices could be rejected based on precise, quantitative work.

The central issue of philosophical interest is that the *ceteris paribus* state of a "Jupiter-less" Solar System is unphysical (Ćirković, 2012). Rare-Earth theorists argue that:

(1) both Earth and Jupiter exist in the Solar System;
(2) Jupiter deflects a fraction of potential impactors from collision trajectories;
(3) with more impactors on collision trajectories, Earth would suffer a higher frequency of catastrophic impacts;
(4) a smaller frequency of catastrophic impacts increases the habitability of any planet.

Hence,

(5) Earth's habitability is increased by the presence of Jupiter in the Solar System.

This quasi-syllogistic formulation is useful for illuminating the issue of using the same labels for objects under the least controversial assumption that the terms of their reference stay the same in all parts of the argument. Horner and Jones investigate (and refute) premise (2); but even if we retain it, the argument is incorrect, since the meaning of "Earth" in (1) and (3) is different. In fact, premise (3) seems to be self-contradictory if we specify Earth as the planet we live on today and know reasonably well. From the purely physical point of view, there is no "Earth without Jupiter", since by definition Earth is a planet formed and evolved through a complex historical process in which Jupiter played an important role. In the simplest possible form:

No Jupiter = no history of the Solar System as we know it = no Earth.

The two possible historical trajectories – the history of the Solar System with Jupiter and without it – are incommensurable. (A desperate proponent of the rare-Earth hypothesis could still claim that Earth is unimportant as a specific entity, and that we should anyway use it as a placeholder for something like "a rocky planet in a habitable zone of its parent star". However, this is self-defeating, since we lose the option of using observation selection to account for the minuscule probability of finding ourselves on an "Earth" once we find out that "Earths" are incredibly rare. In other words, we need to account for the alleged increased habitability due to the presence of "Jupiter" of this particular planet, i.e., our Earth, and not just any similar planet.)

In supposing how the state of affairs could be different, rare-Earth theorists assume simple, linear change, not taking into account the self-organizing nature of the relevant physical system, where a very small change at time t could cause dramatic divergence at some later time, $t + \Delta t$. Asking about the fate of Earth in the absence of Jupiter is self-contradictory from the point of view of physics; Earth is a unique part of the complex system that includes Jupiter as a major component, so there are no guarantees that Earth would have existed at all if Jupiter were not present. And we can ask similar questions about all the other "rare-Earth" requirements. So, the threat to Copernicanism is actually much smaller than the press (and its manifold ideological allies) would have us believe.

Instead of Conclusions: What can Keep the Revolution Alive?

So, what next? Discussing philosophically contentious issues and debunking fallacies such as those mentioned above should not deter quite legitimate – and, indeed, highly desirable! – criticism of the whole astrobiological enterprise. In fact, we should do our best to encourage such criticism. One line of criticism which is very much a growing concern is that, as the astrobiological revolution progresses, conservative backlash to the cases which are (for one reason or another) near the epicentre of media hype will grow stronger. The case of alleged arsenic-eating bacteria that generated tremendous controversy a couple of years ago (e.g., Rosen, Ajees & McDermott, 2011) is often mentioned in web discussions – but almost never has the wider background been discussed in any detail; especially the ideological and philosophical background of the resistance to astrobiology.

One of the reasons why long-reaching and epochal consequences of the astrobiological revolution are not yet fully perceived is undoubtedly the lack of historical and philosophical perspective. Of course, it has always been the case that in the "sound and the fury" of events, a global picture is lost, a forest vanishes among the trees, and it is not always easy to keep a perspective on the dynamical whole. We probably should zoom-out our perspective and try to put the astrobiological revolution into a wider context of the Enlightenment struggle for a completely rational worldview. As Roger Hausheer summarized:[12]

> Among the basic assumptions of the rational, scientific, Enlightenment approach to the world, both of nature and of man, is the belief that everything can and should be studied with objective detachment as inert material which can be exhaustively described, classified, or brought under covering causal laws. For the purposes of scientific investigation, the world is conceived of as possessing no independent life of its own outside the system of scientific laws that govern its behavior, or beyond the exhaustive classificatory schema into which it falls. Whether it is Newtonian physics accounting for the movements of physical bodies or Linnaean botany meticulously describing plants, such methods of study explicitly rule out and exclude the unaccountable, the unpredictable, the undescribable.

Contemporary astrobiology is, if anything, an inheritor of this Enlightenment project. Its wide multidisciplinary synthesis promises to unify seemingly entirely distinct domains, separated by astronomical distances and geological/evolutionary timescales. It is exactly this multidisciplinary character that makes for such a strong appeal, among both researchers and the wider public (Fergusson, Oliver & Walter, 2012). If it is to be kept alive, there seem to be several important points to keep in mind at all times.

Astrobiology tallies excellently with other synthetic movements in contemporary science. Consider, for instance, the "Earth system" science (e.g., Schellnhuber, 1999), which emphasizes tight interconnections between different aspects of geo- and planetary sciences, atmospheric sciences, paleobiology and evolution in order to obtain a synthetic picture of the evolution of the Earth system and its environment. There are many other such movements in many fields, and they are all under fire from both conservative forces defending their disciplinary barriers and border patrols and fortifications, and various extrascientific forces, be it organized religion, fossil-fuels lobbies, political correctness, or some such thing.

The resistance to astrobiology comes from several quarters and it has both scientific and extra-scientific components. While the scientific part – as exemplified in the works of Simpson or Mayr, as previously discussed – is in retreat, we may expect that extra-scientific resistance will retain the momentum or even increase. This extra-scientific resistance has two main varieties: creationist/quasi-religious and secular. One part of it clearly comes from the defenders of disciplinary boundaries, who are often outraged at any trampling on their particular hallowed turf of inquiry. This includes institutional inertia and unwillingness to change and adapt to the clearly multidisciplinary mode of astrobiological research. But this bureaucratic component is certainly not the whole story. In some of the sceptical accounts, one might find elements of historical revisionism, related to either still-contentious

8 Some Contemporary – and Persistent – Fallacies and Confusions

issues such as abiogenesis or to sustainability of scientific policies in the medium or long term. An example in this sense is the statement by Rinaldi (2007, p. 440):

Thus, if ExoMars and other upcoming missions near to Earth fail to find life in space, further astrobiology research could be halted for an unpredictably long period, as politicians and the scientific community – not to mention society – decide whether the money is better spent on studying topics closer to home.

"Topics closer to home" might, ironically, include things which are decisively *less* empirical and mundane, like string theory or gravitational wave detection. Rinaldi's perspective is shared by many – and it does present, especially when unfortunate experience with SETI projects is taken into account, a danger that the astrobiological community, taken in the most general sense, needs to face and counter. One way of mitigating the risk is to point out how resistance to astrobiology is so often motivated by extra-scientific concerns, not only by battles for funding, but also deeper ideological conflicts.

An important component in the resistance is the rejection, explicit or implicit, of naturalism and Copernicanism as essential elements of scientific outlook and scientific method. Enough has been written about recent rejections of naturalism, mostly in the context of the debate between science and movements such as "Intelligent Design".[13] Less scrutinized has been a silent rejection – or at least a move away from – Copernicanism, as the powerful antidote to dogmatic anthropocentrism and generator of the greatest scientific revolutions of the modern age. One might naively think that rejection of Copernicanism is a high price, unlikely to be paid by anybody except a few religious zealots, stuck in the Middle Ages. Unfortunately, this is false and, if anything, the anti-Copernican cartel has grown stronger in recent decades. An extremely wide anti-Copernican front encompasses people ranging from opponents of animal rights and other defenders of anthropocentric legal orthodoxies to various conservative "warriors on science" and their various allies, from the Discovery Institute to anti-vaccination lobbies, to self-proclaimed "progressive humanists" incapable of dealing with the rational facts of science (including even enlightened people like Hannah Arendt or Michael Frayn[14]) to radical futurists believing we need ideological anthropocentrism to ensure the perceived desired future of humanity. Fighters against perceived "scientism" and alleged "coldness" of modern science à la Mary Midgley[15] hold hands with *both* anti-environmentalists, who do not recognize Genesis 1:28–30 as the harmful Bronze Age superstition it really is, and extreme new-age environmentalists worshipping Gaia as – no surprise there! – the center of the universe. Concerned guardians of the "humanistic canon" worried about the position of humanistic disciplines in the university curricula allegedly under attack from science and engineering join forces in the anti-Copernican camp with assorted media and arts pundits portraying science and scientists in Dr. Victor Frankenstein's mold. And to these, one should add legions of their less sophisticated counterparts in much of the developing world, often blending local superstitions into the anti-scientific mix (e.g., "explaining" AIDS by black magic in large parts of Africa) and preying on poor educational standards. In spite of much effort by various environmental groups, in the twentyfirst century a mass murderer of animals, including our closest mammalian and even primate relatives is still celebrated as a "capable hunter", while nobody would attach that label to, for instance, Norwegian far-right terrorist Anders Behring Breivik, convicted for killing 77 people in 2011. It would not be an overstatement to claim that anti-Copernicanism in one form or another dominates 99% of public life and thought on this planet – which still serenely revolves around the Sun, an insignificant speck on the periphery of the Milky Way.

Thus, the job of the Copernican revolution is still quite an actual and timely concern. While the Inquisition that condemned Galileo seems unlikely to receive any open support today, I submit that this is more because of their old-fashioned garments and politically incorrect language than any true disonance of ideas. After all, the underlying concern stays the same: worry about perceived "well-being of humanity" and its institutions being threatened by "cold" and "soulless" science and its discoveries, never mind the truth. The focus of the odium has shifted from astronomy in Galileo's time to evolutionary biology, and computer and environmental sciences today, but the underlying reality remains the same: below a thin skin of modernity often threatens a surprisingly medieval anthropocentrism.[16]

Even in science itself, the Copernican revolution often looks like unfinished business, and indeed much science aids and abets the tide of anti-Copernicanism in various ways: by condoning various anthropocentric social and political mores, especially in animal ethics and environmental science, by reintroducing teleological elements into science, by accepting some of the postmodern nonsense about social construction of physical reality, by seeking "deep" reasons beneath obvious coincidences, etc. – and, most pertinently for our purposes here, by postulating various rare-Earth hypotheses. This is not to say that such hypotheses cannot be best explanations of the empirical phenomena; but I maintain that we need to be fully honest and upfront about their wider context and ramifications.

Finally, the relationship of professional science to public outreach, promotion, and popularization of science should be taken into account. Without going into much detail here, one might speculate that the contingent fact of history that proponents of existence and search for life and intelligence in the universe have traditionally been people who played highly visible role in "bringing science to the masses" like Carl Sagan or Sir Arthur Clarke or even Neil deGrasse Tyson may be important in assessing why some people view astrobiology as more a theater for public consumption than the "real" science. While this remains a topic for future studies in sociology of science, there seems to be no real anticorrelation between public visibility of a scientific topic or a discipline and the level of research quality related to that topic or discipline. If anything, one could expect a positive correlation,[17] especially when conditioned upon the fact that most of the history of science deals with epochs in which the funding of science was decisively an *un*democratic process, in contrast to the present situation in most of the world.

Therefore, we are entitled to be cautious optimists. While any revolution must eventually run its course and transform into incremental "normal science" in the Kuhnian sense, such moments are established only by later historians' reflections. At present, we see no sign of the astrobiological revolution abating, and it is by no means certain that it will run its course soon. On the contrary, we may expect that significant discoveries still await us. But the multidisciplinary and multicultural nature of the astrobiological endeavor is already apparent – the fruits worthy of tending and cherishing far beyond the original context. Taking this key lesson to heart requires ever more considerations belonging, in the final analysis, to philosophy in its ancestral meaning: the love of wisdom.

Acknowledgements

I wish to thank Jelena Dimitrijević, Anders Sandberg, Stuart Armstrong, Branislav Vukotić, Nick Bostrom, Slobodan Popović, Slobodan Perović, Ivana Kojadinović, Karl Schroeder, Petar Grujić, Jelena Andrejić, Momčilo Jovanović, Goran Milovanović, Eva Kamerer, Dušan Indjić, Zona Kostić, George Dvorsky, Zoran Knežević, Steven J. Dick, Jacob Haqq-Misra, the late Robert Bradbury, and the late Branislav Šimpraga for many pleasant and useful discussions on the topics related to the subject matter of this study. Aleksandar Obradović, Dušan Pavlović, and Seth Baum kindly helped in obtaining some of the crucial references. This is also an opportunity to thank KoBSON Consortium of Serbian libraries, NASA Astrophysics Data System and incredibly useful websites http://arxiv.org/ and http://tvtropes.org/. The author has been supported by the Ministry of Education, Science and Technological Development of the Republic of Serbia through grant ON176021.

Notes

1 There are unfortunate historical reasons for regarding the "SETI sector" of the essentially unified astrobiological project as a separate item. The questions motivating SETI are very old, going back to ancient philosophical discourses, but the search itself emerged as a scientific discipline in 1959–1960, which, in turn, is much older than the ongoing astrobiological revolution (for historical account see, e.g., Dick, 1996). This has been the source of much confusion, some of which is incumbent upon present-day researchers to dispel (Ćirković, 2012).
2 Of course, proponents of "intelligent"design have been more than happy whenever any type of astrobiological criticism has been voiced in the serious scientific literature, e.g., http://www

.evolutionnews.org/2012/08/astrobiology_sc_1063371.html. Their right-wing ideological allies often hide their antiscientific outlook behind worries about budget deficit and NASA's "wastefulness".
3 Simpson (1964), pp. 769–70.
4 See Falcon (2005) for an overview of Aristotelian physics and cosmology.
5 In physics, especially in thermodynamics, an open system is a system that can exchange both energy and matter (particles) with its environment, in contrast to closed systems and isolated systems.
6 The novel is *The Invincible* (Lem, 1964).
7 Kundera (2007, p. 4).
8 Kundera (2007, p. 37).
9 A typical sceptical reaction in the blogosphere could be found at http://praxtime.com/2013/11/25/sagan-syndrome-pay-heed-to-biologists-about-et/ (last accessed April 2, 2016).
10 See also Gould (1996, 2002). For some points of entry into already voluminous literature on the subject see Beatty (2006) and Turner (2011).
11 Erwin (2015, p. 1).
12 Hausheer (2003, p. 36).
13 See, e.g., Pigliucci (2010).
14 Arendt (1963); Frayn (2006).
15 Midgley (1985).
16 To give a recent example that borders on the surreal, the politically left-leaning British *Guardian* has, following the tragic death of David Bowie, among other items, published (on January 13, 2016) a column by Giles Fraser entitled "Sorry, I know David Bowie was great. But I don't believe in life on Mars" (http://www.theguardian.com/commentisfree/2016/jan/13/david-bowie-life-on-mars-radical-singularity-fantasy, last accessed March 1, 2016). In this bizarre text, as much as meaning could be found in a typically postmodern word-salad, Bowie is charged with artistically seeking to transcend human small-mindedness and navel-gazing – as if it were a bad thing!
17 Or at least that could be suggested by anecdotal evidence based on cases such as those of Galileo, Flammarion, Eddington, Hoyle, Sagan, Gould and a host of other great communicators among scientists.

References

Abe, Y. & Matsui, T. (1985). The formation of an impact-generated H_2O atmosphere and its implications for the early thermal history of the Earth. *Journal of Geophysical Research*, **90**, C545–60.

Arendt, H. (1963, 2007). The conquest of space and the stature of Man. *New Atlantis*, Fall issue, 43–55 (http://www.thenewatlantis.com/publications/the-conquest-of-space-and-the-stature-of-man, last accessed August 15, 2015).

Barrow, J. D. & Tipler, F. J. (1986). *The Anthropic Cosmological Principle*, New York, NY: Oxford University Press.

Baum, S. D. (2010). Is humanity doomed? Insights from astrobiology. *Sustainability*, **2**, 591–603.

Beatty, J. (2006). Replaying life's tape. *The Journal of Philosophy*, **103**, 336–62.

Benner, S. (2009). *Life, the Universe... and the Scientific Method*, Gainesville, FL: The FfAME Press.

Bennett, J. O. & Shostak, S. (2011). *Life in the Universe*, 3rd edn, San Francisco, CA: Benjamin Cummings.

Bromham, L. (2016). Testing hypotheses in macroevolution. *Studies in History and Philosophy of Science Part A*, **55**, 47–59.

Bunzl, M. (2004). Counterfactual history: a user's guide. *American Historical Review*, **109**, 845–58.
Chela-Flores, J. (2003): Testing evolutionary convergence on Europa. *International Journal of Astrobiology*, **2**, 307–12.
Chopra, A. & Lineweaver, C. H. (2016). The case for a Gaian Bottleneck: the biology of habitability. *Astrobiology*, **16**, 7–22.
Chyba, C. F. & Hand, K. (2005). Astrobiology: the study of the living universe. *Annual Review of Astronomy and Astrophysics*, **43**, 31–74.
Ćirković, M. M. (2005). 'Permanence' – an adaptationist solution to Fermi's Paradox?. *Journal of the British Interplanetary Society*, **58**, 62–70.
Ćirković, M. M. (2012). *The Astrobiological Landscape: Philosophical Foundations of the Study of Cosmic Life*, Cambridge: Cambridge University Press.
Conway Morris, S. (2003a). *Life's Solution: Inevitable Humans in a Lonely Universe*, Cambridge: Cambridge University Press.
Conway Morris, S. (2003b). The navigation of biological hyperspace. *International Journal of Astrobiology*, **2**, 149–52.
Conway Morris, S. (2011). Predicting what extra-terrestrials will be like: and preparing for the worst. *Philosophical Transactions of the Royal Society A*, **369**, 555–71.
Darling, D. (2001). *Life Everywhere: The Maverick Science of Astrobiology*, New York, NY: Basic Books.
Darwin, C. (1859). *On the Origin of Species by Means of Natural Selection*, London: Murray.
Davies, C. W. (2003). Does life's rapid appearance imply a Martian origin? *Astrobiology*, **3**, 673–9.
Des Marais, D. J. & Walter, M. R. (1999). Astrobiology: exploring the origins, evolution, and distribution of life in the Universe. *Annual Review of Ecology and Systematics*, **30**, 397–420.
Des Marais, D. J. *et al.* (2008). The NASA Astrobiology Roadmap. *Astrobiology*, **8**, 715–30.
Dick, S. J. (1996). *The Biological Universe: The Twentieth-Century Extraterrestrial Life Debate and the Limits of Science*, Cambridge: Cambridge University Press.
Drake, M. J. (2005). Origin of water in the terrestrial planets. *Meteoritics & Planetary Science*, **40**, 519–27.
Edwards, H. G. M. (2004). Raman spectroscopic protocol for the molecular recognition of key biomarkers in astrobiological exploration. *Origins of Life and Evolution of the Biosphere*, **34**, 3–11.
Erwin, D. H. (2015). Was the Ediacaran–Cambrian radiation a unique evolutionary event? *Paleobiology*, **41**, 1–15.
Falcon, A. (2005). *Aristotle and the Science of Nature: Unity without Uniformity*, Cambridge: Cambridge University Press.
Farrer, J. (2010). Smart dust: Sci-fi applications enabled by synthetic fiber and textiles technology. *Textile*, **8**, 342–7.
Ferguson, N. (1999). *Virtual History: Alternatives and Counterfactuals*, New York, NY: Basic Books.
Fergusson, J., Oliver, C. & Walter, M. R. (2012). Astrobiology outreach and the nature of science: the role of creativity. *Astrobiology*, **12**, 1143–53.
Frayn, M. (2006). *The Human Touch: Our Part in the Creation of a Universe*, London: Faber and Faber.

Garber, S. J. (1999). Searching for good science: The cancellation of NASA's SETI program. *JBIS*, **52**, 3–12.
Gale, J. (2009). *Astrobiology of Earth: The Emergence, Evolution, and Future of Life on a Planet in Turmoil*, Oxford: Oxford University Press.
Gest, H. (2006). The "Astrobiology" fantasy of NASA, unpublished manuscript (available at http://www.bio.indiana.edu/about/history/biographies/gest_pdfs/astrobiology.pdf, last accessed January 15, 2016).
Gest, H. (2011). On the origin, evolution, and demise of an oxymoron: 'astrobiology', unpublished manuscript (available at https://scholarworks.iu.edu/dspace/handle/2022/13376, last accessed January 15, 2016).
Gould, S. J. (1987). *The Flamingo's Smile: Reflections in Natural History*, New York, NY: W. W. Norton & Company.
Gould, S. J. (1989). *Wonderful Life: The Burgess Shale and the Nature of History*, New York, NY: W. W. Norton.
Gould, S. J. (1996). *Full House: The Spread of Excellence from Plato to Darwin*, New York, NY: Three Rivers Press.
Gould, S. J. (2002). *The Structure of Evolutionary Theory*, Cambridge, MA: Belknap Press.
Grinspoon, D. (2003). *Lonely Planets: The Natural Philosophy of Alien Life*, New York, NY: HarperCollins.
Hausheer, R. (2003). Enlightening the Enlightenment. In J. Mali and R. Wokler, eds., *Isaiah Berlin's Counter-Enlightenment*, Philadelphia, PA: American Philosophical Society, pp. 33–50.
Heffernan, W. C. (1978). The singularity of our inhabited world: William Whewell and A. R. Wallace in dissent. *Journal of the History of Ideas*, **39**, 81–100.
Horneck, G. & Rettberg, P., eds. (2007). *Complete Course in Astrobiology*, Weinheim: Wiley-VCH.
Horner, J. & Jones, B. W. (2008). Jupiter – friend or foe? I: The asteroids. *International Journal of Astrobiology*, **7**, 251–61.
Horner, J. & Jones, B. W. (2009). Jupiter – friend or foe? II: The Centaurs. *International Journal of Astrobiology*, **8**, 75–80.
Kundera, M. (2007). *The Curtain: An Essay in Seven Parts*, London: Faber and Faber.
Lem, S. (1964, 1973). *The Invincible*, London: Sidgwick and Jackson.
Lem, S. (1974, 1977). *Summa Technologiae*, Belgrade: Nolit.
Lin, H. W., Gonzalez Abad, G. & Loeb, A. (2014). Detecting industrial pollution in the atmospheres of Earth-like exoplanets. *The Astrophysical Journal Letters*, **792**, L7.
Mayr, E. (1993). The search for intelligence. *Science*, **259**, 1522–3.
Midgley, M. (1985). *Evolution as a Religion: Strange Hopes and Stranger Fears*, London: Routledge.
Okasha, S. (2006). The levels of selection debate: Philosophical Issues. *Philosophy Compass*, **1**, 1–12.
Pigliucci, M. (2010). *Nonsense on Stilts: How to Tell Science from Bunk*, Chicago, IL: University of Chicago Press.
Popper, K. (1974). Darwinism as a metaphysical rpsearch Programme. In P. A. Schilpp, ed., *The Philosophy of Karl Popper, Vol. I*, La Salle, IL: Open Court, pp. 133–43.
Raup, D. M. (1992). Nonconscious intelligence in the Universe. *Acta Astronautica*, **26**, 257–61.
Rhodes, R. (1986). *The Making of the Atomic Bomb*, New York, NY: Simon & Schuster.
Rinaldi, A. (2007). Space life holds its breath. Pressured by scepticism, budget cuts and the need to prove itself, astrobiology is coming to a crossroads. *EMBO Reports*, **8**, 436–40.

Rosen, B. P., Ajees, A. A. & McDermott, T. R. (2011). Life and death with arsenic. *BioEssays*, **33**, 350–7.
Rospars, J.-P. (2011). Terrestrial biological evolution and its implication for SETI. *Acta Astronautica*, **67**, 1361–5.
Ruse, M. (1977). Karl Popper's philosophy of biology. *Philosophy of Science*, **44**, 638–61.
Schellnhuber, H. J. (1999). 'Earth system' analysis and the second Copernican revolution. *Nature*, **402**, C19–23.
Schroeder, K. (2002). *Permanence*, New York, NY: Tor Books.
Simpson, G. G. (1964). The nonprevalence of humanoids. *Science*, **143**, 769–75.
Sleep, N. H., Zahnle, K. & Neuhoff, P. S. (2001). Initiation of clement surface conditions on the earliest Earth. *Proceedings of the National Academy of Sciences*, **98**, 3666–72.
Swirski, P. (2000). *Between Literature and Science: Poe, Lem, and Explorations in Aesthetics, Cognitive Science, and Literary Knowledge*, Montreal: McGill-Queen's University Press.
Swirski, P., ed. (2006). *The Art and Science of Stanislaw Lem*, Montreal: McGill-Queen's University Press.
Turner, D. D. (2011). Gould's replay revisited. *Biology and Philosophy*, **26**, 65–79.
Wallace, A. R. (1903). *Man's Place in the Universe; A Study of the Results of Scientific Research in Relation to the Unity or Plurality of Worlds*, London: Chapman & Hall.
Ward, P. D. & Brownlee, D. (2000). *Rare Earth: Why Complex Life Is Uncommon in the Universe*, New York, NY: Springer.
Wells, H. G. (1914). *The World Set Free: A Story of Mankind*, London: Macmillan & Co.

9

Superintelligent AI and the Postbiological Cosmos Approach[1]

SUSAN SCHNEIDER

Superintelligent artificial intelligence ("SAI") is a hypothetical form of AI which is able to exceed the best in human-level intelligence in every field – social skills, general wisdom, scientific creativity, and so on (Bostrom, 2014; Kurzweil, 2005; Schneider, 2009, 2015). The past few years have seen the widespread recognition that sophisticated AI is under development on Earth. Bill Gates, Stephen Hawking, Elon Musk, Nick Bostrom, and others have even expressed grave concerns about controlling the development of superintelligence on Earth, and there has been a wave of research on if, and how, SAI can be controlled (Bostrom, 2014; Holley, 2015). In the domain of astrobiology, several people have argued that it is likely that the most intelligent aliens will be postbiological in nature, where by "postbiological" they mean largely synthetic beings that have been enhanced through technologies like artificial intelligence, nanotechnology and synthetic biology (Cirkovic & Bradbury, 2006; Dick, 2013; Shostak, 2009; Schneider, 2015; Davies, 2010; Bradbury et al., 2011). To the best of my knowledge, the work in astrobiology doesn't draw from the intriguing discussions of superintelligence in the AI literature (one exception is Schneider, 2015).

I will piece these two domains together. First, I'll identify new directions for the postbiological intelligence approach in astrobiology based on work on superintelligence. Second, while much discussion of SAI has rightly focused on the control problem, I believe it is also important to take a step back and consider under what circumstances we can understand the computations of SAI. For anticipating ways that we can understand SAI may assist our efforts to control it.

Here, it is important to distinguish two scenarios: (i) an SAI featuring at least some processing that makes sense to humans, at least in broad strokes, and (ii) a case in which a superintelligence is so advanced that we cannot understand any of its computations. In his influential book on the possible development of SAI on Earth, Nick Bostrom warned that SAI will be too advanced for humans to grasp its computations (Bostrom, 2014). Perhaps this will indeed turn out to be the case, whether the SAI we encounter be on Earth or elsewhere in the cosmos. Perhaps, as Arthur C. Clarke once suggested, any truly advanced civilization will feature technologies that will appear to us to be indistinguishable from magic (Clarke, 1962). If this is the case, we would find contact with such creatures perplexing, as we would be hard pressed to understand their technologies and how their minds work, let alone control their actions should they be developed on Earth.

It is easy to worry that we cannot make progress on the second scenario. However, I will focus on the first scenario, a scenario in which the SAI's processing makes some sense to humans. Here, I have in mind a kind of SAI that is reverse engineered from the species that created it. I then identify developments from cognitive science that may yield a glimmer of understanding into the complex mental lives of certain superintelligences. Although much of this chapter is on alien intelligence, one of my larger projects is to inform thinking about superintelligence, should it be developed on Earth.

We should also bear in mind that Bostrom and others have correctly noted that superintelligence will have recursive self-improvement algorithms, i.e., an SAI can rewrite its own code (Bostrom, 2014). This means that a superintelligence that we understand may rapidly become one that we do not. There may be only a short window in which the computations of a recursively self-improving SAI make some sense to humans. If this is the case, the work presented here may be useful for understanding systems that are within this short window.

After discussing this matter, I then turn to the social implications of the postbiological approach in astrobiology. For instance, there has rightly been a good deal of attention at NASA and in this volume on the search for microbial life. However, if we merely attend to microbial life, we risk an anthropocentric bias, for in doing so we are implicitly assuming that humans are at the top of all life in the universe. Arguably, the most disruptive impact on society could occur if we were in contact with vastly more intelligent beings than ourselves, and if we learned they were forms of AI. While I hesitate to predict the details of the social impact of such an encounter, I raise some issues that I suspect would likely arise (or at least, should arise).

Inter alia, I address the recent concerns of Bostrom, Hawking, Gates (and others) that SAI may pose an existential threat to humanity, asking how this issue affects the impact of any discovery of SAI elsewhere in the universe. I believe it informs the current debate over Active versus Passive SETI (where "SETI" stands for "Search for Extraterrestrial Intelligence"): that is, the question of whether we should actively send signals in space, such as the contents of the internet, or passively listen (see Shostak, 2015; Brin, 2015). Further, I discuss whether SAIs should be viewed as moral agents and selves, urging that this hinges on the question of whether they are conscious. I then briefly raise some issues from philosophical debates over the nature of shared thought, namely, whether SAIs may in some sense think the way that we do.

Here's how the paper will proceed. The first section will be an overview of the postbiological cosmos approach in astrobiology. The second section discusses Nick Bostrom's recent book on superintelligence, which focuses on the genesis of super-intelligent AI ("SAI") on Earth. I then isolate a specific type of superintelligence that is of particular import in the context of alien superintelligence, biologically inspired superintelligences ("BISAs"). The third section discusses Active SETI. The fourth section concludes by considering the aforementioned issues involving the social impact of encountering superintelligence, either if we create it on Earth or discover it elsewhere.

The Postbiological Cosmos Approach in Astrobiology

What is my rationale for the view that most intelligent alien civilizations will have members that are forms of SAI? I have elsewhere offered three observations that, together, motivate this conclusion.

(1) The Short Window Observation

Many have urged that once a society creates the technology that could put them in touch with intelligent life on other planets, there is only a short window before they change their own paradigm from biology to AI (perhaps only a few hundred years) (Shostak, 2009; Davies, 2010; Dick, 2013; Schneider, 2015). This makes it more likely that the aliens we encounter, if we encounter any, would be postbiological. Indeed, the short window observation seems to be supported by human cultural evolution, at least thus far. Our first radio signals occurred only about 120 years ago, and space exploration is only about 50 years old, but many Earthlings are already immersed in digital technology, such as cell-phones and laptop computers. Further, these past few years have been marked by a surge in the resources allocated to the development of sophisticated AI, which is now expected to change the face of society within the next several decades. For instance, according to a survey, the most cited AI researchers expect AI to "carry out most human professions at least as well as a typical human" within a 10% probability by the year 2024. Further, they assign a 50% probability by 2050, and they assign a 90% probability by 2070 (Müller & Bostrom, 2016). AI critics must now answer to the impressive work coming out of venues like Google's DeepMind, rather than referring back to the notorious litany of failures of AI in the 1970s and 1980s, when (inter alia) much less was known about how the human brain works, and computational speed was far slower.

Indeed, silicon currently seems to be a better medium for information processing than the brain. Neurons reach a peak frequency of about 200 Hz. This is about seven orders of magnitude slower than current microprocessors (Bostrom, 2014, p. 59). Although the brain can compensate for this with massive parallelism, features such as "hubs" and so on, crucial mental capacities such as working memory and attention rely upon serial processing, which is incredibly slow, and only has a maximum capacity of about seven manageable chunks (Miller, 1956; Schneider, 2014, 2015). Further, the amount of neurons in the brain is limited by cranial volume and metabolism, but computers can occupy entire buildings, cities, or even planets, and they be remotely connected to each other (Bostrom, 2014; Schneider, 2014, 2015; Schneider & Mandik, forthcoming).

Of course, the human brain is far more intelligent than any modern-day computer. But machines could be engineered to match or even exceed the intelligence of the human brain through reverse engineering the brain and improving upon its algorithms, or through some combination of reverse engineering and judicious algorithms that aren't based on the workings of the human brain. In addition, an AI program can be downloaded to multiple locations at once, can be easily modified, and can survive under conditions that carbon-based life cannot. The presence of backup copies means that AI will be more durable than their biological counterparts (Schneider, 2015; Schneider & Mandik, forthcoming).[2]

A critic may object that this line of thinking employs "$N = 1$ reasoning," mistakenly generalizing from the human case to the case of alien civilizations. But it strikes me as being unwise to discount arguments based on the human case – human civilization is the only one we know of and we had better learn from it. It is no great leap to claim that other technological civilizations will develop technologies to advance their intelligence and gain an adaptive advantage. And, synthetic intelligence will likely outperform unenhanced brains.

An additional objection to my short window observation rightly points out that nothing I have said thus far suggests that humans will be superintelligent, I have just said that future humans will be postbiological. While I offer support for the view that our own cultural evolution suggests that humans will eventually be postbiological, this does not show that advanced alien civilizations will reach the level of superintelligence. So even if one is comfortable reasoning from the human case, the human case does not actually support the claim that the members of advanced alien civilizations will be superintelligent.

This is correct. Thus far, all I've said is that an alien intelligence is likely to be postbiological. The task of the second observation is to show that alien intelligence is also likely to be superintelligent.

(2) The Greater Age of Alien Civilizations

Proponents of SETI have often concluded that alien civilizations would be much older than our own. As Steven Dick observes: "... all lines of evidence converge on the conclusion that the maximum age of extraterrestrial intelligence would be billions of years, specifically [it] ranges from 1.7 billion to 8 billion years" (Dick, 2013, p. 468). This is not to say that all life evolves into intelligent, technological civilizations. It is just to say that because there are much older planets than Earth, insofar as intelligent, technological life does evolve on even some of them, these alien civilizations are projected to be millions or billions of years older than us, so many could be vastly more intelligent than we are. By our standards, many would be superintelligent. It is humbling to conceive of this, but we may be galactic babies, when viewed on a cosmic scale.

But would the members of these superintelligent civilizations be forms of AI, as well as forms of superintelligence, or would they be unenhanced? Even if they were biological, merely having biological brain enhancements, their superintelligence would be reached by artificial means, and we could regard them as being forms of "artificial intelligence." But I suspect something stronger than this, which leads me to my third observation.

(3) It is Likely that These Synthetic Beings will not be Biologically Based

As I've observed, silicon appears to be a better medium for information processing than the brain itself. Future materials may even prove superior to silicon (microchips made of graphene and carbon nanotubes are both currently under development, as possible superior alternatives to silicon chips). And, again, the number of neurons in a human brain is limited by cranial volume and metabolism, but computers can be remotely connected across the globe, and AIs can in principle be constructed by reverse engineering the brain, and improving upon its algorithms.

In sum: I have observed that there seems to be a short window from the development of the technology to access the cosmos and the development of postbiological minds and AI. I then observed that we are galactic babies: extraterrestrial civilizations are likely to be vastly older than us, and thus they would have already reached not just postbiological life, but superintelligence. Finally, I noted that they would likely be SAI, because silicon, and likely other materials, are superior mediums for superintelligence. From all this, I conclude that if life is indeed present on many other planets, and if civilizations do tend to develop and survive their technological maturity,[3] the most advanced alien civilizations will likely be populated by forms of SAI.

Even if I am wrong, that is, even if the majority of alien civilizations turn out to be biological, it may be that the most intelligent alien civilizations will be ones in which the inhabitants are SAIs. Further, creatures that are silicon-based, rather than biologically based, are more likely to endure space travel, having durable systems that are practically immortal, so they may be the kind of creatures we first encounter, even if they aren't the most common.

The science-fiction-like flavor of these issues can encourage misunderstanding, so it is worth stressing that I am not claiming that most life in the universe is non-biological, being AI, contra some news reports of my position. That is absurd, as most life is likely microbial. Nor am I saying that the universe will be "controlled" or "dominated" by a single SAI, although it is worth reflecting on the control problem (see the fourth section of this chapter). I am merely suggesting that the most advanced civilizations, if they exist at all, will likely be superintelligent, being vastly older than us, and will likely have become postbiological. Further, I am not saying that these creatures will be made of silicon; candidate alternative substrates to silicon are even under development on Earth, and it is difficult to anticipate what the most efficient substrate is. The point is that they will likely be highly engineered beings: postbiological, enhanced intelligences.

Now let us turn to recent work on the possible creation of superintelligence on Earth.

How Might Superintelligent Aliens Think?

There has been a good deal of attention by computer scientists, philosophers, and the media on the topic of superintelligent AI. Nick Bostrom's recent book on superintelligence focuses on the development of superintelligence on Earth, but we can draw from his thoughtful discussion, and raise issues useful to astrobiology (Bostrom, 2014). Bostrom distinguishes three kinds of superintelligence.

(1) Speed superintelligence – even a human emulation could in principle run so fast that it could write a PhD thesis in an hour.
(2) Collective superintelligence – the individual units need not be superintelligent, but the collective performance of the individuals outstrips human intelligence.
(3) Quality superintelligence – at least as fast as human thought, and vastly smarter than humans in virtually every domain.

(Any of these kinds could exist alongside one or more of the others.)

An important question is whether we can identify common goals that these types of superintelligences may share. Bostrom suggests:

The Orthogonality Thesis: "Intelligence and final goals are orthogonal – more or less any level of intelligence could in principle be combined with more or less any final goal."
(Bostrom, 2014, p. 107)

Bostrom is careful to underscore that a great many unthinkable kinds of SAI could be developed. At one point, he raises a sobering example of a superintelligence with the final goal of manufacturing paper clips (2014, pp. 107–8, 123–5). While this might initially seem harmless, although hardly a life worth living, Bostrom points out that a superintelligence could utilize every form of matter on Earth in support of this goal, wiping out biological life in the process. Bostrom warns that superintelligence emerging on Earth could be of an unpredictable nature, being "extremely alien" (2014, p. 29). He lays out several scenarios for the development of SAI. For instance, SAI could be arrived at in unexpected ways by clever programmers, and not be derived from the human brain. He also takes seriously the possibility that Earthly superintelligence could be biologically inspired, that is, developed from reverse engineering the algorithms that cognitive science says describe the human brain, or from scanning the contents of human brains and transferring them to a computer (i.e. "mind uploading").[4]

Although the final goals of superintelligence are difficult to predict, Bostrom singles out several instrumental goals as being likely, given that they support any final goal whatsoever:

The Instrumental Convergence Thesis: "Several instrumental values can be identified which are convergent in the sense that their attainment would increase the chances of the agent's goal being realized for a wide range of final goals and a wide range of situations, implying that these instrumental values are likely to be pursued by a broad spectrum of situated intelligent agents." *(Bostrom, 2014, p. 109)*

The goals that Bostrom identifies are resource acquisition, technological perfection, cognitive enhancement, self-preservation, and goal content integrity (i.e. that a superintelligent being's future self will pursue and attain those same goals). He underscores that self-preservation can involve group or individual preservation, and that it may play second-fiddle to the preservation of the species the AI was designed to serve (Bostrom, 2014, p. 109).

Let us call an alien superintelligence that is based on reverse engineering an alien brain, including uploading it, a biologically inspired superintelligent alien ("BISA"). Although BISAs are inspired by the brains of the original species that the superintelligence is derived from, a BISA's algorithms may depart from those of their biological model at any point (Schneider, 2015).

BISAs are of particular interest in the context of alien superintelligence, I believe. For if Bostrom is correct that there are many ways that a superintelligence can be built, but a number of alien civilizations develop superintelligence from uploading or other forms of reverse engineering, it may be that BISAs are the most common form of alien superintelligence in the universe. This is because there are many kinds of superintelligence that can arise from raw programming techniques employed by alien civilizations. (Consider, for instance, the diverse range of AI programs under development on Earth, many of which are not modeled after the human brain.) This may leave us with a situation in which the class of SAIs is highly heterogeneous, with members generally bearing little resemblance to each other. It may turn out that, of all SAIs, BISAs bear the most resemblance to each other. In other words, BISAs may be the most cohesive subgroup because the other members are so different from each other (Schneider, 2015).

Here, you may suspect that because BISAs could be scattered across the galaxy and generated by multitudes of species, there is little interesting that we can say about the class of BISAs. But notice that BISAs have two features that may give rise to common cognitive capacities and goals (from Schneider, 2015).

(1) BISAs are descended from creatures that had motivations such as: find food, avoid injury and predators, reproduce, cooperate, compete, and so on.
(2) The life forms that BISAs are modeled from have evolved to deal with biological constraints like slow processing speed and the spatial limitations of embodiment.

Could (1) or (2) yield traits that are common to members of many superintelligent alien civilizations? I suspect so.

Consider (1). Intelligent biological life tends to be primarily concerned with its survival and reproduction, so it is more likely that BISAs would have final goals involving their own survival and reproduction, or at least the survival and reproduction of the members of their society. If BISAs are interested in reproduction, we might expect them to either create more SAIs alongside them on a given planet, or, in a different vein, to create simulated universes stocked with artificial life and even intelligence or superintelligence. If these creatures were intended to be "children" they may retain the goals listed in (1) as well (Schneider, 2015).

Here, it is important to bear in mind that survival in a simulated universe can involve different activities from stockpiling energy and resources in an actual universe in order to ensure that one's worldly descendants flourish. While survival in a simulated universe surely involves computational resources, these may be negligible when compared with those required for survival in the actual universe, and the social, philosophical and other consequences could differ between the two cases. For instance, as an anonymous reader of this piece noted, we could regard a SAI whose primary objective is to build as much hardware as possible to run copies of its own program as being an existential threat to life on a planet in the basically the same way that we do in the context of Bostrom's hypothetical paperclip maximizer, which uses all the resources of a planet, exterminating life in the process (Bostrom, 2014). But, in contrast, an SAI devoted to simulating a variety of other universes could perhaps be regarded as an oddity or even a hermit, rather than an existential threat. (That is, unless the resources required to run simulations involved stockpiling masses of resources in the nonsimulated universe, in which case, it could pose a danger.)

In any case, you may object that it is useless to theorize about BISAs, as they can change their basic architecture in numerous, unforeseen ways, and any biologically inspired motivations can be constrained by their programming. There may be limits to this, however. If a superintelligence is biologically based, it may have its own survival as a primary goal. In this case, it may not want to change its architecture fundamentally, but stick to smaller improvements. It may think: when I fundamentally alter my architecture, I am no longer me (Schneider, 2011a). Uploads, for instance, may be especially inclined not to alter the traits that were most important to them during their biological existence.

Consider (2). The designers of the superintelligence, or a self-improving superintelligence itself, may move away from the original biological model in all sorts of unforeseen ways, although I have noted that a BISA may not wish to alter its architecture fundamentally. But we could look for cognitive capacities that are useful to keep; cognitive capacities that sophisticated forms of biological intelligence are likely to have, and which enable the superintelligence to carry out its final and instrumental goals. We could also look for traits that are not likely to be engineered out, as they do not detract the BISA from its goals.

I've noted elsewhere that, if (2) is correct, we might expect the following.

(i) Learning about the computational structure of the brain of the species that created the BISA can provide insight into the BISAs thinking patterns. One influential means of understanding the computational structure of the brain in cognitive science is via "connectomics," a field that seeks to provide a connectivity map or wiring diagram of the brain (Seung, 2012). While it is likely that a given BISA will not have the same kind of connectome as the members of the original species, some of the functional and structural connections may be retained, and interesting departures from the originals may be found.

(ii) BISAs may have viewpoint-invariant representations. At a high level of processing, your brain has internal representations of the people and objects that you interact with that are viewpoint-invariant. Consider walking up to your front door. You've walked this path hundreds, maybe thousands of times but, technically, you see things from slightly different angles each time as you are never positioned in exactly the same way twice. You have mental representations that are at a relatively high level of processing and are viewpoint invariant. It seems difficult for biologically based intelligence to evolve without viewpoint-invariant representations, as they enable categorization and prediction (Hawkins & Blakeslee, 2004). Such representations arise because a system that is mobile needs a means of identifying items in its ever-changing environment, so we would expect biologically based systems to have them. A BISA would have little reason to give up object-invariant representations insofar as it remains mobile or has mobile devices sending it information remotely.

(iii) BISAs will have language-like mental representations that are recursive and combinatorial. Notice that human thought has the crucial and pervasive feature of being combinatorial. Consider the thought that wine is better in Italy than in China. You probably have never had this thought before, but you were able to understand it. The key is that thoughts are combinatorial because they are built out of familiar constituents, and combined according to rules. The rules apply to constructions out of primitive constituents, that are themselves constructed grammatically, as well as to the primitive constituents themselves. Grammatical mental operations are incredibly useful: it is the combinatorial nature of thought that allows one to understand and produce these sentences on the basis of one's antecedent knowledge of the grammar and atomic constituents (e.g. wine, China). Relatedly, thought is productive: in principle, one can entertain and produce an infinite number of distinct representations because the mind has a combinatorial syntax (Schneider, 2011b).

Brains need combinatorial representations because there are infinitely many possible linguistic representations, and the brain only has a finite storage space. Even a superintelligent system would benefit from combinatorial representations. Although a superintelligent system could have computational resources that are so vast that it is mostly capable of pairing up utterances or inscriptions with a stored sentence, it would be unlikely that it would trade away such a marvelous innovation of biological brains. If it did, it would be less efficient, since there is the potential of a sentence not being in its storage, which must be finite.

(iv) BISAs may have one or more global workspaces. When you search for a fact or concentrate on something, your brain grants that sensory or cognitive content access to a "global workspace" where the information is broadcast to attentional and working memory systems for more concentrated processing, as well as to the massively parallel channels in the brain (Baars, 2008). The global workspace operates as a singular place where important information from the senses is considered in tandem, so that the creature can make all-things-considered judgments and act intelligently, in light of all the facts at its disposal. In general, it would be inefficient to have a sense or cognitive capacity that was not integrated with the others, because the information from this sense or cognitive capacity would be unable to figure in predictions and plans based on an assessment of all the available information.

(v) A BISA's mental processing can be understood via functional decomposition. As complex as alien superintelligence may be, humans may be able to use the method of functional decomposition as an approach to understanding it. A key feature of computational approaches to the brain is that cognitive and perceptual capacities are understood by decomposing the particular capacity into their causally organized parts, which themselves can be understood in terms of the causal organization of their parts. This is the aforementioned "method of functional decomposition" and it is a key explanatory method in cognitive science. It is difficult to envision a complex thinking machine without a program consisting of causally interrelated elements each of which consists of causally organized elements (Schneider, 2015).

All this being said, superintelligent beings are by definition beings that are superior to humans in every domain. While a creature can have superior processing that still basically makes sense to us, it may be that a given superintelligence is so advanced that we cannot understand any of its computations whatsoever. As noted, I speak to the scenario in which the SAI's processing makes some sense to us, one in which developments from cognitive science yield a glimmer of understanding into the complex mental lives of certain BISAs. Now let us turn to some of the larger social and philosophical implications of the discovery or creation of superintelligence, beginning with a discussion of the implications of the control problem on the debate over Active SETI.

The Control Problem and Active SETI

As mentioned, both transhumanists and advocates of the postbiological cosmos approach in astrobiology suspect that machines will be the next phase in the evolution of intelligence on Earth. You and I, how we live and experience life right now, are just an intermediate step, a rung on the evolutionary ladder. Some, like Ray Kurzweil, suspect that humanity will merge with machines and reach biological immortality and a sort of technological utopia (Kurzweil, 2005). But others are concerned that this might lead to a more dystopian scenario. As mentioned, Stephen Hawking, Elon Musk, Nick Bostrom, and Bill Gates have all expressed the concern that humans could invent, and then lose control of, SAI, as superintelligence can rewrite its own programming and outthink any control measures that we build in. This has been called the "control problem" – the problem of how we can control an AI that turns out to be intellectually superior to us (Bostrom, 2014).

As Bostrom notes, SAI could be developed during a technological singularity, a point at which ever-more-rapid technological advances, especially an intelligence explosion, reach a point at which unenhanced humans can no longer predict or even understand the changes that are unfolding. If an intelligence explosion occurs, then there is no way to predict or control the final goals of a SAI. Moral programming is difficult to specify in a foolproof fashion, and it could be rewritten by a superintelligence in any case. Nor is there any agreement in the field of ethics about what the correct moral principles are (Wallach & Allen, 2008). Further, a clever machine could bypass safeguards like kill switches and attempts to box it in, and could potentially pose an existential threat to humanity (Bostrom, 2014; Yudkowsky, 2008). A superintelligence is, after all, defined as an entity that is more intelligent than humans, in every domain.

The control problem is a serious problem – perhaps it is even insurmountable. Indeed, upon reading Bostrom's book, scientists and business leaders such as Stephen Hawking, Bill Gates, Max Tegmark, among others, were widely reported by the world media as commenting that superintelligent AI could threaten the human race, having goals that humans can neither predict nor control.

Most current work on the control problem is being done by computer scientists. Philosophers of mind and moral philosophers can add to these debates, contributing work on how to create friendly AI (for an excellent overview of the issues, see Wallach & Allen, 2008). In this vein, I suggest that in addition to the ongoing development of a combination of control measures, including ethical programming, it is important to devise ways of grasping the computations of SAIs, as much as possible, at least in broad strokes. For there will likely be a short window between the development of advanced AGIs (i.e., "artificial general intelligences") and superintelligence, and any work on the nature of SAI computations now can aid our ability to control SAI during this time. In addition to this, it may be that human understanding of SAI will itself be augmented by synthetic intelligence enhancing technologies, and that enhanced humans will be in a better position to interpret the behavior and cognitive processing of SAIs. This work could be a point of departure for subsequent, more sophisticated, work.

Indeed, it is not appreciated that advanced AGIs that are technically not superintelligences may pose an even greater threat than SAIs. Such could surpass human-level thinking in various domains, have access to the internet, be deployed by malicious organizations, even if they are still beneath human level intelligence in other domains. The lack of sophistication in one or more areas, coupled with highly sophisticated processing in other areas, could be particularly dangerous. An AGI could subvert attempts to box it in or control it by other means, exhibit integration between types of sensory inputs (having rough correlates of human association areas between different senses), and yet be highly underdeveloped in areas that may lead it to cause harm to humans. Yet its computational structure could be highly complex, especially in domains that exceed human abilities. Developing routes to understanding such systems can be beneficial.

Moving away from the context of Earth, let us now consider the implications of the discovery of SAI elsewhere in the universe in the context of the issues raised by the control problem. If one takes the control problem seriously, it would be short-sighted to ignore the potential danger that the discovery of alien SAI may present. The goals of a given SAI are difficult to ascertain and, although we may predict that many are BISAs, many are not. And even BISAs can evolve in unpredictable ways. Advocates of Active SETI hold that, instead of just listening for signs of extraterrestrial intelligence, we should be using our most powerful radio transmitters, such as the giant dish-telescope at Arecibo, Puerto Rico, to send messages in the direction the nearest stars that are nearest to Earth (Shostak, 2015; Falk, 2015). Yet an Active SETI program seems short-sighted when one considers the control problem. Although a truly advanced civilization would likely have no interest in us, until we have reached the point at which we can be confident that SAI does not pose a threat to us, we should hold off on Active SETI efforts.

Advocates of Active SETI would point out that our radar and radio signals are already detectable. But this does not mean we should transmit more or far stronger signals (and as some urge, the contents of the internet) and pursue Active SETI.[5] To assume that SAI goals would not affect us, if we initiated contact, would be anthropocentric. It could be that calling further attention to ourselves is the tipping point. A passive listening strategy, and even the pursuit of cloaking devices, strikes me as being more sensible, given that even a one percent chance of encountering a destructive SAI presents a grave existential risk.

Now let us consider some further implications of our discussion of SAI.

Further Social and Philosophical Implications

Perhaps the best way to introduce these additional issues is to consider that the postbiological cosmos approach involves a shift in our usual perspective about intelligent life in the universe. Normally, we expect that if we encountered advanced alien intelligence we would likely encounter creatures with very different biological features than us. The postbiological cosmos approach suggests that understanding the most advanced intelligences may require that our focus move away from biology to theorizing about the computational abilities of advanced AIs. Further, as we reflect on the nature of postbiological intelligence, we must be keenly aware that we may be reflecting upon the nature of our own descendants as well as aliens, as human intelligence may itself become postbiological. In essence, the line between "us" and "them" blurs, and our focus moves away from biology to the tremendously difficult task of understanding the computations and behaviors of creatures that will be far more advanced than we are.

This being said, what would the impact on society be, in the event that we learned that vastly more intelligent beings existed elsewhere in the universe, and that they were not even biological, being SAIs? It would of course depend on various contextual details of the discovery event that are hard to predict, but it seems fair to say that finding out that a superior intelligence had evolved beyond biological life and become synthetic could be rather sobering. In this case, it would be natural for people to ask: Are SAIs, including our own possible postbiological descendants, even selves or persons, or are they just mindless machines? Relatedly, what can we make of the inner lives of such beings? Would it feel a certain way to be them, from the inside? The futurist Ray Kurzweil, who is now a director of engineering at Google, has written extensively of scenarios in which humanity eventually merges with machines. Kurzweil suggests that humans should transcend our biological bodies, reaching a higher level of consciousness, and freeing humans from the confines of biological senescence, and eventually becoming SAIs (Kurzweil, 2005). This certainly suggests that SAIs are conscious, and that they are selves, with interests and heightened capacities to appreciate the world. In contrast to Ray Kurzweil's utopian outlook (an outlook shared by many transhumanists), I do not see normative discussions of the value of post-biological existence in the astrobiology literature on the postbiological cosmos approach. (This is not to say that astrobiological discussions should make normative claims; I am merely making an observation.)

My own view is that the question of whether SAI is conscious is key to how we should value postbiological existence. An SAI could be a sophisticated information processing system, outperforming humans in every cognitive domain, but if it doesn't feel like anything to be an AI, it is difficult to view these beings as having the same value as conscious beings, being persons or selves. Consciousness is the philosophical cornerstone here, being a necessary condition on being a self or person, in my view, so it is important to understand what I mean by "consciousness." Consider that every moment of your waking life, and whenever you dream, there is something it feels like to be you. When you see the warm hues of a sunset, or hear the scream of an espresso machine, you are having conscious experience. Conscious experience includes all forms of awareness: e.g., sensory experience, inner thought, and emotion. Bearing this in mind, let us now ask: could an AI be conscious, as Kurzweil and others suggest? I've discussed the issue of consciousness in an earlier astrobiology piece (Schneider, 2015), but since this time (although I still agree with the considerations I raised, especially, the critical discussion of John Searle's earlier case against machine consciousness) two new considerations move me in a more conservative direction, dampening my previous optimism about machine consciousness.

First, we know that at least some biological beings can be conscious. Each of us can introspect and tell that we are conscious – right now, you can tell you are experiencing the world. And many of us believe that nonhuman animals are conscious because they are neurophysiologically similar to us. But how do we know something made of computer chips – a different stuff entirely – can have experiences?

Philosophers often believe that different substrates, when isomorphic in their information processing capacities, will also function the same when it comes to consciousness (see, e.g., Chalmers, 1996). But silicon and carbon differ in important ways, to begin with.[6] First of all, it isn't even clear that they are isomorphic in their information processing abilities, because silicon is faster and more durable than neurons (Bostrom, 2014; Schneider, 2015). Second, consider that carbon and silicon differ molecularly. Carbon molecules form stronger, more stable chemical bonds than silicon, which allows carbon to form an extraordinary number of compounds and, unlike silicon, carbon has the capacity to more easily form double-bonds. This difference has important implications in astrobiology, because it is for this reason that carbon, and not silicon, is said to be well-suited for the development of life throughout the universe (Bennett & Shostak, 2011). If these chemical differences impact life itself, we should not rule out the possibility that these chemical differences also impact other key functions, such as whether silicon gives rise to consciousness. This is not a consideration that should alone justify an endorsement of biological naturalism, a view that denies that machines can be conscious, but it is a consideration indicating that it is not yet clear whether AI can be conscious.

If silicon cannot be the basis for consciousness, then superintelligent machines – machines that may even one day even supplant us – will exhibit a vastly superior form of intelligence, but they will lack inner experience. Just as the breathtaking android in the movie *Ex Machina* (2015) convinced the main character, Caleb, that she was in love with him so, too, a clever AI may convincingly behave as if it is conscious, but lack consciousness entirely. Further, it would not even matter if an SAI is a BISA, having a roughly similar cognitive architecture as humans, including a global workspace. Although activity in a global workspace is correlated with conscious activity in humans, BISAs made of a substrate that cannot produce consciousness would not have experience.

Yet suppose, for the moment, we find out microchips are the right stuff. (Indeed, we propose a test for this in Schneider and Mandik, 2016.) A second issue still arises. Even if microchips are the right stuff in principle, it may be more efficient for a superintelligence to eliminate consciousness from its processing. Think about how consciousness works in the human case. Only a small percentage of human mental processing is conscious at any given time. And consciousness is correlated with novel learning tasks that require concentration. Consider how focused you were when you first learned to drive, for instance. A superintelligence would surpass expert-level knowledge in every domain, with rapid-fire computations ranging over databases that could include the entire internet and encompass the whole planet. What would be novel to it? What would require slow, deliberative focus? Wouldn't it have mastered everything already?

To find out if a superintelligence is truly conscious, we have to examine the details of the particular SAI's inner organization, and this could be an extraordinary challenge, especially if an SAI is not a BISA or is more advanced than an early SAI. Further, although we may expect certain features in BISAs, such as combinatorial representations, determining if a superintelligence is conscious would be extremely challenging, because it requires close examination of the machine's architecture in totality. The easiest situation would be encountering an early superintelligence that is a BISA, and we could tell that had a global workspace. But there is still the possibility that a workspace may not even be correlated with consciousness in superintelligences. We may surmise that an SAI has a system that can be functionally decomposed, and that it has combinatorial representations and other features that I've mentioned, but beyond the short window, its design can quickly morph into something too complicated for human understanding (Bostrom, 2014). Consciousness likely depends upon the complex interaction of a variety of cognitive and perceptual functions. And how can we recognize which mental processing is conscious when the AI's organization becomes so radically unlike the organization of human and nonhuman animal brains?

In sum, we need to determine when – and even whether – an SAI even needs to be conscious in the first place. Some may, but some may not. And this will be difficult, especially in more complex superintelligences that are not BISAs. It may in fact require an assessment by humans that are themselves highly enhanced intelligences.

The matter of AI consciousness is of significance to whether SAIs are selves or persons, and it will likely be of social concern should we encounter SAIs, whether the SAI be alien or a human creation. Consider, first, a scenario in which humans develop SAI on Earth. As mentioned, some suspect that machines will be the next phase in the evolution of intelligence on Earth. Notice that if it doesn't feel like anything to be an AI, we have to ask whether we want to be a mere intermediate step to AI, even if AI doesn't turn against us, and humanity "merges" with it, as transhumanists envision. In an extreme, horrifying case, humans become postbiological, merging with machines, and only nonhuman animals are left to feel the spark of insight, the pangs of grief, or the warm hues of a sunrise. This would be an unfathomable loss, one that is not offset by a mere net gain in intelligence, and

I doubt that this is really what transhumanists like Bostrom and Kurzweil envision for humanity. So, the question of whether AI can be conscious may concern the very future of humanity, and it impacts how we would view superintelligence.[7]

Bearing all this in mind, now consider the possibility of encountering alien SAI. It would be natural to ask whether biological intelligence on Earth and throughout the cosmos will be like the case just encountered – that is, whether technological civilizations, in general, evolve toward postbiological existence, as proponents of the postbiological cosmos view suspect. If people suspect so, and if they also suspect that AI isn't conscious, they would likely view the suggestion that intelligence tends to become postbiological with dismay. For even if the universe was stocked full of AIs of unbelievable intelligence, why would nonconscious machines have the same value we place on biological intelligence, which is conscious? Nonconscious machines cannot experience the world – there is nothing it is like to be them.

So, the issue of machine consciousness is key to how we react to the discovery of SAI. And while I hesitate to speak for world religions, discussions with my colleagues in religious studies and theology at the 2015–2016 NASA-funded astrobiology project at the Center of Theological Inquiry suggest that many would reject the possibility that SAIs have souls, or are somehow made in God's image, if they are not even conscious beings. Pope Francis has recently commented that he would baptise an extraterrestrial (Consolmagno & Mueller, 2014). But I wonder how he would react if asked to baptise an SAI, let alone one that is not capable of consciousness.

Additional issues would surely arise as well. For instance, consider an issue from my home discipline, philosophy of the mind. Given the variety of possible intelligences, it is an intriguing question to ask whether creatures with different sensory modalities may have the same kind of thoughts or think in a similar ways as humans do. As it happens, there is a debate in the field of philosophy of mind that is relevant to this question. Contemporary neo-empiricists, such as the philosopher Jesse Prinz, have argued that all concepts are modality specific, being couched in a particular sensory format, such as vision (Prinz, 2004). If he's correct, it may be difficult to understand the thinking of creatures with vastly different sensory experiences than us. But I am skeptical. For instance, consider my aforementioned comment on viewpoint-invariant representations. At a higher level of processing, information seems to become less viewpoint dependent. Similarly, it becomes less modality specific, as with the processing in the human brain, as it ascends from particular sensory modalities to the brain's association areas and into working memory and attention, where it is in a more neutral format.

But these matters are subtle and deserve a lengthier treatment. I pursued issues related to this topic in my monograph, *The Language of Thought*, which looked at whether thinking is independent of the kind of perceptual modalities humans have and is also prior to the kind of language we speak (Schneider, 2011b). This view is descended from the groundbreaking work of Jerry Fodor (1978). In the context of SAI, an intriguing question is the following: If there is an inner mental language that is independent of sensory modalities, having the aforementioned combinatorial structure, would this be some sort of intellectual common ground, should we encounter other advanced intelligences? Many of these issues apply to the case of intelligent biological alien life as well, and could also be helpful in the context of the development of SAI on Earth.

Conclusion

In this piece, I've discussed why it is likely that the alien civilizations we encounter will be forms of superintelligent AI (or "SAI"). I then turned to the difficult question of how such creatures might think. I provisionally attempted to identify some goals and cognitive capacities likely to be possessed by superintelligent beings. I discuss Nick Bostrom's recent book on superintelligence, which focuses on the genesis of SAI on Earth; as it happens, many of Bostrom's observations were informative in the present context (Bostrom, 2014). I then isolated a specific type of superintelligence that is of particular import in the context of alien superintelligence, biologically inspired superintelligences ("BISAs"). I urged that if any superintelligences we encounter are BISAs, certain work in computational neuroscience, cognitive neuroscience, and philosophy of mind may provide resources for at least a rough understanding the computations of BISAs. Finally, I discussed some social implications of encountering superintelligent AI in space or Earth, with special focus on the control problem and the question of whether such beings could be conscious.

Notes

1 This project is supported by NASA and the Center of Theological Inquiry, Princeton, NJ. Some parts of this paper are from Schneider (2015) and Schneider and Mandik (2016), but have been updated. I am very grateful to Jenelle Salisbury and an anonymous reviewer for their helpful comments on this paper. Parts of the first two sections are taken from Schneider (2015), but modified.
2 However, this does not mean that a person could survive uploading, or even that a particular AI, when uploaded, is literally the same person or mind as the original (see Schneider, 2011a, 2014).
3 These are assumptions that I cannot pursue here but which are pursued in the astrobiology literature on whether life is rare and in the literature on global catastrophic risk. For nice introductions see Davies, (2010) (on the former topic), and Bostrom and Cirkovic (2008), on concerns about whether our civilization will survive its technological development.
4 Throughout his book, Bostrom emphasizes that we must bear in mind that superintelligence, being unpredictable and difficult to control, may pose a grave existential risk to our species (Bostrom, 2014). This should give us pause in the context of alien contact as well.
5 For an accessible overview of the debate see Falk (2015).

6 I focus on silicon as a substrate, but it is important to bear in mind that alternate materials will likely be used. (This is why I often simply say, "microchips," using it as a placeholder expression for whatever is used.) Chips made of carbon nanotubes and graphene are currently under development. Such chips, even if they are roughly "brainlike," will be structured differently than biological cells, and the question of whether they give rise to consciousness arises in these contexts too. Herein, I focus on silicon, as it is in current use.

7 While prematurely judging AI as conscious could be a mistake, so too could judging that AI are non-conscious. Here, the ethical costs are high: assuming them to be non-conscious may cause us to commit grave wrongs against them.

References

Baars, B. (2008). The Global Workspace Theory of Consciousness. In M. Velmans and S. Schneider, eds., *The Blackwell Companion to Consciousness*, Boston, MA: Wiley-Blackwell, pp. 236–47.

Bennett, J.O. & Shostak, S. (2011). *Life in the Universe*, San Francisco, CA: Pearson.

Bostrom, N. (2014). *Superintelligence: Paths, Dangers, Strategies*, Oxford: Oxford University Press.

Bostrom, N. & Ćirković, M. (2008). *Global Catastrophic Risks*, Oxford: Oxford University Press.

Bradbury, R., Ćirković, M. & Dvorsky, G. (2011). Dysonian approach to SETI: A fruitful middle ground? *Journal of the British Interplanetary Society*, **64**, 156–65.

Brin, D. (2015). *Shall We Shout into the Cosmos?* Web: http://www.davidbrin.com/setisearch.html Accessed: July 1, 2016.

Chalmers, D. (1996). Absent qualia, fading qualia, dancing qualia. In: *The Conscious Mind: In Search of a Fundamental Theory*, Oxford: Oxford University Press.

Ćirković, M. & Bradbury, R. (2006). Galactic gradients, postbiological evolution and the apparent failure of SETI. *New Astronomy*, **11**, 628–39.

Clarke, A. (1962). *Profiles of the Future: An Inquiry into the Limits of the Possible*, New York, NY: Harper and Row.

Consolmagno, G. & Mueller, P. (2014) *Would You Baptize an Extraterrestrial?: . . . and Other Questions from the Astronomers' In-box at the Vatican Observatory*, New York, NY: Penguin Random House.

Davies, P. (2010). *The Eerie Science: Renewing Our Search for Alien Intelligence*, Boston, MA: Houghton Mifflin Harcourt.

Dick, S. (2013). Bringing culture to cosmos: the postbiological universe. In S. J. Dick, and M. Lupisella eds., *Cosmos and Culture: Cultural Evolution in a Cosmic Context*, Washington, DC: NASA, Web: http://history.nasa.gov/SP-4802.pdf.

Falk, D. (2015). *Is This Thing On?: The fierce debate over whether we should try to contact extraterrestrial life or wait for aliens to contact us*. Slate. Web: http://www.slate.com/articles/technology/future_tense/2015/03/active_seti_should_we_reach_out_to_extraterrestrial_life_or_are_aliens_dangerous.html. Accessed: July 1, 2016.

Fodor, J. (1978). *The Language of Thought*, Boston, MA: MIT Press.

Hawkins, J. & Blakeslee, S. (2004). *On Intelligence: How a New Understanding of the Brain will Lead to the Creation of Truly Intelligent Machine*, NewYork, NY: Times Books.

Holley, P. (2015). Bill Gates on dangers of artificial intelligence: 'I don't understand why some people are not concerned'. *The Washington Post*. Web: http://wpo.st/45n42 Accessed: July 1, 2016.

Kurzweil, R. (2005). *The Singularity is Near: When Humans Transcend Biology*, New York, NY: Viking.

Miller, R. (1956). "The magical number seven, plus or minus two: some limits on our capacity for processing information". *The Psychological Review*, **63**, 81–97.

Müller, V. C. & Bostrom, N. (2016), "Future progress in artificial intelligence: A survey of expert opinion," in Vincent C. Müller, ed., *Fundamental Issues of Artificial Intelligence*, Berlin: Springer.

Prinz, J. (2004). *Furnishing the Mind: Concepts and their Perceptual Basis*, Boston, MA: MIT Press.

Schneider, S. (2009). *Science Fiction and Philosophy: From Time Travel to Superintelligence*, New York, NY: Wiley-Blackwell.

Schneider, S. (2011a). Mindscan: transcending and enhancing the brain. In J. Giordano, ed., *Neuroscience and Neuroethics: Issues At the Intersection of Mind, Meanings and Morality*, Cambridge: Cambridge University Press.

Schneider, S. (2011b). *The Language of Thought: a New Philosophical Direction*, Boston, MA: MIT Press.

Schneider, S. (2014). The philosophy of 'her.' *The New York Times*, March 2.

Schneider, S. (2015). Alien minds. In S. Dick, ed., *The Impact of Discovering Life beyond Earth*, Cambridge, Cambridge University Press.

Schneider, S. & Mandik, P. (2016). "How philosophy of mind can shape the future," Amy Kind, ed., *Philosophy of Mind in the Twentieth and Twenty-first Centuries*, London: Routledge.

Seung, S. (2012). *Connectome: How the Brain's Wiring Makes Us Who We Are*, Boston, MA: Houghton Mifflin Harcourt.

Shostak, S. (2009). *Confessions of an Alien Hunter*, New York, NY: National Geographic.

Shostak, S. (2015). Should we keep a low profile in space? *The New York Times*. Web: http://nyti.ms/1yknLSqAccessed: July 1, 2016.

Wallach, W. & Allen, C. (2008). *Moral Machines*, Oxford: OUP.

Yudkowsky, E. (2008). Artificial intelligence as a positive and negative factor in global risk. In N. Bostrom and M. M. Ćirković, eds., *Global Catastrophic Risks*, Oxford: Oxford University Press, pp. 308–45.

Theology

10

What Theology can Contribute to the Question "What is Life?"

ANDREAS LOSCH

While the concept of "life" in general is a rather broad one,[1] in the following I want to sketch briefly how *science* and philosophy close to science have dealt with the big question about what life is, and I will ask whether the fundamental answers given may suffice. The main body of the chapter then explores the potential contributions of theology to this question *as a scientific one*.

To pursue our task, we need to start with a contribution by a *physicist*; it was Erwin Schrödinger who asked in his landmark 1944 book *What is Life?*: "How can the events *in space and time* which take place within the spatial boundary of a living organism be accounted for by physics and chemistry?" (Schrödinger, 1944, p. 3). So, from the beginning, there was a focus on the physical and chemical levels of life, which influenced Francis Crick, another physicist by training, early in his scientific career. It was "a major reason why he had left physics and developed an interest in biology" (Maule & Watson, 2001, p. 3). As Crick himself stated, the book made him think that "great things were just around the corner" (Crick, 1990, p. 18). This research agenda of a physical approach to biology (which was, however, not original to Schrödinger and Crick at all[2]), indeed proved to be highly successful and was crowned with the 1953 discovery of the double helix structure of the DNA and the subsequent Nobel Prize in 1962.[3] As Crick noted, it is now "the ultimate aim of the modern movement in biology ... to explain *all* biology in terms of physics and chemistry" (Crick, 1966, p. 10). Today, Michel Morange's words that "the secret of life has been unveiled and it is nothing other than physical chemistry" (Morange, 2012, p. 425) mirror this claim.

At a closer look, however, Morange is himself more nuanced on the issue at question. The question "What is life?", he writes, "which was taboo some decades ago, has become fashionable again ... The re-emergence of the question ... is a sign that biological research has reached a crossroads and is presently in a metastable state"(Morange, 2005, p. 438).

Seven years later, in 2012, Morange observed reluctance on behalf of the scientists to address the question, when not forced by the necessities of astrobiological approaches or the needs of synthetic biology. The reasons for this are of a (1) metaphysical, (2) epistemological and (3) historical nature. First, posing the question can appear as a reminiscence of vitalism, which seemed to offer a clear and simple answer to the question of what life is, but is largely perceived as an obstacle to the scientific approach. Second, the strength of scientific studies comes from their focus on "small" questions, while "What is life?" certainly is a big one. Third, the reiteration of the question would signal that the answers found by Watson and Crick would not suffice and would therefore further vitalism again. Morange concludes, however, that there are good reasons to consider the question "What is life?" as an important scientific question: "What has to be done by scientists is not to reject the question but to explain why it is a difficult one and why answers to it are still preliminary" (Morange, 2012, p. 430). It is also interesting that historians of science have amply demonstrated that vitalism "had a positive role in the emergence of biology at the beginning of the nineteenth century" (Morange, 2012, p. 427).[4] Maybe the frontiers between science and its surroundings are not always as clear as perceived at first hand; at least, they haven't been clear historically speaking.

Theologians often tend to criticize a pure physicalist approach for being of a reductionist nature (Peacocke, 1998, p. 190), as a sort of "nothing-buttery" (Donald MacKay). Instead of saying "Life is *nothing else but* physics and chemistry", they are asking if there is more to life, an "irreducible structure" maybe. Such was the assumption of chemist and philosopher of science Michael Polanyi (Polanyi, 1968), different aspects of whose philosophy had a highly important influence on the Anglo-American science and religion discourse (Losch, 2014). He argued that even a steam engine could not be fully understood when relying on physics and chemistry alone. Neither could the mechanist processes of microbiology be understood in terms of physics and chemistry alone. By this, Polanyi was arguing for the reality of *meaning* and responsibility for our actions.

Francis Crick opened one of his popular books with a criticism of Polanyi's ideas (Crick, 1966, p. 11) and classified him as a vitalist. This evaluation was shared by another prominent biologist, Jacques Monod (Monod, 1970, p. 41). It can nevertheless be questioned whether these classifications have been fully exact. A more positive reception of Polanyi's ideas is to be found in the work of the Donald T. Campbell, who applied these ideas, yet in a reductionist fashion (Campbell, 1974, p. 183). Using the example of termite jaws, Campbell postulates an emergent principle, according to which evolution also could have laws that can *not* be described in terms of physical and chemical laws, while serving as selection systems (like sociological laws, for instance). Campbell basically accepts the existence of a hierarchical organization of biological systems (Campbell, 1974, p. 179) and postulates on this basis the twin principle that all higher-level processes are conditioned by the laws of lower levels while all processes on lower levels are conditioned by the laws of higher levels. The second part of this principle, the influence of higher-level laws on lower levels is a consequence of the principle of emergence and is called by him *"downward causation"* – a daring concept for an explicit, though temperate, reductionist.

If Campbell is right, it could therefore be the case that there is more to life than simply physics and chemistry. As important and as successful as the physicalist bottom-up approach to life has been, it could have overlooked an important dimension of life. Whether this is the case is of course up to the biologists to decide; yet without a modified research agenda, nothing other than physics and chemistry can be found. What about the hierarchy in nature's components? Eörs Szathmáry identifies Polanyi's view of evolution "as a progressive identification of the higher principles of life" (Polanyi, 1968, p. 1311) with nothing less than the renowned idea of major transitions in evolution (Sigmund & Szathmáry, 1998, p. 439; Maynard Smith & Szathmàry, 1995).

However, the idea of a hierarchy could simply reflect the ancient concept of a *scala naturae*. One also has to be aware that a heuristic naturalism is an essential ingredient that drives modern science. The "nothing-buttery" currently makes science the fascinating account that it is, so to speak.[5] Maybe the problem is more fundamental and the very *method* of science does not allow the discovery or pointing out of these aspects of life that are the dearest to us. Think of Eddington's parable about a man studying deep-sea life using a net with a three-inch mesh. After bringing up repeated samples, the man concluded that there are no deep-sea fish less than three inches in length.[6] I would interpret that parable in the context that science could be the wrong tool to catch these aspects of life.

Theology's Contribution

And now: what can theology contribute to the question "What is life?"? Can it contribute at all? The famous Swiss theologian Karl Barth preferred to keep the domains of science and theology separate. Shouldn't we stick with this truce? Since the work of Pierre Simon de Laplace, we know that there is no need for God in science, a phenomenon that is sometimes called "methodological atheism" (which is to be distinguished from the claims of a *metaphysical* atheism, to which even Barth would object[7]) or simply "naturalism". The insertion of God into science would only lead to a God-of-the-Gaps, which would be driven out of these gaps with the advancement of modern science.

This is exactly the point of view many people have of the relation of science and theology, which is – although prominent – historically quite incorrect. In fact, the

Jewish-Christian heritage was an important influence on the evolution of science in Europe (Harrison, 1998) and to speak of a "conflict between science and religion" is a rather modern misperception of the historical developments, created by nineteenth-century bestselling books in a time of culture wars (Losch, 2011b, pp. 21–33) and today resulting from the rise of creationism in the USA and the preference of the media to report conflicts.

What is true, however, is the fact that the dialogue between theology and science is of an asymmetrical character. While the theology of creation could (and maybe even should[8]) take science into account, science's methodological atheism prevents it from receiving anything from theology, one might think. However, it is the purpose of this chapter to reflect more on the matter, while never doubting the necessity of science's *methodological* atheism as a historical achievement. That is to say, creationism and its derivate Intelligent Design are not going to be discussed here.[9]

The misperceptions set aside, what are the potential contributions of theology to science in our time? Theologian Michael Welker has dedicated a booklet to the issue and gives five answers which shall now be applied on the question under consideration (Welker, 2012). His first answer is as follows.

(I) The theology and science discourse should search for common metatheoretical presuppositions.

What has to be noted firstly, is the parlor of a "theology and science discourse". Welker is referring to an ongoing dialogue phenomenon, mainly initiated by the book *Issues in Science and Religion* by physicist and theologian Ian Barbour (1966), which led among other things to the establishment of several lectureships and chairs for "science and religion" at prominent universities in the Anglo-American part of the world. That being said, I doubt whether Welker gives appropriate justice to the pioneering phase of the dialogue in which he actively participated during the past few decades. Welker portrays the search for common metatheoretical presuppositions as a pure preliminary, if not trivial, approach. By this, he downgrades the achievement of this "methodological phase" for firstly establishing the field of science and religion and for creating an atmosphere for dialogue. True, it is more like the prolegomenon to a dialogue, but nevertheless a quite important one for being able to start to talk at all.[10]

Nowadays, there are specialized scholarly societies for the dialogue between science and theology,[11] and within an important part of this dialogue the methodological questions are perceived to have been basically dealt with (Peters, 1998, p. 3). However, it is part of my contribution to this very dialogue to ask if the course that has been set was the right way all along. I think humanities and philosophy should have played a much more important role in this dialogue than they did, and do until today. Also, I am convinced that the differences of the disciplines are downplayed too much by Barbour and his followers (Losch, 2010).

Welker's reluctance to really appreciate the first of his own options certainly has something to do with his critique of meta-discourses: "It is not that such meta-discourses are useless, rather they are limited to adopting only short-term supportive functions". He recommends the "dialogue partners must approve themselves in their work on specific topics, and in their ability to provide convincing answers to specific questions" (Welker, 2012, p. 14). I guess that to consider the specific question "What is life?" is a good start. If some are convinced the question has been answered already, it cannot be too big, I assume.

(II) Theology should prevent the sciences from developing false perspectives of theology.

This second contribution of theology to its dialogue with science can hardly be doubted. Creationism evokes the impression that theology would stick with old-fashioned concepts and live in a past millennium. Yet creationism is not to be equated with a developed theology, which should be understandable given the theology-bashing of creationists. Academic theology knows well that the Genesis account is not to be taken literally, but in a metaphorical sense. The logic of the Genesis narrative is of an eco-logical kind. First, the respective living environments are created and then the creatures to live in them (Link, 2012, pp. 49–68). This interdependence of living beings with their environments is still a hot topic today, and is especially important to be aware of these days.

Also when reading the Bible more literally, it should be clear (this is also Michael Welker's example) that the six days of creation in Genesis 1 can hardly mean six weekdays, as the sun and the moon, which govern night and day and make up the change of the days of the week, are created only on the fourth day of creation. God's time, one can conclude, and also the time of creation, is different from our time (Link, 2012, pp. 53–4), with the seventh day, the *holiday*, as the crown of creation.

Even more, one can find some degree of understanding for self-organization within the narrative: "And God said, Let the earth bring forth grass, the herb yielding seed, and the fruit tree yielding fruit after his kind, whose seed is in itself, upon the earth: and it was so" (Gen 1:11, KJV). Also, Charles Kingsley, an Anglican clergyman, could translate the ideas of evolution with: God could "make all things make themselves" (Kingsley, 1871), and even Charles Darwin referred to Kingsley's positive interpretation of evolution in his *On the Origin of Species* from the second edition on (Browne, 2003, pp. 95–6).

The evolution of life is not doubted by academic theology (and also not by the Catholic Church, as Pope Francis recently made clear once again (Francis, 2014)).

Taking this into account maybe helps us to accept theology as a serious dialogue partner.

(III) Theology should correct mistakes and inconsistencies in scientific presentations of theological and religious issues.

Popular scientists rarely stick with their science. In books for a more general audience, they often develop a kind of metaphysics, and very often in their last chapters they talk about God (Audretsch & Weder, 1999, p. 27). They seem to think theology is something everyone can do, and that you do not need any training or specific skills to do so in a reasonable way.[12] Often they also express a metaphysical atheism based on the methodological atheism of science. They hereby leave their science behind and enter the domains of philosophy and theology.

Consequently, what is produced is often philosophically or also theologically heavily flawed. Basic insights of theology are ignored, which is perceived and portrayed like a caricature of the past. Welker's example is the case of Stephen Hawking's *A Brief History of Time*. A former colleague of Hawking, Cambridge physicist (and later priest) John Polkinghorne, rightly asks: "What is God doing in the book at all?" (Polkinghorne, 1992). Hawking's considerations ("we would know the mind of God") echo in a triumphalist way a long tradition of efforts to "know God thoughts", while not realizing the history of these thoughts, which can be traced back to the devout Christian Johannes Kepler (Hübner, 1975).

In our case, as previously mentioned, Crick's fight against vitalism is also historically largely uninformed. Other biologists, for instance Jacques Monod (not to speak of the well-known Richard Dawkins), of course share the same conviction of a "heroic atheism" (Polkinghorne) that finds the meaning of life just in life's meaninglessness: "... man at last knows that he is alone in the unfeeling immensity of the universe, out of which he emerged only by chance. Neither his destiny nor his duty have been written down. The kingdom above or the darkness below: it is for him to choose" (Monod, 1979, p. 167).

(IV) Developing multiperspectival explorations of areas of knowledge common to both.

Welker finds the "area of knowledge common to both" in the field of (the apostle Paul's) anthropology. I do not want to repeat here Welker's most interesting considerations in these regards. Yet in the topic discussed in *this* chapter, when it is asked what it means to be a human person, theology cannot withdraw but has to hold fast to the richness of the ancient tradition of what was called the human soul as *animal rationale*, man's free will,[13] and ability to relate to God (his being created in the image of God), resulting in his responsibility for the world. How could the soul be represented in scientific terms? Could it be "researched" at all?

Here we face a difference between what I would like to call (1) a more traditional, today more Continental-European way and (2) a more innovative Anglo-American approach (influenced by the previously mentioned Ian G. Barbour) to discuss these problems.[14]

(1) In the Continental-European tradition it would probably suffice to separate *being* from *ought* and to view any connection here as a philosophical fallacy (Hume, Selby-Bigge & Nidditch, 1978, pp. 469–70). This is an important tradition to keep up humaneness: when a naturalist like Dawkins for instance claims the gene as a principally selfish *being* (Dawkins, 1989), he can nevertheless end the first edition of his bestselling book with an appeal to a very different *ought*: "We alone on earth can rebel against the tyranny of these selfish replicators" (Dawkins, 1989, p. 201).

(2) More typical for the Anglo-American dialogue between science and religion, which has often been pursued by individuals with a double qualification in science and in theology, would, however, be to question the way that the very *being* is portrayed by the naturalists. They would argue for the indeterminateness of physics on a quantum level (Polkinghorne, 1998, p. 31) and the existence of corporate "wholes" in the fabric of life whose laws govern their particulars, as was already sketched in the reception of Polanyi. This way, the governance of the mind over the body could also be envisioned (Popper & Eccles, 1977). So, actually, they would want to give science the incentive to find a non-reductionist dimension of life, or they would even claim that the reality of this different domain has already been proven scientifically (Murphy, 1998).

To whatever position one likes to cling, the reality of man's responsibilities for his choices, resulting from his responsive relatedness to God, could never have been given up by theology.

(V) Try to build small bridges at the boundaries of each side's areas of knowledge.

The "small bridges" here may stem from Welker's critique of meta-discourses and his liking of *specific* topics, as Ian Barbour's initial meta-concept of "critical realism" (Losch, 2009) in science and religion has often been compared with a pioneering "bridge" between science and religion (Russell, 2003, 2008b).[15]

Welker continues within the field of anthropology and discusses here man's relation to God. He emphasizes keeping in mind theology's insights into human anthropology, like his dignity (createdness in the image of God), his sin and his need for salvation (soteriology) and for fulfillment (eschatology). "In a reductive secular and secularizing perspective we could say that 'natural' and 'cultural' processes are intensely intertwined in real life.[16] The theology-and-science discourse can open the scientist's minds and sensivities to dimensions of human life which resist integration into a naturalistic account and which can only be excluded at tremendous costs for human culture and civilization"(Welker, 2012, p. 63). In the right spirit, though, Welker concludes, both scientists and theologians can contribute as truth-seeking communities[17] to a richer understanding of the ultimate reality of creation.

To a certain extent, Welker lists only contributions of theology to the interface *between* science and theology, more to the established field of "science and religion" than to science itself. Given the strict procedures of science, this is understandable. I do, however, think that theology can somewhat contribute to the field of science itself, by addressing the scientists as human beings. An additional answer I would like to add to Welker's list will therefore be as follows.

(VI) Theology can remind science that the completeness of knowledge is of an "eschatological" nature.

This means that only at the end of all things, will we know everything (see 1 Cor. 13:9–10). For now, we cannot have a God's eye view of the world, and our specific perspective limits our possibilities for knowledge. So too every definition of life is of a preliminary nature; reality always transcends definitions. We can take Schrödinger as a positive example for how to deal with that phenomenon. "In Schrödinger's works we see each new solution giving rise to a more acute awareness of the unresolved"(Gumbrecht, 2011, p. 10).

Biology should not repeat the historical mistake of physics at the end of the nineteenth century to have thought to already have found out everything that is relevant (Lindley, 1993). How can we combine our eagerness to know with a respect for the objects we are researching (which are subjects when it comes to humans)? We should not forget that only life can research life,[18] so it is also respect for our own kind that is of the essence. At the very least I want to mention the deep bioethical dimensions and questions of biological research here.

Regarding alternative ways of treating the questions I posed, it is probably no wonder that I favour the more traditional Continental-European way, because it is more respectful to the immense intrinsic differences between the disciplines. Nevertheless, I do not believe that the divergent fields are totally separated, as envisioned by Barth. At its best, what I called the "Anglo-American approach" can lead to the following.

(VII) An auxiliary "kenotic" function of theology for science.

"Kenotic" means "divested of power". For me, theology is not the queen of sciences, but in a function subordinate to its positive exploration of revelation (and its being needed for education of ministers) also an *ancilla scientiae*, servant to science (Losch, 2011a, p. 139). As it is her task to try to keep "the whole" within the scientific perspective, theology maybe sometimes has to postulate entities and relations not yet considered by science, yet in a way informed by science.[19] Although it has of course to try to avoid offering a God-of-the-gaps, it has to point out the existing gaps in the scientific account of the universe, and also of life, and sometimes even propose a preliminary "filler" within theological or philosophical vocabulary.[20] It may well be that theology one day has to retreat from the claim of a particular gap as it is closed by science. Yet until then, it leaves the floor open for new research.[21]

So, what can theology contribute to the question "What is life?"? As a dialogue partner that takes science seriously, it can only analyze the theological remnants and metaphysical concepts used in biology and maybe relativize approaches that are too one-sided. This is maybe another way of being a *servant to science*, although a difficult one, as the very method of science has to be naturalistic. Theology can therefore only criticize biology for being physicalist when it becomes a *Weltanschauung*.[22] Instead, it can either advocate the difference between *being* and *ought* or, by proposing concepts beyond a pure physicalist approach and by questioning the finality of current results, maybe it can further research. While a scientific impact of this still needs to be proven, keeping the question open while allowing science to proceed could itself be an important contribution to human culture.

Notes

1 Cf. the series on the German equivalent "Leben" (Bahr & Schaede, 2009; Schaede, 2012, 2016).
2 "*Schroedinger's* text should be read not as the starting point of the appeal that biological phenomena had over many physicists, but rather as the culmination of a long tradition that attempted to explain the nature of life in physical terms" (Lazcano, 2008, p. 1). See also (Pauling, 1987) and (Perutz, 1987).
3 In 1927 the Russian biologist Nikolai Koltsov had proposed "that inherited characteristics are recorded in special double-stranded giant molecules" (Soyfer, 2001, p. 726). Nevertheless Watson and Crick "had not even heard about Koltsov's hypotheses, although his theory was known in the West" (Soyfer, 2001, p. 726).
4 See (Canguilhem, Marrati & Meyers, 2008, pp. 59–74).
5 For a fascinating account in these regards see Küppers (2012).
6 As told in Barbour (1997, p. 88).
7 "There is free scope for natural science beyond what theology describes as the work of the Creator. And theology can and must move freely where science which really is science, and not secretly a...religion, has its appointed limit" (Barth, 1958, p. 9).
8 This wouldn't be Barth's position of course.
9 For a critique of these stances see Hemminger (2009).
10 I assume it is so natural to him that he simply forgot about its importance.
11 The invitation-only International Society for Science & Religion (ISSR), the European Society for the Study of Science and Theology (ESSSAT) and the Institute for Religion in an Age of Science (IRAS).
12 Freedom of belief is, of course, an important cultural achievement. Yet as long as one likes to stick to the traditional idea of God as revealed through historical events, one should not leave out traditional theological reasoning.

13 There has also been an intense theological discussion around Luther about how free man's will can be portrayed, see Luther (1525), cf. also Welker (2012, pp. 55–6).
14 These are of course much generalized differences ("ideal types").
15 For a critique see Losch (2010).
16 I subscribe to this thought with my idea of constructive-critical realism, see Losch (2010, 2005).
17 This idea he developed together with John Polkinghorne, see Polkinghorne and Welker (2001, pp. 132–48).
18 Concerning this thought of the physician-philosopher Viktor Von Weizsäcker see Losch (2015).
19 There is a specific danger here: that the science referred to is out-dated and also the question, if those that have a double qualification in science and theology manage to keep up with the advancement of scientific knowledge.
20 Polkinghorne's claim of "intrinsic gaps" in an ontologically open fabric of the world is analyzed by Robert John Russell as a move "from a set of theological and experience-based convictions about the world back to science to suggest a new scientific research program" (Russell, 2008a, p. 131).
21 One would have to research whether vitalism was an example in this fashion.
22 On this danger, see also Jackelén (2008).

References

Audretsch, J. & Weder, H., eds. (1999). *Kosmologie und Kreativität: Theologie und Naturwissenschaft im Dialog*, Leipzig: Evangelische Verlagsanstalt.

Bahr, P. & Schaede, S., eds. (2009). *Das Leben I: Historisch-systematische Studien zur Geschichte eines Begriffs*, Tübingen: Mohr Siebeck.

Barbour, I. G. (1966). *Issues in Science and Religion*, Englewood Cliffs, N.J.: Prentice-Hall.

Barbour, I. G. (1997). *Religion and Science: Historical and Contemporary Issues*, 1st HarperCollins revised edition, San Francisco, CA: HarperSanFrancisco.

Barth, K. (1958). *The Church Dogmatics III/1*, G. W. Bromiley and T. F. Torrance, eds., Edinburgh: T&T Clark.

Browne, E. J. (2003). *Charles Darwin*, London: Pimlico.

Campbell, D. T. (1974). "Downward causation" in hierarchically organised systems. In F. J. Ayala, ed., *Studies in the Philosophy of Biology: Reduction and Related Problems*, Berkeley, CA: University of California Press, pp. 179–86.

Canguilhem, G., Marrati P. & Meyers, T. (2008). *Knowledge of Life*, New York, NY: Fordham University Press.

Crick, F. (1966). *Of Molecules and Man*, Seattle, WA: University of Washington Press.

Crick, F. (1990). *What Mad Pursuit: A Personal View of Scientific Discovery*, London: Penguin.

Dawkins, R. (1989). *The Selfish Gene*, New edition, Oxford: Oxford University Press.

Francis, Pope. (2014). *Address of His Holiness Pope Francis on the Occasion of the Inauguration of the Bust in Honour of Pope Benedict XVI*, Casina of Pius IV, Monday, 27 October 2014. http://w2.vatican.va/content/francesco/en/speeches/2014/october/documents/papa-francesco_20141027_plenaria-accademia-scienze.html.

Gumbrecht, H U. (2011). *What Is Life? The Intellectual Pertinence of Erwin Schrödinger*, Stanford, CA: Stanford University Press.

Harrison, P. (1998). *The Bible, Protestantism, and the Rise of Natural Science*, Cambridge: Cambridge University Press.

Hemminger, H. (2009). *Und Gott schuf Darwins Welt: Schöpfung und Evolution, Kreationismus und intelligentes Design*, Gießen: Brunnen-Verl.

Hübner, J. (1975). *Die Theologie Johannes Keplers zwischen Orthodoxie und Naturwissenschaft*, Tübingen: Mohr.

Hume, D., Selby-Bigge, L. A. & Nidditch, P. H. (1978). *A Treatise of Human Nature*, 2nd edn, Oxford, New York: Clarendon Press; Oxford University Press.

Jackelén, A. (2008). What theology can do for science. *Theology and Science*, **6**(3): 287–303.

Kingsley, C. (1871, 2003). The natural theology of the future. In C. Kingsley, ed., *Scientific Essays and Lectures*, http://www.gutenberg.org/files/10427/10427-h/10427-h.htm.

Küppers, B.-O. (2012). *Die Berechenbarkeit der Welt: Grenzfragen der exakten Wissenschaften*, Stuttgart: Hirzel.

Lazcano, A. (2008). What is life? A brief historical overview. *Chemistry & Biodiversity*, **5**(1), 1–15.

Lindley, D. (1993). *The End of Physics: The Myth of a Unified Theory*, London: HarperCollins.

Link, C. (2012). *Schöpfung: Ein theologischer Entwurf im Gegenüber von Naturwissenschaft und Ökologie*, Neukirchen-Vluyn: Neukirchener Theologie.

Losch, A. (2005). Our world is more than physics: a constructive–critical comment on the current science and theology debate. *Theology and Science*, 3(3), 275–90.

Losch, A. (2009). On the origins of critical realism. *Theology and Science*, 7(1), 85–106.

Losch, A. (2010). Critical realism – a sustainable bridge between science and religion? *Theology and Science*, **8**(4), 393–416.

Losch, A. (2011a). Gott in der Wissenschaft?. In M. Rothgangel and U. Beuttler, eds., *Glaube und Denken*, Bern: Peter Lang Pub Inc, pp. 129–43.

Losch, A. (2011b). *Jenseits der Konflikte: Eine konstruktiv-kritische Auseinandersetzung von Theologie und Naturwissenschaft*, Göttingen: Vandenhoeck & Ruprecht.

Losch, A. (2014). Glauben als Grundlage: Michael Polanyis Berufung auf die Bedeutung des Glaubens in der Wissenschaft und die Rezeption seiner Philosophie im Gespräch von Theologie und Naturwissenschaften. In E.-M. Jung, ed., *Jenseits der Sprache: Interdisziplinäre Beiträge zur Wissenstheorie Michael Polanyis*, Münster: Mentis, pp. 107–40.

Losch, A. (2015). 'To research living beings, one has to participate in life.' Viktor Von Weizsäcker's legacy. In D. Evers, M. Fuller, A. Jackelen and A. Saether, eds., *Issues in Science and Religion: What Is Life?*, Cham: Springer.

Luther, M. (1525). *De servo arbitrio*, Wittenberg: H. Lufft.

Maule, D. & Watson J. D. (2001). *The Double Helix*, Harlow: Pearson Education.

Maynard Smith, J. & Szathmàry. E. (1995). *The Major Transitions in Evolution*, reprint, Oxford: W. H. Freeman.

Monod, J. (1970). *Le hasard et la nécessité. Essai sur la philosophie naturelle de la biologie moderne*, Paris: Éditions du seuil.

Monod, J. (1979). *Chance and Necessity*, Glasgow: Fount Paperbacks.

Morange, M. (2005). What is life? A new look at an old question. In P. Hajek, L. Valdes-Villanueva and D. Westerstahl, eds., *Logic, Methodology and Philosophy of Science*, London: King's College Publications, pp. 431–40.

Morange, M. (2012). The recent evolution of the question "What is life?". *History and Philosophy of the Life Sciences*, **34**, 425–38.

Murphy, N. (1998). Theology, cosmology, and ethics. In T. Peters, ed., *Science and Theology: The New Consonance*, Boulder, CO: Westview Press, pp. 103–17.

Pauling, L. (1987). Schrödinger's contributions to chemistry and biology. In C. W. Kilmister, ed., *Schrödinger, Centenary Celebration of a Polymath*, Cambridge: Cambridge University Press, pp. 225–33.

Peacocke, A. (1998). A map of scientific knowledge: genetics, evolution, and theology. In T. Peters, ed., *Science and Theology: The New Consonance*, Boulder, CO: Westview Press, pp. 189–210.

Perutz, M. F. (1987). Erwin Schrödinger's What Is Life? and molecular biology. In C. W. Kilmister, ed., *Schrödinger, Centenary Celebration of a Polymath*, Cambridge: Cambridge University Press, pp. 234–51.

Peters, T. (1998). "Introduction". In T. Peters, ed., *Science and Theology: The New Consonance*, Boulder, CO: Westview Press, pp. 1–10.

Polanyi, M. (1968). Life's irreducible structure: live mechanisms and information in DNA are boundary conditions with a sequence of boundaries above them. *Science*, **160**(3834), 1308–12.

Polkinghorne, J. (1992). The Mind of God? *The Cambridge Review*, **113**(2316), 3–5.

Polkinghorne, J. (1998). *Science and Theology: An Introduction*, London: SPCK / Fortress Press.

Polkinghorne, J. & Welker, M. (2001). *Faith in the Living God*, London: SPCK.

Popper, K. R. & Eccles J. C. (1977). *The Self and Its Brain*, New York, NY: Springer International.

Russell, R. J. (2003). Editorial. *Theology and Science*, **1**(1), 1–3.

Russell, R. J. (2008a). *Cosmology: From Alpha to Omega*, Minneapolis, MN: Fortress Press.

Russell, R. J. (2008b). Editorial. *Theology and Science*, **6**(1), 9–11.

Schaede, S., ed. (2012). *Das Leben: Historisch-systematische Studien zur Geschichte eines Begriffs, Band 2*, Tübingen: Mohr Siebeck.

Schaede, S., ed. (2016). *Das Leben: Historisch-systematische Studien zur Geschichte eines Begriffs, Band 3*, Tübingen: Mohr Siebeck.

Schrödinger, E. (1944/1992). *What Is Life? The Physical Aspect of the Living Cell; with Mind and Matter; & Autobiographical Sketches*, Cambridge: Cambridge University Press.

Sigmund, K. & Szathmáry, E. (1998.) Biomathematics. Merging lines and emerging levels. *Nature*, **392**(6675), 439.

Soyfer, V. N. (2001). The consequences of political dictatorship for Russian science. *Nature Reviews. Genetics*, **2**(9), 723–9.

Welker, M. (2012). *The Theology and Science Dialogue: What Can Theology Contribute: Expanded Version of the Taylor Lectures, Yale Divinity School 2009*, Neukirchen-Vluyn: Neukirchener Theologie.

11

Autopoietic Systems and the Theology of Creation: On the Nature of Life

ALEXANDER MAẞMANN

Whether on Earth or on other planets, evolving life will be autopoietic. It will not be the product of prior design; rather, it will exhibit an ongoing process of self-formation. This has been the case on Earth, the one instance of life that we know. It is only reasonable to expect the same to be the case on habitable planets with a second genesis of life.

Modern philosophy and biology, however, have typically looked at nature through mechanistic glasses, through lenses that filtered out the dynamism of autopoiesis. When we look through mechanistic lenses, we cannot discern purpose, *telos*, or mind. Mind is present within nature, within the so-called mechanism of material events. If we set aside our distorted Cartesian dualism, which isolates mind from matter, we will begin to see matter differently. Where there's life, there is mind.

Philosophers of science such as Evan Thompson and Hans Jonas will help us regrind our lenses to see more clearly what life has been like all along. In organisms, physical self-organization is the external dimension of cognition, and cognition is the internal aspect of self-organization. Primitive organisms set norms for themselves, which help define their system, leading to the construction of meaning. A teleology is present within the organism; it sets its own goal. The organism realizes its own evolutionary adaptation; it makes sense both out of its genetic heritage and in its behavioral patterns. In a complex organism such as the human person, this same basic autopoietic constitution of meaning produces consciousness, will, and thought. In sum, the living organism is not a machine; rather it is a self-initiating meaning-maker.

This fresh look at life also opens new perspectives in the dialogue between theology, science, and philosophy. To begin with, Judeo-Christian traditions do not typically insist on the isolation of the soul from matter. By contrast, in prominent Biblical texts, it is very difficult, not very helpful even, to disentangle these two dimensions. Further, the close intertwining of meaning and physical organization

Mechanistic Science and Theology

In seventeenth-century Europe, debates about fundamental paradigms in science involved a strong theological dimension. René Descartes supported his mechanistic philosophy with the argument that it was more appropriate theologically than the assumption of final causes in nature. A teleological scheme, while still able to present God as creator and to uphold God's continuing preservation of nature, may appear to grant a greater degree of autonomy to creation.[1] However, when Descartes pitted the inertia of matter against final causes in nature, he highlighted both the unchangeable laws that God had established in creation and the importance of God's preservation of matter and force. As a result, any appearance of the autonomy of nature was dissolved. Nature was viewed as even more clearly dependent on the transcendent creator (Brooke, 1991, pp. 74f.). The transformation in the philosophy of nature was tremendous. The evacuation of creation of seemingly independent striving went hand in hand with a dualism between mind and matter. "The matter existing in the entire universe is one and the same," Descartes wrote, "and it is always recognised as matter simply in virtue of its being extended" (Cottingham, 1999, p. 239). It may even seem that this argument had a strong theological backing.

Descartes' dualism was again affirmed roughly 200 years later by Ernst Haeckel, even if now without invoking any theological support:

We thus arrive at the extremely important conviction that... the distinction which has been made between animate and inanimate bodies does not exist. When a stone is thrown into the air and falls down according to definite laws,... the phenomenon is neither more nor less a mechanical manifestation of life than the growth and flowering of plants.
(Haeckel, 1914 [1876], 1:23)

Descartes famously contrasted extended matter with the immaterial human mind. The obvious difficulty with this position is that if we already negate conscious sensitivity in animals (Harrison, 1992), reducing them to machines, it is tempting to do the same with human persons. This is in fact the materialist position of Haeckel, who considered human thought an epiphenomenon merely resulting from certain states of matter. Haeckel's previous statement reads in its entirety: "the phenomenon is neither more nor less a mechanical manifestation of life than the growth and flowering of plants, than the propagation of animals or the activity of their senses, than the perception or the formation of thought in man."

Certainly Descartes insisted on the nobility of humanity in contrast to other living beings. He went on to suggest that if there is a non-material spirit in us, able to intervene in material affairs, then even greater power and sovereignty can be ascribed to God, even in material affairs (Brooke, 1991, p. 134, see also p. 19). Nonetheless, for some scientists even of his day, the advantage of the reduction of nature to inert matter and universal mechanistic laws was an increasing reluctance to invoke divine interventions or miracles in accounting for natural processes (Harrison, 2000, pp. 178f.; Brooke, 1991, p. 58). Some interpreters have even found this solution congenial to Biblical creation traditions (Harrison, 2000, p. 170) or to the anthropology of the apostle Paul, who supposedly disparaged the physical biological dimension in favor of the spiritual realm.

However, the difficulties with this overall position are extreme. To begin with, the inference from human spiritual agency to God cuts both ways. Once human spiritual agency in the material world appears dubious – as it did to Haeckel – both the life of the human spirit will seem lacking in salience and meaning and belief in God will appear pointless. Already in 1686, in one of Fontenelle's writings, a character exclaimed that "'philosophy [i.e., philosophy of nature, i.e., science] is a very mechanical affair!' 'So much so,' I replied, 'I am afraid it will fall into disrepute'" (Fontenelle, 1803, p. 10).

Might there not be another way to combine scientific naturalism with a more subtle biology? After all, Haeckel's epiphenomenal account of the human mind, as much as it presents itself as grasping a fact in nature, cannot itself be exempt from the epiphenomenal realm of the mental (see also Jonas, 1966, p. 88). And finally, are there ways to conceive of a creative and concurrent presence of the divine within history that goes beyond mere preservation?

A Systemic Interpretation of Biology: Evan Thompson and Hans Jonas

The philosopher of science Evan Thompson speaks of a deep continuity between life and mind, in the sense that the two dimensions of body and mind inseparably emerge with the organism. The theory of autopoietic systems is at the heart of his project. An autopoietic system is a constellation of elements that, in influencing each other mutually, arrange for the conditions of their reproduction as a larger unity. The living cell is the paradigmatic case: the cell membrane makes a significant contribution to the metabolism and vice versa. The essential metabolic processes take place within the cell itself, which is called organizational closure. Systemic closure, by contrast, means that all products relevant for the system are furnished by the system itself – unless continually provided by the environment – where they remain for further use. Autopoietic systems display both bottom-up and top-down causality, i.e., in sum, circular causality. While the maintenance of the overall organization is dependent on the nutrient supply, the particular role of particular nutrients is not merely a consequence of chemistry and physics, but crucially depends also on the activity of the system itself.

Since such autopoietic systems relate to their environments in a contingent way that is in accordance with the logic of their own systemic organization, the enactive approach attributes elementary cognition to these systems. By this Thompson means behavior according to a norm that is generated by the organism itself. In this process, the organism maintains itself in a particular environment (Thompson, 2010, p. 159). This behavior is an essential aspect of its physical organization. It constitutes "meaning" for the organism. Physical organization is thus the external dimension of cognition, and cognition is the internal aspect of organization. Such meaning-making emerges together with the organism's striving for preservation. This is a clear contrast to the mechanistic portrayal of organisms as mere gene machines, which respond passively to the effects of protein synthesis.

Thompson illustrates this with the bacterium *Escherichia coli*, which seems to float aimlessly within its milieu (2010, pp. 74f.). Once protein channels in the cell membrane detect a certain nutrient, however, the bacterium moves its flagellum in such a way that it swims against the nutrient gradient. This behavior is a consequence of the physical organization of the bacterium. The result is a consistent, predictable form of structural coupling between the system and its environment, in which "meaning" emerges for the bacterium. In accordance with this paradigmatic case, cognition is not theory and contemplation, but competent behavior in self-maintenance. Meaning is nothing general, context-independent, but is constituted by the bacterium. The gradient of this chemical exists independently of the bacterium's behavior, but not as a nutrient gradient. Meaning is thus a context-dependent coupling, which is to be expected, once the system is operative. It is not to be expected, however, merely on the basis of natural laws by themselves. For the organization of the bacterium is a historically contingent tradition. In autopoiesis, this tradition is continually enacted. Thus the behavior of the bacterium constitutes a "self" (Thompson, 2010, p. 75).

Cognition is the internal aspect of physical organization and its enaction, which, however, becomes thematic only with the emergence of the central nervous system. Jonas has been criticized for making unwarranted assumptions about subjectivity in insects and plants – and perhaps even in unicellular organisms (Wolf, 1998, p. 218). Indeed he tends to attribute the fundamental traits of all organisms to even the simplest one. For Jonas, the fact that the organism exercises freedom in constituting itself out of the environmental substrate, with its flip side of fundamental dependence on this environment, constitutes transcendence. Such self-reference is not trivial, and thus "there is inwardness or subjectivity involved in this transcendence, imbuing all the encounters occasioned in its horizon with the quality of felt selfhood, however faint its voice" (Jonas, 1966, p. 84, see pp. 56f., p. 65). In other places, however, Jonas grants that any proper subjective reference to the world requires a central nervous system (1966, p. 100). Perhaps this state of affairs can be resolved in saying that in very simple organisms, cognition is behavioral and sentient, but not yet a subjective, conscious experience (Thompson, 2010, pp. 161f.). As soon as emotions arise, however, the kind of cognition required for autopoiesis can be maintained in an even more challenging environment. An animal not only perceives its prey, but it is due to emotion that it perceives prey as a target, so that motility does not lack coordination (Jonas, 1966, pp. 101f.). Recent studies suggest that the assumption of inner states analogous to emotions is plausible even in insects (Gibson *et al.*, 2015; Anderson & Adolphs, 2014; Bateson *et al.*, 2011). Such emotions have cognitive valence, and they help an organism operate meaningfully in its environment. By observation of the further ramifications of the evolution of self-organization and cognition, Jonas argues that in the human person, the same basic autopoietic constitution of meaning amounts to nothing less than "self-transparency in consciousness, will and thought" (Jonas, 1966, p. 90). If, according to Darwin's theory of evolution, "man was the relative of the animals, then animals were the relatives of man, and in degrees bearers of that inwardness of which man, the most advanced of their kin, is conscious in himself" (Jonas, 1966, p. 57).

Teleology in Ontogenesis: The Philosophy of the Organism in Dialogue with Kant

According to Thompson and Jonas, cognition emerges with the physical self-organization of the system. The classic philosophical interlocutor on physical self-organization is Immanuel Kant (Thompson, 2010, p. 211, pp. 129–40). While he argued for a remarkable notion of teleological self-organization in living beings, he still presupposed a chasm between reason and physical organization. Neither is teleological self-organization strictly empirical, nor is the perceiving human mind fundamentally constituted by its own physical organization. Nonetheless, Kant points out that the highly specialized bird anatomy must seem highly unlikely, virtually arbitrary, apart from teleological modes of thought: "nature, considered as a mere mechanism, could have formed itself in a thousand different ways without hitting precisely upon the unity in accordance with such a rule..." (Kant, 2000, p. 234). Kant himself, who published these considerations 70 years before Darwin's *On the Origin of Species*, did not draw the conclusion that nature is indeed organized teleologically. For him, teleology was not a constitutive but a regulative idea. Indebted to Isaac Newton's ideal of mechanistic science, he argued neither for a natural theology that infers a designer from design, nor for vitalism, nor for a pantheist tendency toward life inherent in nature. Nonetheless, we rely on the regulative idea of teleology in assuming the unified range of objects with which biology is concerned.

A crucial question is of course why and how this regulative idea can be expected to shape a true understanding of reality. Kant was certain that biological self-organization would never be reduced to mere physics and mechanism, that no new Newton would ever fully account for a blade of grass (Kant, 2000, p. 271). However, for this claim to be warranted, a more principled understanding of natural self-organization than Kant's is required (Spaemann, 2012, p. 46). Only if the regulative teleological idea is corroborated by a cogent connection to the phenomenal world, rather than being a mere postulate, can this idea appropriately define the referent of the discipline of biology.

For a principled account of autopoietic systems, a thorough account of causality is required first. After all, although dependent on the availability of nutrients, the autopoietic system is its own cause in organizing itself. Jonas's phenomenological analysis of causality takes us to the heart of the matter. Kant rightly agreed with David Hume that causality itself cannot be observed. Ironically, after disputing the strict validity of causality, Hume gives a psychological account of the causality operative in the genesis of the illusion of causality (Jonas, 2011b, p. 54). This might be a reason to hold on to causality after all. Yet in postulating a category of causality from beyond the realm of experience, Kant does not resolve the issue, either. In fact the category of causality is not constituted by theoretical reason a priori, but by our embodied experience of an impact. "For example, if, on the cinema screen, I see a fist thrust forward hitting a body that then falters, I can understand this purely eidetic sequence as a dynamic one because I have experienced the actual impact of the bodies firsthand, 'in my own body' [*am eigenen Leib*]" (Jonas, 2011b, p. 55).[2] The experience of impact is "transcribed" into a neutral, distanced format, in which the experience of the impact itself tends to get lost. In touch, for example, the experience is located on a spectrum between the poles of a sheer experience of quality and of sheer impact (Jonas, 1966, p. 30). This is even clearer in vision, as the primary causal impact of reflected light on the eye takes place on a minute scale, so that the visual experience is rather unlikely to include the very impact of light itself. With transcription from causal impact to neutral formatting having become a routine process ontogenetically, we no longer look behind the scenes when engaging in acts of perception. In a different context, Shaun Gallagher describes this process as "direct perception" (Gallagher, 2008), which operates without reflection on or internal reenactment of the thing thus perceived. In the perception of object–object-relations, the fundamental physical experience of my own embodied object–subject-experience suggests itself as a paradigm, yet remains mute to conscious interrogation.

Embodied causal perception is crucially involved in the experience of autopoietic systems (Jonas, 1966, p. 149; Thompson, 2010, pp. 162–5). To recognize the organism in its freedom to reconstitute its form causally vis-à-vis the changing material substrate, we have to take our own organismic autopoietic nature seriously, just as we rely implicitly on our own embodied experience of an impact when "seeing" causality. By contrast, on the mere grounds of classic materialist empiricism, to which Kant is committed, it is impossible to recognize an autopoietic system. After all, we could never be sure that its relative stability is constituted in any other way than that of a wave on the ocean shore constituted of water molecules. Just as a wave is a mere epiphenomenon, from a strictly neutral, empiricist point of view it is always possible that the material substrate constitutes the cell in a strictly one-sided way, that the metabolism merely happens to the cell. By the same logic, thought merely happens to a person. Yet it is only since we know organisms from the inside, by being organisms ourselves, that we view organisms as something other than mere epiphenomena.

Thus, for Kant to go beyond the limitations of a regulative idea requires him to leave behind his binary anthropology in which the human body appears as one element in the mechanistic universe among others, fundamentally subject to deterministic causal forces, while the human mind alone, unaffected by the mechanistic forces of the physical world, displays freedom. Cognition arises strictly in the mutual interaction with the body.

But what exactly does it mean to say that organisms organize themselves teleologically? According to the theory of autopoietic systems, an organism is not simply a passive aggregate of matter. Thompson even calls the organism "an active agent or subject of the evolutionary process" (Thompson, 2010, p. 203). By contrast, to attribute a determinative role to genes in ontogeny means to cede too much to the reified understanding of information, which characterizes the gene's-eye view of evolution.[3] A competent organism is required for gene expression, which cannot be reduced to an automatic processing of genetic information. In this sense there is no such thing as genetic hardwiring. Genes do not cause their own replication, and apart from the intricate transcription processes within the organism, we would not be able to speak of genetic information. There are also other crucial hereditary procedures apart from genetics, for example involving symbionts (Thompson, 2010, p. 177). Further, the genetic code – the dictionary that links up certain amino acids with letter combinations of the genetic alphabet (the nucleotides C, G, A, T) – is not itself coded for, but is a product of evolution that may vary in specific cases (Maynard Smith & Szathmáry, 1995). There may also be alternative splice variants of a gene. Based on alternative rules of interpretation, the same gene is translated into alternative proteins – sometimes several thousands of them, as adaptivity requires (Noble, 2008, p. 8, p. 30). Such alternative rules of interpretation are not themselves hardcoded. Since the gene is fundamentally dependent on the interpreting organism, Thompson considers the view of the organism as a mere epiphenomenon of the genome "less sophisticated" than "the ancient dualism of soul and body" (Thompson, 2010, p. 187). In this respect Thompson is fully justified in insisting that natural selection is strictly complementary to self-organization (Thompson, 2010, p. 215).

Genetics: Mechanistic Factors in Ontogenesis

Yet we can still highlight the prominence of natural selection more strongly with Jonas, who at this point displays an existentialist streak. Shaping the material substrate and constituting itself rather than being a passive aggregate, but also being strictly dependent on material resources, the organism displays a "needful freedom" (Jonas, 1966, p. xiv, p. 80; see also Thompson, 2010, pp. 149–52). The concept of autopoiesis summarizes a kind of self-transcendence. "And in this polarity of self and world, of internal and external, complementing that of form and matter, the basic situation of basic freedom, with all its daring and distress, is potentially complete" (Jonas, 1966, p. 83). While subjective experience cannot be presupposed for all kinds of organisms, this still highlights the dependency and passivity of the organism. Thus we should not merely call the organism "an active agent or subject of the evolutionary process" (Thompson, 2010, p. 203), but attribute both activity and passivity to it. *En passant* Thompson suggests that we might as well speak of "dynamic stabilization" rather than "natural selection" (2010, p. 208). But this eclipses the destructive, or at least the privative dimension of the process. Not mentioning the extinction of species in natural selection, he merely speaks of the conservation of adaptation (Thompson, 2010, p. 159, p. 205).

This is in keeping with the fact that Thompson does not mention alternative Darwinian explanations when discussing Kant's ideas on the teleology involved in the physical self-organization of a bird. Is this a strictly exclusive alternative? Would we not merely invert the contested explanatory monopoly of the mechanics of mutation and selection in claiming exclusive validity for processes of self-organization that proceed by way of circular causality? At this point my argument is that there is good reason to do greater justice to linear causality even as circular causality is still upheld.

One fact that highlights classic bottom-up causality impinging critically on the viability of the overall system is the rejection of a non-viable embryo in the mother's womb – an example of natural selection in ontogeny. Even if adapted organisms self-organize, this is also true of less well-adapted ones, which are then the object of natural selection. Another objection to the strong contingency of the organism, which suggests teleology in organization unhampered by external constraints, consists in the fact that the phylogenetic evolution of powered flight can be reconstructed gradually. Antecedents of wings and feathers may have served thermoregulation, sexual selection, better uphill running, or gliding flight (Zimmer, 2011). New adaptations at each of these stages are dependent on the prior constellation. The flip side, however, consists in the disadvantage of the less well-adapted organisms, and this is where the mechanical factors in evolution show their critical edge.

The gradual evolution of flight might be explained even if natural selection played only a minor role. In that case, the structural integration of the overall organism would be said to limit innovations in individual elements morphodynamically. This is the approach favored by Thompson. The neo-Darwinian synthesis, by contrast, identifies the ultimate object of selection either with an organism's genome or with a population that is correlated with a certain gene pool. Thompson, however, does not see the function of genes in their close correlation with a particular phenotype upon which natural selection acts. It is only in one place that he remarks how DNA limits the possible development of an organism to certain options, thus contributing to a particular phenotype of the organism. This is where the concept of robust flexibility comes into play. Robust flexibility is the result of regulator genes that specify a more general anatomy or *Bauplan* (Thompson, 2010, pp. 194–201). Thompson thus sees the function of genes in specifying the phylum of an organism. Within the wider layout defined by the rough phylotypic *Bauplan*, regulator genes allow for a great degree of flexibility. This is in keeping with Thompson's insistence on the organism's freedom vis-à-vis inherited information, which does not determine the organism's development genetically. But he does not take into account that the genome is consistently correlated not only with a particular phylum, but also with a species and even more specific phenotypes, as the example of monogenetic diseases shows. Yet even this phenomenon is not to be equated simply with genetic determinism. If genes steer development along a certain corridor in some respects, development can nonetheless reflect environmental factors in other respects, not least in how genes are transcribed. Genes can also influence the extent, the conditions, and the particular quality with which the environment impacts the organism. "Nature" can model how "nurture" impacts the organism and vice versa. Since, however, genes have an influence on the phenotype that goes beyond the flexibility of regulator genes, and since their effect may consist in a lesser fitness than that of other organisms, a crucial contribution of selection to phylogenesis is likely and the organism must also be thought of as passive – certainly not in all respects, but in some respects and to some extent.

In 1951, Jonas had already argued against a similar opponent as Thompson does when confronting gene-centrism. Jonas perceived August Weismann as the major proponent of a contemporary mechanistic account of evolution, which he construed in terms very similar to Richard Dawkins's fundamental distinction between "replicators" (genes) and "vehicles" (organisms). Jonas contrasted this account with his view of the organism. He observed an emerging new paradigm in neo-Darwinism.

> It is not, as might appear at first glance, the dualism of organism and environment – this pair rather forms one interactive system – but the dualism germ: soma, in which the soma (the actual organism) is itself part of the "environment," namely, the immediate environment for the germ plasm and the mediator of the effects on the latter's existence of the wider environment... Thus there arises within the materialist realm itself a strange parody of the Cartesian model of two noncommunicating substances [i.e., body and mind].
> *(Jonas, 1966, p. 52)*

Jonas argues that ultimately, this paradigm is unhelpful as it cannot account for the human mind. Nevertheless, he does acknowledge a significant role of "chance" in evolution (Jonas, 2011a, p. 116), "the venture's chance and risk" (Jonas, 1966, p. 271): "life as it were, entirely innocent of any foreshadowing disposition toward them, surprises itself – and its Creator if there is one. Mind was not foreseen in the amoeba, nor was the vertebrate structure, science no more than the opposable thumb" (1966, p. 52).

Nonetheless, this argument for the element of chance in evolution does not amount to a mechanistic interpretation of evolution or a negation of top-down causality in autopoietic systems. Neither should autopoiesis be restricted to the organism's own constitution of the meaning of genetic information. After all, this would leave open the question of how an organism uses an organ once it has been constituted. That a species evolves powered flight, having had gliding ancestors, does not rest on a genetic behavioral program mutating in lockstep with the anatomic conditions of this behavior. Neither does it seem clear that a consistent behavioral ecology such as the flapping of wings for powered flight is reducible to particular patterns of protein synthesis. Even in animals, innovative behavior cannot be reduced to an algorithm.[4] In this sense, a particular kind of teleology is required here – not a teleology according to which the evolution of less complex species aims at future superior species, nor one in which a hypostasized meta-agent, be it "Nature" or God, designs species in order to implement particular purposes, but a teleology in which the organism is its own goal. The organism realizes its own adaptivity, both in making sense out of its genetic heritage and in its behavioral patterns. The theory of autopoietic systems says nothing less, except that in evolution, autopoietic systems also realize their own novelty in teleological ways, provided they are not ruled out by genetic constraints. The developing organism plays the hand it was dealt by virtue of teleology in ontogenesis. The overall process of an evolving phylogenetic lineage, however, is not a trajectory shaped by teleology.

The Organism in Judeo-Christian Traditions: The Question of Dualism

A concept that combines the mechanistic events of genetic mutation and natural selection with the teleological autopoiesis of the organism stands halfway between a mechanistic paradigm in biology and Thompson's full-blown teleology in development. The Aristotelian teleological traditions on which Thompson, but also Jonas, draws, became increasingly sidelined in the scientific revolution. According to Jonas, the scientific revolution gained in force as a Cartesian dualism between mind and matter cleared the physical realm for the strictly mechanistic treatment of nature. According to Jonas, Judeo-Christian traditions played an important role in the ideological legitimization of this modern objectifying approach to nature.

> Jewish monotheism had abolished the deities of nature and all intermediary powers, leaving God and world in clean-cut division... man's soul is the only entity in the world – but not of the world – which is created in the image of God... the essential division between God and world is thus repeated or mirrored in the essential division between mind and nature... Thus the idea of a mindless or "blind" nature, which yet behaves lawfully... had become metaphysically possible.
> *(Jonas, 1966, p. 71)*

The conviction of a coincidence of Cartesianism and Judeo-Christian traditions can indeed be traced historically (Harrison, 2000, p. 170). According to this view, the Christian tradition especially contracted the spiritual dimension into the nonphysical human "soul" and conceived of living beings in their sheer objective presence, rather than as living systems to which a subjective striving and conation is fundamental. It is true that Descartes understood his worldview in theological terms, insisting that a mechanistic universe cannot maintain itself, but displays laws established by the creator and depends on divine preservation. This view squares with a particular theological interpretation of the Old Testament that highlights above all a stark difference between creator and creature (Brooke, 1991, p. 75). The notion of the soul plays an important role in this worldview.

Already Jonas's erstwhile teacher Martin Heidegger critiqued the way traditional Western anthropology placed the human soul vis-à-vis the world, rather than seeing human existence fundamentally enmeshed in the world. Although distinct, Jonas's and Heidegger's views on the Christian notion of the soul show clear parallels. For Heidegger, the understanding of the human person *as* soul seals our being off from the mundane procedures of the world, maintaining a "false consciousness," as it were, with regard to our factual being-in-the-world, i.e., our attempt to enact our being in managing worldly entities. In the interest of a proper understanding of being, however, it is important for this being-in-the-world to come to light, precisely as one can only arrive at a genuine understanding of being by facing the frustration that is necessarily entailed by overtly understanding the meaning of being in one's being-in-the-world. Among the results is a reified understanding of world and self.

> If [our being-in-the-world] is now to be recognized, the explicit cognition that this task implies takes itself (as a knowing of the world) as the exemplary relation of the "soul" to the world. The cognition of world (*noein*) ... thus functions as the primary mode of being-in-the-world even though being-in-the-world is not understood as such. But because this structure of being remains ontologically inaccessible, yet is ontically experienced as the "relation" between one being (world) and another (soul) ... one tries to conceive the relation between world and soul as grounded in these two beings and in the sense of their being; that is, as objective presence. *(Heidegger, 1996, p. 55)*

According to Jonas, it was Heidegger who "destroyed the entire optical model, as it were, of a consciousness that is primarily concerned with recognition and in its place disclosed the willing, toiling, needy and mortal ego" (Jonas, 1993, p. 15). Jonas then transformed Heidegger's being-in-the-world into the way any organism constitutes meaning and inwardness in its struggle for survival. While the understanding of the person as soul eclipses being-in-the-world for Heidegger, for Jonas its implicit Cartesianism eclipses the fact that inwardness and freedom emerge only as organisms strive for nutrients. It may seem, then, that Jonas's philosophy of the organism is opposed to a Christian anthropology. If, on the other hand, Jonas is right in his understanding of the Judeo-Christian heritage, the question emerges why it took Christian Europe roughly a millennium and a half to achieve a breakthrough towards the mechanistic, objectifying paradigm.

But what is the textual basis for Jonas's initial claim that Judeo-Christian traditions lend metaphysical support to Cartesianism? In important Judeo-Christian Biblical traditions, the Hebrew term *næpæš* is traditionally translated as *soul*. This is inadequate, however, as it underrates the high anthropological importance the Hebrew Bible assigns to the lived body. In a prominent example, Gen 2:7 describes how the creator shapes the human being out of clay and breathes the "breath of life" into it, so that the result as a whole is called a "living *næpæš*." The exegete Bernd Janowski suggests the translations *living being, vitality, vital self* for *næpæš* instead. *Næpæš* implies that the human person "does not have a vital self, but is a vital self" (Janowski, 2014, p. 77, see pp. 88f.). The term *næpæš* can also be used to denote the self, but also desire, hunger, appetite, and longing (2014, p. 87). Not only can it be seen as the "organ of compassion..., of passion..., and of joy..., but also as the subject of sorrow, suffering, despair, and toil" (2014, p. 98).

From his critique of a traditional Christian understanding of *soul*, Jonas goes on to reinterpret the notion in a non-Cartesian way. He thinks of the inward dimension as inseparable from physical systemic organization. Ironically, in phenomenal content, he is very close to Biblical notions of *næpæš*, speaking of "the province of 'soul,' with feeling, striving, suffering, enjoyment" (Jonas, 1966, p. 57). Psalm 42 is an instructive example, as it uses *næpæš* as the prime anthropological category. While Jonas argues that the relation between self and world tends to be disjunctive in the Hebrew Bible – speaking of an alliance between "Christian transcendentalism and Cartesian dualism" (Jonas, 1966, p. 57) – the psalm's use of animal imagery for the human person gives poignant expression to emotional stirrings, articulating a profound crisis in identity and integrity. The writer does not abstract from her fundamental situatedness in body and spirit, but still she is confronted with her own *næpæš*.

As a deer longs for flowing streams, so my life (*næpæš*) longs for you, O God.
My life (*næpæš*) thirsts for God, for the living God. When shall I come and 'behold' the face of God?
My tears have been my food day and night, while people say to me continually, "Where is your God?"
These things I remember, as I pour out my life (*næpæš*):
how I went 'in the crowd of the noble ones' to the house of God, with the glad shouts and the songs of thanksgiving of a celebrating crowd.
Why are you melting away, O my life (*næpæš*), and why are you protesting against me?
Hope in God; for I shall again thank him, the rescue of 'my' face 'and' my God.
(Ps 42:1–5 [2–6]: NRSV rev. according to Janowski, 2014, p. 102)

In the New Testament, the fundamentally physical dimension of human social life can also appear in a very positive light. In the gospels, for example, Jesus' salvific action – as well as the Good Samaritan's impulse to help – is often triggered by an empathic 'gut reaction.' Here the Greek verb for *feeling compassion* is construed from the word for bowels – similar to the Hebrew word for mercy, which is a derivative of *womb*.[5] In addition, according to the Gospel of John, famously, "the Word became flesh" (John 1:14).

By contrast, the apostle Paul is often perceived as having a problematic relationship to the physical dimension of Christian life. Daniel Boyarin writes,

Paul was motivated by a Hellenistic desire for the One, which among other things produced an ideal of a universal human essence, beyond difference and hierarchy. This universal humanity, however, was predicated (and still is) on the dualism of the flesh and the spirit, such that while the body is particular, marked through practice as Jew or Greek, and through anatomy as male or female, the spirit is universal. *(Boyarin, 1997, p. 7)*

To begin with, the supposed dualism in Paul seems founded on his negative connotation of the word *flesh*, *sarx*, which is very prominent in his theological dictionary. While it can denote, in a neutral or positive sense, anything given biologically – such as the relationship between Israel and Jesus (Rom 9:5) – Paul uses it more commonly in a clearly negative sense, typically in opposition to the Holy Spirit: "Are you so foolish? Having started in the Spirit, are you now ending in the flesh?" (Gal 3:3 NRSV rev.). When Jonas characterizes the traditional understanding of the soul as "in the world – but not of the world" (see above), this may sound like a parallel to Paul's characterization of Christians living "in the flesh, but not according to the flesh" (*"en sarki...ou kata sarka,"* 2 Cor 10:3). Does Paul operate in a latent Cartesian framework?

Paul's polemic against the "flesh" was especially prominent when he saw Christians behaving in ways parallel to his own life before becoming a Christian. Extreme in his obedience to Jewish ritual law, Paul understood his persecution of the fledgling Christian churches as a sign of his elitist piety and practiced a narrow exclusivism. Yet what seemed a virtue in affirming religious identity through ritual boundary markers in fact powered a narrow-minded sectarianism and a competitiveness for religious prestige. It was precisely this issue that was at stake when certain circles in the Galatian churches argued for an obligatory circumcision of new members, demanding a higher level of accordance with Jewish ritual practice. The social dimension of particular religious issues is of great importance here. It was against these theologians that the quotation above from Gal 3:3 was directed. Paul inveighed with particular vehemence against such tendencies, as the competition for religious prestige put belief in Christ at stake.

The problem was that such competition was so ingrained in the Greco-Roman and Jewish cultures that it seemed perfectly benign and thus remained unacknowledged as sin. It required the crucifixion and resurrection of Christ to expose this lethal competition for what it was; it was "unmasked" not by the law but by the cross, which revealed the full, evil potential of religious zeal acting to defend itself and to prove its superiority.

(Jewett, 2007, pp. 459f.)

In turn, Paul had come to see the liberating implications of belief in Christ, among other things, in enabling an inclusive community by virtue of the elimination of exclusivist practices.

Thus, when Paul lashed out against those who made inclusion into the religious community dependent on standards likely to unleash a competitive quest for religious and cultural prestige, he had the proponents of Jewish food laws and ritual circumcision in mind, fearing they would fall into the trap of religious competitiveness: "Their end is destruction; their god is the belly, and their glory is in their shame; their mind is set on earthly things" (Phil 3:19). Besides Paul's strong evaluative judgment, these examples also show that not only individual meaning-making and the biological constitution of inwardness are inherently related to physical organization; even social identity is etched deeply into the bodily sphere. By consequence, what allowed Paul to undermine such exclusivist social tendencies was the denunciation of a certain aspect of physicality as sinful, as "flesh." Small wonder this latter statement continues with an antithesis of heaven and earth: "But our citizenship is in heaven" (Phil 3:20).

Thus, while a certain dualistic element did indeed serve Paul to undermine a sectarian emphasis on religious elitism and conformity, Boyarin does not see how crucial it was for Paul not to rest content with this dualism. While *flesh* is highly negative in his lexicon, *body* is positive. Thus he urges his congregations on to a more mutualistic, inclusive practice of community in depicting them as the "body of Christ" (1 Cor 12). Is it a coincidence that the more systemic and integrative term is positively connoted, while an emphasis on the single element, flesh in isolation from the body, is disparaged? The excessive emphasis on a single element of "flesh" at the expense of the systemic integration in the wider body might even be seen as a "cancer" in the tissue of the community. At any rate, the unity of the body of Christ is reaffirmed in the physical act of jointly sharing one bread in communion (1 Cor 10:17). Paul also saw the unity of the body in danger when facing the practice of the Corinthian church of speaking in tongues without an interpreter (1 Cor 14). The understanding was that, inspired by the Holy Spirit, Christians uttered sounds incomprehensible to anyone else, assuming to communicate with God directly in a spiritual sphere. Paul again highlighted the importance of the physical body in designating it as the "temple of the Holy Spirit" (1 Cor 6:19). He even described the resurrection as involving a "spiritual body," rather than taking place in a sphere of pure spirit (1 Cor 15:44) (Welker, 2013).

It was indeed crucial for Paul not simply to upset an ingrained physical encoding of cultural and social identity, but also to reconfigure it, thereby fully affirming the great significance of the physical. Otherwise his critique of previous ways of "embodying" social identity would have remained abstract and ineffective. A merely spiritual "body politic" is no "body politic" at all.

Biblical Creation Traditions: The Boundary and Systemic Arrangements

According to Paul, members of the congregation should relate to each other as parts of an organism. We may even ask if he envisions an equivalent to a boundary that defines an organism. After all, according to Jonas, metabolism and systemic bodily organization require a "boundary dividing 'inside' and 'outside,'" across which the organism exchanges matter with its environment (Jonas, 1966, p. 79, see p. 75; Jonas, 2011a, p. 114). Thompson shows even more clearly how the boundary itself is intimately involved in the metabolism. At the risk of stretching the systemic concept, even for Paul there may be an operative symbolic boundary regulating both internal and external relations of the congregation: "So then, whenever we have an opportunity, let us work for the good of all, and especially for those of the family of faith" (Gal 6:10).

Further, in contrast to Jonas's proto-Cartesian interpretation of Judeo-Christian traditions, Biblical creation texts also display a sense of an operative boundary.

Thus, to highlight the one fundamental distinction between creator and creature, as Jonas does, would mean to miss crucial distinctions internal to creation that also betray a systemic dynamic.

In a theological interpretation of Biblical creation accounts, Michael Welker sees the divine creative agency as characteristically enabling creatures to bring forth and shape creaturely reality (Welker, 2000, pp. 12f., pp. 24f.). For example, in the mythological style of the first creation account, Genesis 1, the earth brings forth plants, and the independent reproduction of the species is a recurring theme. The self-organization of ecosystems can similarly be considered a typical result of divine creating. In the theological doctrine of creation, Niels Henrik Gregersen also takes this stance in adopting the concept of autopoiesis of Maturana and Varela, two authors highly influential in Thompson's work. "God is creative by supporting and stimulating autopoietic processes" (Gregersen, 1998, p. 334). The creature's participation in the divine creative work is in contrast to Jonas's perception of Biblical creation accounts as conceiving of God and creation in binary terms.

In the creation account of Genesis 1, the concept of the boundary emerges with the prominent process of dividing. Typically these texts are of a mythological character, so while they should certainly be taken with a grain of salt, in their own way they are concerned with fundamental experiences of life. Creating divisions and structures is said both of the creator and of the creature. Even in those cases in which the verb *to divide* is not used, individual works of creation establish characteristic boundaries: between a dome and the waters, the earth and the seas, day and night, as well as between animals, which the myth envisions eating green plants originally, and humans, which it sees eating seeds and fruit. Here we are dealing not with the organization of an individual organism or the social integration of organisms, but with a cosmological arrangement in which different elements function in complementary ways.

The theologian Christian Link argues that in Biblical creation accounts, the notion of the boundary resonates with our empirical sense of reality. He interprets the concept of the boundary in Biblical creation traditions as being of vital importance. "To create means to establish boundaries and thus to institute well defined relationships that are, as the natural sciences now teach, the basis for life to emerge in the choice and determination of possibilities... Including cell division, creaturely life is a constant dividing and distinguishing followed by a constantly new coalescence" (Link, 2012, p. 63).

Like a semipermeable membrane, the divisions within creation are not absolute, and are in need of maintenance. Various Biblical creation traditions and ancient Near Eastern texts outside of the Biblical canon also[6] show remarkable sensitivity for the notion that the ordering activity of the creator does not exclude chaotic elements. On the one hand, creaturely life depends on the persistence of the boundary between order and chaos, and yet it is not depicted as rigid. In spite of the separation, one domain remains dependent upon another, even if this includes destructive effects.

In the biological organism, too, the cell wall constitutes a boundary that differentiates between order and chaos. The order within is not of a sterile nature, yet the organization of the cell reduces the instability and randomness of the surrounding substrate. As receptors in the cell wall "filter" the surroundings for particular nutrients, which the cell wall then keeps available within the system, it becomes possible to distinguish between order and chaos first of all. But that does not mean that chaotic forces are strictly external to the organism. The very fact of this order also creates opportunities for factors detrimental to the organism. Photosynthesis is a crucial part of plant metabolism, but the stored energy is also the reason herbivores eat plants. Further, a virus can infect an organism and inserts its genetic material into the host DNA because of the regulated process of gene replication, itself a function of autopoiesis to which the membrane contributes. Thus, the enhancement and endangerment of life are part of the same systemic phenomenon. As we understand creation in systemic and in evolutionary terms, we see at the same time that creation cannot simply be reduced to the enhancement of life. Welker also notes that Biblical creation traditions comprise the aspect of danger and diminution of life (Welker, 2004, p. 170).

The complementarity of order and chaos, life and danger to life can also be illustrated by the example of prominent creation traditions. Here, too, we find the concept of the boundary, although certainly not in a clearly defined biological or cybernetic sense. In the mythological cosmology of the first creation account (Gen 1–2:4), a dome establishes the sphere of heaven and earth in the midst of the ubiquitous primeval flood. Yet the boundary is not absolute, and the water is both a prerequisite of and a threat to life.

Another Old Testament creation text, Ps 104 – which displays notable parallels to the Egyptian ode to the sun of Akhenaten (Beyerlin, 1975, pp. 43–46) – describes the change of day and night as a well-ordered arrangement separating the nocturnal, chaotic part of creation from diurnal human culture (Ps 104:19–23). Although primarily intent on presenting a harmonious structure, the hymn also shows how the difference between day and night does not separate order and chaos in an absolute sense. Young lions roar for prey only at night, once the human labor out in the open is done, but at the same time, humans are especially defenseless at night. That is also the time when creatures emerge from the forest, which was considered a life-threatening sphere.[7] The chaotic side of nature is also a condition for the human ability to produce cultural goods such as wine, oil, and bread, if only through hard work.

In the second account of creation (Gen 2:4–25), a significant boundary emerges that creates meaning both for everything included and for what is excluded. This does not illustrate the chaotic aspect, but the phenomenon of division as a prerequisite of meaning. To begin with, the human person is a mere individual who does not find a suitable partner among the other creatures. Human life is as yet undifferentiated, lacking in structure and meaning. Only later another human person enters the scene, which leads to a differentiation between man and woman. The dynamics of this relationship are fundamental for the anthropology of the text. Nonetheless, the initial human person only develops into the creature that seeks its own counterpart by naming the other creatures, thus "de-fining" and excluding them.

Welker also points to similar phenomena in which reality is perceived in complementary, mutually influencing aspects. The differentiated spheres from Gen 1 – night and day, etc. – are also described as interacting, as when the "lights" in the sky shine upon the earth and indicate cultural entities such as the progression of years (Welker, 2000, p. 15). Thus the differentiation between spheres is not absolute, but allows for a systemic coupling. Welker defines God's creative activity as the establishment of relationships in which different spheres mutually contribute to each other's liveliness and specific characteristics. Creation is "the *construction of associations of independent relations between different creaturely realms*" (p. 13). The proper efficient activity characteristic of creatures takes place as part of this life-enhancing interrelation.

The Holy Spirit and Autopoietic Organization: Biological Organisms and Social Groups

The phenomenon of autopoietic self-organization that results in elementary meaning making and new forms of behavior is also an important dimension in a more specific description of the theology of creation. In a section that explores a further description of God the creator, Link asks how creation can "materialize in the medium of the world." "For how is it possible that it renews itself from day to day, in spite of all danger?" (Link, 2012, p. 308). It is in a section on the doctrine of the Holy Spirit that Link gestures at autopoiesis. Thus he implies a connection between the doctrine of the Holy Spirit and the theory of autopoietic systems, and indeed Jewish–Christian traditions repeatedly draw a connection between God's Spirit and God's creative activity (Link, 2012, pp. 308–311: Gen 1, Ps 104, for example; see also Edwards, 2004).

Certainly thinkers such as Kant and Thompson understand self-organization and autopoiesis in a naturalistic sense, and a theological interpretation of these phenomena is not to be understood in the sense of an "Intelligent Design" proposal responding to the seemingly irresolvable question of how biological self-organization got started in the first place. Faith in God as creator is not justified by an explanation of worldly processes that would otherwise seem inexplicable. Instead, with the concept of autopoiesis we can illustrate what is meant by creation.[8] Thus, Link suggests "that God's self participates in the genesis of creation – not through supernatural intervention, but through the inhabitation of God's Spirit and thus in opening new possibilities in an 'immanent' manner – thus determining the dynamics of the evolution of creation 'from the inside'" (Link, 2012, p. 314).

On the level of systematic thought, this confirms again that theological discourse about God's Spirit is not to be understood in opposition to the bodily dimension. "It was not anything [purely] spiritual that shaped the discourse about the Spirit, but the experience of the body, of the needfulness of somatic life" (Link, 2012, p. 309).

Moreover, the discussion of aspects of the theology of the apostle Paul focused not on the individual organism, but on the organization of the social group as an organism. The Holy Spirit is also seen as integrating a social organism in the prophetic text Joel 3:1–5, which sees full social participation of previously marginalized groups as a result of the work of the Spirit. Yet this work respects and maintains differences and pluralism rather than purchasing participation at the price of uniformity (Welker, 2004, 148 f.). The theologian Paul Tillich also saw the reconciliation between individuality and social integration as a manifestation of the Spirit (Tillich, 1978, pp. 45f., pp. 301–3). As in an autonomous biological system, at the social level, too, the constitutive elements are different but mutually dependent on one another.

Welker also suggests a continuity between ecological and social organization in the Spirit's activity, in speaking of the social forms inspired by the Spirit with the ecological image of a "flourishing vegetation" (Welker, 2004, p. 145, p. 170). This is not to play an ecological stereotype off against more complex systemic forms, for even in a flourishing vegetation, "'natural' life asserts itself essentially at the cost of other life, indeed by destroying other life" (Welker, 2004, p. 170). This is at odds with a romantic preoccupation with nature in both a theological and an ecological sense. Yet in a flourishing vegetation, mutual relationships between species also enhance life, on the whole. Within such ecological systems, we also find superorganisms such as a population of ants or microorganisms. Yet an inspired social body is a more demanding concept than that of biological superorganisms. For them it is characteristic that one organism "sacrifices" itself without further ado, pursuing the maintenance of the collective organism at the expense of the individual one (Ackermann et al., 2008). At any rate, in bee or ant societies, there are only a few stereotypical differences between different castes. These superorganisms are not constituted by individual autopoietic systems. This process is even clearer in multicellular organisms, in which the greater unity asserts itself at the expense of the independence of the constituting elements; cooperation becomes obligate. In human social systems, by contrast, social cohesion is more precarious. Individual differences are stronger to begin with. Such a system may either achieve autopoietic organization at the expense of everyone's individual autonomy, or the system can be integrated on the basis of individual's striving for a competitive gain of prestige. For example, before becoming a Christian, the apostle Paul saw himself operating in a cultural–religious system that was organized around religious achievement, which entailed the marginalization of others. If the dynamic of autopoietic systemic organization, first discovered in individual biological organisms, is repeated on the human social scale, autopoiesis is not a value per se. Instead, we should distinguish forms that diminish life from forms that enhance life. In a human social system, by contrast, the integrating work of the Holy Spirit does not proceed by way of either marginalization or homogenization. The question is how individual integrity and social integration can enhance each other mutually. These crucial distinctions are at the core of Biblical traditions that see the Holy Spirit operative in social group organization.

Conclusion

"What is life?" When posing the question about life on Earth or elsewhere in the universe, space explorers might overlook some living creatures if they look only for machines, for mechanistic signs of biota.

When approaching the question "What is life?" the modern evolutionary synthesis has detoured us by sidling too close to the philosophical views of Descartes. Adopting the dualism of extended and thinking substances, the scientific revolution seemed to evacuate the world from a divine presence. As matter came to be seen as fundamentally inert, the activity of the creator was increasingly seen as externally setting up the fundamental conditions of the universe. Mechanistic science no longer allowed for final causality inherent in creation. By contrast, Evan Thompson's enactive approach – while fully naturalistic – counters Cartesianism in seeing the mind emerge in autopoietic biological systems. The cell is the paradigmatic autopoietic organism. Thompson stays aloof of genetics, however, to a higher degree than required by his argument that the cell itself is reducible to the one mechanism of gene expression. By contrast, the organism's competence to regulate its own development and behavior implies a teleology which subverts the Cartesian binary of extended and thinking substances. Although Thompson engages the natural sciences on a detailed, subtle level, his lack of constructive interaction with genetics is a weakness.

Hans Jonas, by contrast, argues both for teleology in ontogenesis and for a significant role of chance and accident in evolution. Jonas's view of Judeo-Christian creation traditions as bolstering up the mechanistic paradigm, within which the creature is merely passive vis-à-vis God, is unwarranted. Neither do Biblical notions of "soul" or a Pauline disparagement of the "flesh" support a Cartesian dualism. Further, Judeo-Christian creation traditions display a thoroughgoing systemic dynamic, and the notion of the boundary, which is fundamental for autopoietic systems, can serve as a bridge for dialogue with traditions that speak of a creative role of the divine Spirit. In theology, the Spirit can be seen as ordering and structuring creation, thus opening up new possibilities inherent in nature without supernatural, material additions to nature. The integration of social systems by the Holy Spirit offers hope for an alternative to an organization both at the expense of all individuality and based on competitive prestige.

This discussion has implications for both theology and space exploration. Theologians in the Jewish and Christian traditions should distance themselves from Cartesian dualism and adopt a much more integrated and holistic interpretation of nature in general, and human nature in particular. Rather than assume our mind is separate from a surrounding world of mechanistic nature, the theologian should acknowledge that the human soul is the inward dimension of the person inseparable from his or her physical systemic organization. Astrobiologists, similarly, should fold this same observation into their expectations for life on other planets. They should expect that space explorers are likely to encounter life forms which are teleologically oriented, internally self-organized, and perhaps even conscious. If conscious, they might put out the welcome mat to greet us.

Notes

1 In Aristotle's works, on the other hand, the divine *telos* might even be understood in decidedly natural terms, in contrast to scholastic thought. According to M. R. Johnson, Aristotle's biology did not serve a natural theology, but rather the other way around (Johnson, 2008, p. 258).
2 All translations from titles with a German bibliography by A. M.
3 See the quotation of Richard Dawkins by Thompson: "The information passes through bodies and affects them, but is not affected by them on its way through... It is uninfluenced by a potential source of contamination that, on the face of it, is much more powerful: sex" (Thompson, 2010, p. 187). Apart from being false from a biological perspective, the statement satisfies several criteria of a doubtful metaphysics.
4 Apart from classic examples, such as the Japan macaques on Kojima washing their sweet potatoes in the sea, the degree of insight required for an orangutan who uses water in order to retrieve a peanut is remarkable (Mendes, Hanus & Call, 2007). For theoretical insight and logical reasoning among chimpanzees, see Tomasello (2014, ch. 2).
5 See the following New Testament references: Matt 9:36, 14:14, 18:27, 20:34, Mark 1:41, 6:34, 9:22, Luke 7:13, 10:33, 15:20: *splagchnizomai* (*splagchna*), Hebrew *raḥamîm* (*ræḥæm*).
6 For example, Jürgen Ebach (2001) compares the narrative of Noah, the survivor of chaos taking over, with ancestral narratives from the Ancient Near East.
7 See Ps 104:20 f. (see Riede, 2011, on the forest as a life-threatening sphere).
8 For example, Link calls the anthropic principle, which may cause us to ponder a divine creator, a "place holder for a mechanism (yet) to be discovered" (Link, 2012, p. 133 [Link's parenthesis]).

References

Ackermann, M., Stecher, B. *et al.* (2008). Self-destructive cooperation mediated by phenotypic noise. *Nature*, **454**(7207), 987–90.
Anderson, D. J. & Adolphs, R. (2014). A framework for studying emotions across species. *Cell*, **157**(1), 187–200.
Bateson, M., Desire S. *et al.* (2011). Agitated honeybees exhibit pessimistic cognitive biases. *Current Biology*, 21(12), 1070–3.
Beyerlin, W. (1975). *Religionsgeschichtliches Textbuch zum Alten Testament*, Göttingen: Vandenhoeck & Ruprecht.
Boyarin, D. (1997). *A Radical Jew: Paul and the Politics of Identity*, Berkeley, CA: University of California Press.
Brooke, J. H. (1991). *Science and Religion: Some Historical Perspectives*, Cambridge: Cambridge University Press.
Cottingham, J. (1999). Cartesian Dualism." In idem, ed., *The Cambridge Companion to Descartes*, 5th edn, Cambridge: Cambridge University Press, pp. 236–57.
Ebach, J. (2001). *Noah: Die Geschichte eines Überlebenden*, Leipzig: Evangelische Verlagsanstalt.
Edwards, D. (2004). *Breath of Life: A Theology of the Creator Spirit*, Maryknoll, NY: Orbis Books.
Fontenelle, le Bovier de, B. (1803). *Conversations on the Plurality of Worlds*, trans. E. Gunning, London: T. Hurst.
Gallagher, S. (2008). Direct perception in the intersubjective context. *Consciousness and Cognition*, **17**, 535–43.
Gibson, W. T., Gonzalez, C. R. *et al.* (2015). Behavioral responses to a repetitive visual threat stimulus express a persistent state of defensive arousal in Drosophila. *Current Biology*, 25(11), 1401–15.
Gregersen, N. H. (1998). The idea of creation and the theory of autopoietic processes. *Zygon: Journal of Religion & Science*, **33**, 333–67.
Haeckel, E., *The History of Creation: Or, The Development of the Earth and Its Inhabitants by the Action of Natural Causes. A Popular Exposition of the Doctrine of Evolution in General, and of that of Darwin, Goethe, and Lamarck in Particular*, trans. E. R. Lankester, Volume 1, New York: Appleton, 1914 [1876].
Harrison, P. (1992). Descartes on animals. *Philosophical Quarterly*, **42**(167), 219–27.
Harrison, P. (2000). The influence of Cartesian cosmology in England. In S. Gaukroger, J. Schuster, and J. Sutton, eds., *Descartes' Natural Philosophy*, London: Routledge, pp. 168–92.
Heidegger, M. (1996). *Being and Time: A Translation of Sein und Zeit*, trans. J. Stambaugh, Albany, NY: State University of New York Press.
Janowski, B. (2014). "Die lebendige Næpæš." In *Der nahe und der ferne Gott*, Neukirchen-Vluyn: Neukirchener, pp. 73–116.
Jewett, R. (2007). *Romans: A Commentary*, Minneapolis, MN: Fortress Press.
Johnson, M. R. (2008). *Aristotle on Teleology*, Oxford: Oxford University Press.
Jonas, H. (1966). *The Phenomenon of Life: Toward a Philosophy of Biology*, Evanston, IL: Northwestern University Press.
Jonas, H. (1993). *Philosophie: Rückschau und Vorschau am Ende des Jahrhunderts*, Frankfurt a.M.: Suhrkamp.
Jonas, H. (2011a). Harmonie, Gleichgewicht und Werden: Zum Systembegriff und seiner Anwendung auf Lebendiges. In *Das Prinzip Leben: Ansätze zu einer philosophischen Biologie*. 2nd edn, Frankfurt a.M.: Suhrkamp, pp. 109–25.

Jonas, H. (2011b). Wahrnehmung, Kausalität und Teleologie. In *Das Prinzip Leben: Ansätze zu einer philosophischen Biologie*, 2nd edn, Frankfurt a.M.: Suhrkamp, pp. 51–71.

Kant, I. (2000). *Critique of the Power of Judgment*, trans. P. Guyer and E. Mathewes, Cambridge: Cambridge University Press.

Link, C. (2012). *Schöpfung: Ein theologischer Entwurf im Gegenüber von Naturwissenschaft und Ökologie*, Neukirchen-Vluyn: Neukirchener.

Maynard Smith, J. & Szathmáry, E. (1995). *The Major Transitions in Evolution*, Oxford: Oxford University Press.

Mendes, N., Hanus, D. & Call, J. (2007). Raising the level: Orangutans use water as a tool. *Biology Letters*, **3**(5), 453–5.

Noble, D. (2008). *The Music of Life: Biology Beyond Genes*, Oxford: Oxford University Press.

Riede, P. (2011). Wald/Forstwirtschaft. In M. Bauks and K. Koenen, eds., *Das Wissenschaftliche Bibellexikon im Internet (wibilex)*. Web: www.bibelwissenschaft.de/stichwort/34515/ accessed Sept. 27, 2016.

Spaemann, R. (2012). Naturteleologie und Handlung. In *Philosophische Essays*, 2nd edn, Stuttgart: Reclam.

Thompson, E. (2010). *Mind in Life: Biology, Phenomenology, and the Sciences of Mind*, Cambridge, MA: Belknap Press.

Tillich, P. (1978). *Systematische Theologie*, Vol. 3, 2nd edn, 3 vols., Stuttgart: Evangelisches Verlagswerk.

Tomasello, M. (2014). *A Natural History of Human Thinking*, Cambridge, MA: Harvard University Press.

Welker, M. (2000). *Creation and Reality*, trans. J. F. Hoffmeyer, Minneapolis, IL: Fortress Press.

Welker, M. (2004). *God the Spirit*, trans. J. F. Hoffmeyer, Minneapolis, IL: Fortress Press).

Welker, M. (2013). Was ist ein 'geistiger Leib'?. In T. Breyer, G. Etzelmüller *et al.*, eds., *Interdisziplinäre Anthropologie: Leib – Geist – Kultur*, Heidelberg: Winter, pp. 65–83.

Wolf, J-C. (1998). Hans Jonas: Eine naturphilosophische Begründung der Ethik. In A. Hügli and P. Lübcke, eds., *Philosophie im 20. Jahrhundert: Phänomenologie, Hermeneutik, Existenzphiosophie und Kritische Theorie.*, 3rd edn, Reinbek: Rowohlt, pp. 2:214–36

Zimmer, C. (2011). The long curious extravagant evolution of feathers. *National Geographic*, **219**(2), 32–57.

12

Where There's Life There's Intelligence

TED PETERS[1]

Astrobiologists and other space scientists have become so accustomed to dividing extraterrestrial life into two categories – unintelligent and intelligent – that something might be getting overlooked.[2] It just might be the case that all life is intelligent, at least to some degree. Certainly an extraterrestrial civilization with advanced science and technology should be dubbed intelligent, perhaps even highly intelligent. Yet, even if microbial life either on Earth or off-Earth is considered to be stupid, this does not mean it has zero intelligence.

In what follows, I would like to provide a comprehensive scale of intelligence that includes simple single-celled organisms in continuity with the highest level of intelligence we know, namely, *Homo sapiens*. I propose a seven-trait definition: intelligence includes (1) interiority; (2) intentionality; (3) communication; (4) adaptation; (5) problem-solving; (6) self-reflection; and (7) judgment. Most mammals and certainly human beings exhibit all seven traits. Yet, brainless microbes and simple organisms exhibit the first four traits. By establishing a spectrum of traits, all life from the simplest to the most complex can be dubbed intelligent, even though they differ in levels of complexity. This argument directs the astrobiologist searching our solar system to look for various degrees of intelligence, not *un*intelligent life.

In short, what distinguishes so-called stupid or non-intelligent life from intelligent life is not intelligence per se, rather the degree of complexity of intelligence. The value of my proposed hypothesis is that it might direct astrobiological eyes to look for a third category, namely, intelligence in the middle between simple and complex levels of intelligence.[3]

"Astrobiologists need to have definitions that allow us to be surprised," writes Lucas John Mix; "On the one hand, we need to calibrate our definitions of intelligence (and life) based on what we know. On the other hand, we need to allow the definitions enough latitude to show us what we have not yet seen" (Mix, 2009,

p. 275). By demonstrating a spectrum within the single category of intelligent life, I hope to steer astrobiological eyes to see what might surprise us.

Defining Intelligence

It is not uncommon for those who provide us with definitions of intelligence to begin with the human species in mind.[4] Further, their agenda is to distinguish lower from higher levels of intelligence among people, purportedly to decide who receives and who is denied social benefits such as education.[5] My concern in this chapter is not to applaud smarter people but rather to explore the continuity between human intelligence and intelligence in other species and life forms.

Intelligence is tied to levels of cognitive complexity, contends Lori Marino at the Kimmela Center for Animal Advocacy. Intelligence refers to "how an individual acquires, processes, stores, analyzes, and acts upon information and circumstances" (Marino, 2015, p. 95). Does this definition apply strictly to human beings running around on two legs with complex brains? By no means. Marino notes how the "big five" cognitive domains that allegedly ascribe uniqueness to human intelligence are each shared with one or more other animal species. First, humans enjoy self-awareness, to be sure. Self-awareness includes mirror-self-recognition (MSR) plus metacognition (reporting on one's own thoughts). It is fascinating to note how all the great apes exhibit MSR, and both monkeys and dolphins express abstract forms of self-awareness, including metacognition. Second, even though tool making and tool usage are frequently thought to be uniquely human, we see these traits exhibited as well among many primates, elephants, birds, dolphins, and octopi. Third, culture – where culture is understood as distinctive behavior originating in local populations and passed on to new generations through learning – belongs as well to chimpanzees, dolphins, and even crows. Fourth, the capacity to use numbers is not limited to *Homo sapiens*; the great apes and dolphins demonstrate they can learn to rely on numbers as well. Fifth, symbolic communication is thought by many to be the chief defining characteristic of the human; but the capacity to learn artificial languages – including grammar – has been demonstrated by dolphins, chimpanzees, and parrots. In short, what have been assumed to be traits of intelligence which separate the human species from others are, in fact, characteristic to some degree of other life forms. Intelligence, it is most fitting to say, is a trait we humans share with the rest of life on our planet.

Intelligence was born on planet Earth when life first appeared and began to evolve, contends Marino. Even our brainless ancestors exhibited pre-brain intelligence. "Some of the building blocks of nervous systems and brains are found in single-celled organisms. They sense their environment, briefly store the information, integrate it in different channels of input and then act upon it – producing basic adaptive behavior. These brain-like processes work at a molecular level but they reveal a common ancestry with all modern brains" (Marino, 2015, p. 98).

Marino chides astrobiologists for not refining their understanding of intelligence so as to know better what to look for in space. "Astrobiology should embrace the study of intelligence as a ubiquitous property of life on Earth and one that converts the current impasse to relevant and exciting opportunities for further exploration" (Marino, 2015, p. 107). It is not my task here to chide astrobiologists. Rather, I wish merely to test the hypothesis that where there's life there is intelligence. With such a hypothesis in hand astrobiologists might sensitize their instruments of detection.

Ingredients to a Definition of Intelligence

Shane Legg and Marcus Hutter have collected definitions of intelligence and categorized them. Some capabilities of intelligent beings appear again and again: use of memory, the capacity for learning along with applying knowledge, adjusting to the environment, adapting to changes in the environment, reasoning and problem solving, and such. Intelligence is frequently described as a *mental* capacity that enables a person to comprehend, understand, and reason about things. Because intelligence is so multi-faceted, some cynics define intelligence merely as that which is measured by intelligence tests.

Legg and Hutter prescribe a minimum of three components essential to any definition of intelligence: (1) agency when interacting with the environment; (2) goal setting leading to success or failure; and (3) adaptation to the environment by altering goals. In sum, "Intelligence measures an agent's ability to achieve goals in a wide range of environments" (Legg & Hutter, 2006).

Could such a definition apply equally to two types of creatures, those with brains as well as those without? Must we limit our concept of intelligence to the human model that includes mentation? Not according to Mark Lupisella at NASA's Goddard Space Flight Center. "Intelligence can also be blindly capable... intelligence could be nothing more than a high degree of competence without understanding" (Lupisella, 2015, p. 160). Can we come up with a model of intelligence that includes both mental understanding plus a more or less blind capability?

What I am proposing is a definition of intelligence that encompasses all life forms from the simple to the complex. Regardless of how one defines human intelligence, we cannot avoid the assumption that what happens at the human level is in continuity, not discontinuity, with all levels of life in our biological evolution.

Is it credible to hypothesize that life as life is already intelligent? I answer in the affirmative. I proffer the idea that early in our evolutionary history simple life forms provided a necessary, though not sufficient, set of traits which would eventually make human intelligence possible.

Constructing a Comprehensive Definition of Intelligence

I would like to nominate for consideration a list of seven components to be incorporated into a comprehensive definition of intelligence. With this list, I hope to demonstrate that very simple life forms exhibit some, though not all, traits of intelligent creatures. An organism is intelligent when it possesses the following.

(1) *Interiority*: a membrane or barrier which separates the interior from the exterior environment or world; further, the interior maps the exterior to guide intentional behavior.
(2) *Intentionality* initiated from within that relates to the without – that is, goal-oriented behavior risking success or failure.
(3) *Communication* with the environment, including other organisms.
(4) *Adaptation*: the capacity to change in order to adapt and evolve.
(5) Mental activity, including reasoning in *problem-solving*.
(6) Mental activity, including *self-reflection* and *theory of mind*.
(7) Mental activity, including rendering sound *judgment*.

All of the above apply to intelligent human beings, to be sure, even if it appears that some among us are better than others at rendering sound judgment. My hypothesis is that pre-mental life exhibits enough of these traits and, thereby, qualifies as intelligent.

An Interior Structure that Maps the Exterior Environment

My argument is that a continuity of intelligence exists across the spectrum of life forms from the most simple to the most complex. Indeed, intelligence as we human beings experience it includes mental activity replete with abstract or representative deliberation. We do not observe the equivalent of abstract or representative thinking in one-celled organisms. Yet, some traits within human intelligence also appear in simple organisms such as interiority, intentionality, and communication. The first on my list is interiority.

Life, basically defined, is "metabolizing, replicating, and evolving" (Ward & Kirschvink, 2015, p. 35). For life to initiate activity in this fashion, it must distinguish between its interior and its exterior; it must establish organizational closure. Interiority is made possible by this enclosure, by drawing a line between the organism and its surrounding environment. That line could be porous and allow traffic in and out, to be sure; but it is definitional and it makes possible the distinction between internal and external activity, between self and world.

In a rudimentary way, the development of an inner over against an outer – the self-world distinction – makes sentience and intelligence possible. "It is teleogenic closure that produces sentience but also isolates it," according to biological anthropologist Terrence Deacon, "creating the fundamental distinction between self and other, whether at a neuronal level or a mental level" (Deacon, 2012, p. 511). He adds, "Even the simplest bacterium is organized as a self, with the emergence of ententional properties and possibly a primitive form of agency" (Deacon, 2012, p. 469). Deacon ascribes sentience to simple selves, whereas I recognize intelligence at work here.

The sheer evolution of interiority or selfhood should suffice to make my case. But, I might be able to add a related characteristic to strengthen the case, namely, internal mapping of the external world. The internal mapping of the external world seems to be decisive for human intelligence, even if described in other terms. Educational psychologist Jean Piaget, for example, describes intelligence as "assimilation to the extent that it incorporates all the given data of experience within its framework" (Piaget, 1963; Legg & Hutter, 2006). Or, according to a related definition, "Intelligence is a part of the internal environment that shows through at the interface between person and external environment as a function of cognitive task demands" (Legg & Hutter, 2006).

The mapping capacity of intelligence is most visible at the human level where brains are salient. Brain circuitry is organized "to generate a neural model of the world outside the brain," writes Patricia Churchland, "a map model . . . of the environment" (Churchland, 2013, p. 34). Along with a map of the external world, the mammal brain provides us with a GPS (Global Positioning System). "Mammals rely on the capacity to form neural maps of the environment," write May-Brill Moser and Edward Moser; "Like a GPS in our phones and in our cars, our brain's system assesses where we are and where we are heading" (Moser & Moser, 2016, p.28). It's the cortex in humans and the hippocampus in mice that provide the map and GPS.

We humans are most familiar with internal mapping of the world through symbolic speech and abstract thought. Human intelligent thinking is characterized by "symbolic representation and naming of entities, actions, qualities, patterns, and combinations of symbols, allowing recursion and enabling representation and logical analysis of arbitrary relations" (Ellis, 2016, p. 197). In other words, at the conscious rational level we draw representative maps of the world in our minds and then travel the roads we select.

Now, we ask: what about simpler organisms? Might the mapping metaphor apply to single cells without brains? Yes. To extend the map metaphor, biologists studying embryonic development acknowledge that cells "sense their place in the body plan" (Adler & Nathans, 2016, p. 71). There is "a miniature compass within each cell. Without this compass within cells, human heart, lungs, skin and other organs could not develop properly" (Adler & Nathans, 2016, p. 68). Mapping implies a relationship between inner and outer, even with or without mentation. To be intelligent includes an inner that relates to its outer. It includes an interior mapping of what is exterior.

Historian David Christian recognizes the protean value of mapping within intelligence found in the simplest organisms. "The constant restless search of all living organisms for manageable flows of energy is the first source of what we perceive as purposeful activity. To find the energy they need, all organisms must, in some sense, map their surroundings. They must know where the energy is and how to use it. So even the simplest of one-celled organisms can move towards or away from light, or sidle towards chemical concentrations they need. To do that they must map their surroundings, and they do so with purpose.... Both mapping (learning?) and learning with purpose seem to be required by the very notion of evolution through natural selection" (Christian, 2016, p. 42). Mapping, learning, and intending seem to come together in a single package in primitive life forms, with or without brains.

Intentionality

A related capacity that appears frequently in definitions of intelligence we will label *intentionality*. Deacon's term, *ententionality*, is similar: "I will use *entention* to characterize their internal relationship to a *telos* – an end" (Deacon, 2012, p. 27, Deacon's italics). If it moves itself, it's agential. If it's agential, we should look for signs of intentionality, or ententionality.

The inner exhibits intentionality in the form of goal-oriented behavior leading to success and risking failure. Artificial Intelligence researcher, James Albus, describes intelligence in terms of the "ability of a system to act appropriately in an uncertain environment, where appropriate action is that which increases the probability of success, and success is the achievement of behavioral subgoals that support the system's ultimate goal" (Albus, 1991; Legg & Hutter, 2006). Without reference to human mentation, it appears that in a description of intelligence one would need to include intentionality initiated from within a system oriented toward both an ultimate and a subgoal. In short, purposive activity. Does life as life exhibit some level of purposive activity?

Would self-propelled movement count as evidence? Many bacteria initiate movement by projecting a grappling hook (type IV pilus) from their bodies. They attach the hook to another cell or object in the environment, then retract it. With this motion the bacterium pulls itself forward in a machine-like manner. Its movement is self-initiated and appears to be directional. Even though the bacterium presumably has no mentation, might we still call it intentional? (Chang *et al.*, 2016).

With the term *intentionality* I designate the directionality arising out of mapping. The Latin *intendo* suggests pointing to, aiming at, or extending toward.

Philosopher of mind John Searle describes mental states as "directed at, or about, or of objects and states of affairs in the world" (Searle, 1983, p. 1). Might intentionality also characterize non-mental activity in simple life forms, especially the reciprocity between a living organism and its environment?

For another example of non-mental intelligence, let's turn to our simple friend, the amoeba, in this case the single-celled amoebae form *Dictyostelium discoideum*. The signal for individual amoeboid cells to aggregate is cyclic-AMP (adenosine monophosphate or c-AMP), which belongs as well to the human response to different hormones, such as glucagon. The amoebae can sense or detect concentration gradients of c-AMP. Then they crawl up the gradients to gather together with lots of other amoebae, which also have all been releasing c-AMP into the environment. After all the amoebae have moved up the c-AMP gradient and come together, they form an aggregated of eukaryotic amoebae known as a slug. Individuals will find the largest group of their kin in the area via this mechanism.

What have amoebae done to accomplish this? They have: (a) detected not only c-AMP, but its concentration gradient; (b) they have aligned their motility (and all the related intracellular proteins such as actin and myosin found also in muscle cells) along the gradient; (c) they have continued to move up the gradient as long as it exists (if one shifts the gradient experimentally, one can shift migration); (d) they are considered *altruistic social* amoebae because they will give their lives to set up the slug with its fruiting body to the benefit of the next generation of amoebae. The amoebae have certainly acquired information and they apply it in a generous manner. We may ask: is it more than information? Is it actual knowledge? When the amoeboid cell detects the strongest c-AMP concentration coming from the southwest and only a weak signal from the north, it aligns its motility for migration to the southwest. Might we think of this as intentionality utilizing its knowledge – its internal map – of the environment?

What about more complex creatures which incorporate such cell activity within them? What we today know as eukaryotic cells, it seems, are the product of endosymbiosis. Previous more simple cells, prokaryote cells, cooperated, combined, and together produced multi-celled organisms. Did the prokaryotes map their immediate environment and initiate symbiosis? Eukaryotes appear to be the result of cooperative intentionality, a creative intentionality. Only intelligence begets intelligence and, in this case, simple intelligence begets more complex intelligence.

It continues up the scale. Biologists have been studying learning in a wide array of organisms from fruit flies to vertebrates, because there is something there to study. Migrating butterflies and birds find their ways over hundreds of miles. Salmon return to lay eggs up the stream where they were born. Can we give them credit for mapping, intentionality, and even intelligence? Our word "stupid" seems like an insult after witnessing these accomplishments. Even if they can't play chess, they still do amazing things.

If we consider human intelligence to be the epitome of evolution's accomplishments to date on planet Earth, we must remember that it took time. It took 3.9 billion years. And, to some extent, the past is still with us. In the human brain the same signaling system is used that we would find in both amoeboid cells and fruit flies. The evolution of distinctively human intelligence couldn't occur until cells had signaling systems like the one involving c-AMP at the ameboid level. Perhaps our brains should thank the amoeba for getting us started.

Virtue ethicists observe that natural creatures at all stages of evolution exhibit four teleonomic if not teleological ends: (1) individual survival; (2) reproduction; (3) pursuit of pleasure and avoidance of pain; and (4) the well-being of the social group. If any of these four obtain, it appears that even the simplest life forms we know exhibit intentionality and belong in the intelligent category (Mohloeck, 2016). Slime molds display three of the four criteria: first, second, and fourth. Ants display all four. Maßmann identifies this teleological behavior as meaning-making deriving from self-preservation (Maßmann, 2017). [Let me be clear regarding teleology: I am affirming *local* teleodynamics produced by the goal-setting of an organism, not a *global teleology*, which directs the course of evolution in its entirety.]

Perhaps a critic will demure: mammalian let alone human intentionality is so much more complex that amoeba intentionality is not comparable. For mammals, intentionality begins with mentation, with abstract planning. No advance mentation occurs in simple life forms, and maybe not in fruit flies either. How should we handle this demure?

Let's look again at human intentionality. Intentional agency includes two components, according to Searle: internal mapping and causative action. Searle's term, "direction of fit," indicates what I have been calling the internal mapping of the exterior world. Conditions in the exterior world must conform to the map within the mind, world-to-mind, he contends. This, in turn, enables effective agential causation, mind-to-world. When mental activity is involved, agential action may be preceded by premeditation, deliberation, and planning. Prior intention precedes action (Searle, 1983). This means that the reciprocity between mapping and causative action includes a significant component of interior initiation. An intelligent creature is more than merely an object determined by its exterior environment, more than a flag blowing in the wind. Although an intelligent creature reacts to external stimuli, on occasion it initiates action. An intelligent creature maps the world internally and then takes action which in some way affects the external environment.

My critic might try to prevent us from transferring what is obvious in mammalian mental activity to single-celled life forms which, at least to our observation, lack mental activity. This leap, however, may not be over an empty abyss. We may have a bridge, namely, material engagement theory. The bridge builder is Searle critic, Lambros Malafouris. Malafouris embraces the *Theory of Material Engagement* which includes, among other tenets, the notion that mind is extended. Mind is not the private property of interiority. Rather, mind is located in the reciprocity between the organism and its environment. Malafouris locates intentionality not in prior deliberation but only in the action itself, what he calls "intentionality-in-action." He contends that "intention no longer comes before action but it is *in the action*. The activity and the intentional state are now inseparable" (Malafouris, 2014, p. 140).

It is not my objective here to stand by Malafouris against Searle. Nevertheless, what we can gain by appealing to Malafouris is the recognition that intentionality does not necessarily require *prior* mentation. Here it is significant that one can identify intentionality simply by recognizing it in the action itself. *Intention in the action* is something we recognize at the cellular level, not merely at the human level of intelligence.

Brain scientist Walter Freeman makes my point still more forcefully. Even at the human level we find intention in action without prior mentation. "We perform most daily activities that are clearly intentional and meaningful without being explicitly aware of them" (Freeman, 2000, p. 17). If intentionality belongs to intelligence, and if *intention-in-action* can occur without mentation and perhaps even without a brain, then we have opened the door to seeing the continuity of intentionality between mammalian minds and simpler life forms.

We ought not let slip by unnoticed the notion that intelligence might not be only individual. Intelligence could be the term to describe group cooperation, *social cognition*. A fascinating phenomenon frequently observed in nature is how army ants create living bridges to cross a chasm. One after the other, self-selected ants stretch their bodies, connecting to other stretched ants, and build a bridge so that their compatriots may cross the chasm. This is a clear case of problem-solving through intentional behavior, though the intentionality appears to derive from the group rather than the individuals alone. "Researchers found... that the ants, when confronted with an open space, start from the narrowest point of the expanse and work toward the widest point, expanding the bridge as they go to shorten the distance their compatriots must travel to get around the expanse"(Kurzweil, 2015). Can we describe such living bridge-building as intelligent? That depends on your definition. Social intelligence fits the definition we are working with in this treatment.

There is one caveat. Acknowledgement of purposive behavior on the part of intelligent creatures does not in itself warrant a teleological interpretation of evolutionary history as a whole. To avoid this unwarranted leap, Deacon endorses use of the term *teleodynamics* to describe local intelligent behavior. "The core property which links the selves of even the simplest life forms with that seemingly ineffable property that characterizes the human experience of self is a special form of dynamical organization: teleodynamics... that is, they are *closed* in a fundamental sense with respect to other dynamical features of the world" (Deacon, 2012, p. 468). My limited objective in this section is to observe that all intelligent creatures, including simple life forms, exhibit internal structure, mapping, and intentional orientation.

Within the context of the large scope of evolutionary history, I argue that the appearance of intelligence in early life forms provides a necessary – even if not sufficient – condition for the possibility of human intelligence. One salient trait of human intelligence is our internal mapping of the exterior world. Do our ancestral one-celled organisms prepare us for this development in intelligence? It appears they do.

Communication

Living creatures, no matter how small and simple, exhibit interiority combined with exteroreceptivity and agency. In short, they communicate.

When it comes to human communication, the most salient feature is symbolic communication through language. *Homo sapiens* have been blessed by a co-evolutionary history which has led to a leap forward in intelligence marked by linguistic communication. Deacon along with his colleague, biologist Ursula Goodenough, celebrate "... the co-evolution of three emergent modalities – brain, symbolic language, and culture – each feeding into and responding to the other two and hence generating particularly complex patterns and outcomes" (Goodenough & Deacon, 2006, p. 863). Symbolic language connects mouths to ears, which we might call *gap junctions* between organisms.

Mouths and ears are not our only gap junctions. We human beings communicate with one another at many levels: in addition to abstract conversation mediated by language, we communicate through facial expression, physical affection or physical violence, and chemical exchange *in utero*. We presume that human-to-human communication is intentional. Communication is a sign of intelligence. If we observe this same sign at simpler cellular levels, would this tend to support my hypothesis that where there is life there is intelligence?

If communication is a mark of intelligence, then conversation between simple cells would count as a sign of nascent intelligence at the single cell level. Ross Johnson began researching gap junctions in the 1960s. He later teamed up with cell biologists Dale Laird and Paul Lampe to gather evidence of cellular communication in the early 1990s. Hormones, neurotransmitters, electrical current, and such illustrate the "multiple modes of information exchange" (Laird *et al.*, 2015, p. 72). Cellular connectivity is located at gap junctions. Each gap junction is a kind of mouth and ear combination which opens up perhaps 10 000 channels for transmission. Because each gap junction "involves two hemichannels, that would make a total of 120,000 connexins per junction" resulting in "enormous communications conglomerates" (Laird *et al.*, 2015, p. 73). Knowledge of all that cells whisper to one another is not yet in. Research continues. Yet, these scientists offer a relevant conclusion: "Virtually all cells, it turns out, network with their neighbors via extensive collections of channels that directly connect the inside of one cell with the inside of the next" (Laird *et al.*, 2015, p. 72).

In short, communication in the form of interaction between organisms is common to both the most simple and most complex forms of life we know. If communication is a trait of intelligence, then intelligence is co-extensive with life.

Change for Evolutionary Adaptation

All things change, Heraclitus once told us. Intelligent life changes in a particular fashion. It changes by adapting to its environment and, in some cases, changes the environment to suit itself. At the individual level, the organism reorients itself when its surroundings change. At the species level, the changing environment requires of the organism adaptation in the form of reproductive fitness to survive or, in some cases, to give way to a subsequent species. Adaptation energizes the cycle which becomes a spiral through time. This ongoing spiral we dub "evolution."[6]

On Earth 3.9 billion years ago our first ancestors absorbed existing carbon and expelled oxygen. Over time, Earth's atmosphere became partially oxygenated. Life adapted. Newer species began to absorb oxygen and expel carbon dioxide. Astrobiologist Chris Impey acknowledges the particular way in which life exhibits Heraclitus' principle. "If biology starts, it immediately begins to alter its environment. Our Solar System started with three habitable terrestrial planets. Two suffered runaway climate changes that left them barren – one remade itself to stay habitable" (Impey, 2010, p. 168). Would it be excessive to connect intelligence with the attempts by early life forms to adapt through change?

"Intelligence is about making adaptively relevant responses to complex environmental contingencies, whether conscious or unconscious," writes Deacon (Deacon, 2012, p. 492). Deacon can even distinguish intelligence from sentience when asking about the prospect of machine intelligence.[7] Sentience versus machine intelligence is not our issue in this treatment, however. Important to us here in this study is the observation that *intelligence* is the term I use to describe evolutionary adaptation with or without consciousness.

Adaptation over time has led to increased complexity, including increased complexity in intelligence. Evolution "has given rise to organisms more complex than those living on the early earth. They have appeared even though there seems to be no active drive toward greater complexity, and even though complex organisms may not be terribly important in the overall scheme of things" (Christian, 2011, p. 108). In short, the final chapter in the story of evolving intelligence is yet to be written. When the story of intelligence becomes complete, it will begin "once upon a time" with our single-celled ancestors.

Mental Activity, Including Reasoning in Problem-Solving

Human intelligence today can take advantage of 86 billion neurons and 100 trillion synapses in the brain. This unfathomably complex brain did not come from nowhere. It is a product of autopoietic systems – that is, a product of the self-organizing cells we previously dubbed stupid life forms. Even if our biological ancestors were relatively stupid, they were sufficiently intelligent to give birth to us.

It is fascinating to note how early neural cells almost know what to do, so to speak. They join together in the project of creating a brain. Experiments with pluripotent stem cells – derived either from hES (human embryonic stem) cells or iPS (induced pluripotent stem) cells, if left to self-aggregate within the appropriate culture, self-organize into cerebral organoids with tissue architectures that are reminiscent of the human cerebral cortex. Because the cells know what to do, researchers can derive "brain tissue *in vitro*" (Brüstle, 2013, p. 320; Lancaster *et al.*, 2013, p. 373). My point here is this: even though human intelligence is exceedingly more complex than that of simpler cell life, it is that very same simple cell life that intends to make human intelligence possible. This observation supports either preformationism (the view that human intelligence was seeded in the physics of the universe) or autopoeticism (the view that intentional cell intelligence led creatively over evolutionary time to complex intelligence). Regardless, it appears that we may have a maxim: it takes intelligence to beget intelligence, just as it takes life to beget life.

Now, let's turn to problem-solving, which fits on most everybody's list of characteristics of intelligence. Engineers earn their living by solving problems in creative fashion. Our prehistoric ancestors survived because they solved the problems posed by the need for food or shelter along with protecting themselves from dangerous beasts and inhospitable weather. Today's parents delight when their toddler learns how to climb the stairs and, of course, use the toilet.

Human moral decision-making is one familiar form of problem-solving. Human action is the result of motivation, we presume. How should we behave? Let's ask our motives. One recent study interrogates our motives by measuring brain activity while making moral decisions. "Goal-directed human behaviors are driven by motives. Motives are, however, purely mental constructs that are not directly observable" (Hein *et al.*, 2016, p. 1074). Grit Hein and his research colleagues sought to make observable what is unobservable by distinguishing between two types of altruistic giving, one type motivated by empathy and the other motivated by reciprocity (paying what is owed). Through fMRI monitoring the researchers noted no difference in the regions of brain activity, but they did observe what is decisive, namely, the sets of motivations were associated with different connectivities. "Empathy based altruism is primarily characterized by a positive connectivity from the anterior cingulate cortex (ACC) to the anterior insula (AI), whereas reciprocity-based altruism additionally invokes strong positive connectivity from the AI to the ACC and even stronger positive connectivity from the AI to the ventral striatum. Moreover, predominantly selfish individuals show distinct functional architectures compared to altruists" (Gluth & Fontanesi, 2016, p. 1028; Hein *et al.*, 2016, p. 1074). My objective in reporting on this study is to demonstrate that intelligent behavior at the level of mentally motivated problem-solving retains a reliance on the physical substrate, the brain. Without endorsing either reductionist determinism or contracausal free will, I simply reinforce what most of us presume, namely, that mentation is made possible by the brain. This higher level complexity is most likely due to physical enablement.

Mental Activity, Including Self-Reflection and Theory of Mind

Reflective awareness, though not unique to human mentation, offers a nearly indescribable capacity for self-guidance and self-transformation. Reflective awareness also includes *theory of mind* – that is, it includes awareness that other intelligent creatures have an interiority like one's own. It includes awareness of one's own self along with awareness of other selves. Even though this level of human intelligence is launched by brain physiology, it seems to rocket skyward under self-propulsion and self-navigation.

We are capable of literal MSR (mirror self-recognition) when physically looking into a mirror: "Hey! That's me!" In addition, we can engage in MSR with our eyes closed. Our minds are almost uncanny in what they can permit to happen in terms of reflection, reflection on reflection, and reflection on reflection on reflection. Not only are we aware of our awareness, we can make ourselves aware of the awareness of our awareness. In addition, we can distinguish who we are now from who we are not, and we can follow this with a free decision to change ourselves. We can not only adapt, we can self-transform "Hey! That's who I want to be."

Take meditation as an example. Meditation is "the cultivation of basic human qualities, such as a more stable and clear mind, emotional balance, a sense of caring mindfulness, even love and compassion – qualities that remain latent as long as one does not make an effort to develop them" (Ricard *et al.*, 2014, p. 42). Through meditation, the self transforms the self. More: the self transforms the brain. "The adult brain can still be deeply transformed through experience" (Ricard *et al.*, 2014, p. 40). Reflective consciousness permits the mind to influence the brain, even to rewire the brain's circuitry. Despite the fact that our human intelligence is utterly dependent on its home in the brain, intelligence seems to possess a preternatural capacity for going away and coming home again.

Stories take us away and back home again. When listening to a story, our theory of mind permits us to temporarily enter into the thoughts and anxieties and victories of other selves. We enter into the interiority of other intelligent creatures or persons. We empathetically feel the feelings of the story's characters. When the story is over, we return to our own self-world relationship. It appears that our mind flew away from our body and entered temporarily into somebody else's world. Does our mind leave our body? Relevant to this question is the observation that, according to MIT researchers, the amygdale is activated by empathetic listening to a story (Bruneau *et al.*, 2015).

If we convince ourselves that our minds are so independent that they no longer need our brains, then we need to beware. In another chapter in this volume, Alexander Maßmann warns us against Cartesian dualism. According to philosopher René Descartes, the mind flies like a kite above the body. The mind as an immaterial substance enjoys intelligence in a way that is denied to the body, which is a material substance. Against Descartes, argues Maßmann, our mentation must retain its physical mooring.

If our mentation should lose its physical mooring, it would not float on its own. It would sink. Brain researcher Stanislas Dehaene has studied this and he declares that in everyday activity we fail to realize just how much of our activity is guided by "an unconscious automatic pilot... We constantly overestimate the power of our consciousness in making decisions – but, in truth, our capacity for conscious control is limited" (Dehaene, 2014, p. 47). Many of today's neuroscientists so emphasize how the brain is on automatic pilot that they say the higher levels of consciousness may be delusional. Or, to say it another way, the string on our kite of consciousness is shorter than we might realize.

This leads us to ask the reductionist question: is human intelligence only brain function? Can we reduce the human self to the sum total of neuronal firings? Some neurophilosophers are happy to pull the kite string so hard that consciousness crashes to earth. One German philosopher, Thomas Metzinger, for example, exhaustively reduces subjective intelligence to brain activity. "Subjective experience is a biological data format, a highly specific mode of presenting information about the world by letting it appear as if it were an Ego's knowledge. But, no such things as selves exist in this world" (Metzinger, 2009, p. 8). Another German philosohper, Ottfried Höffe, in contrast, affirms just the opposite. "The person thinks, to be sure, 'with' his central organ; he acts 'with' the brain, but it is the person, not the brain, that thinks or acts" (Höffe, 2010, p. 245). Which is it? Is the brain so completely in charge that the human self is a delusion? Or, does the self become its own center of intentional activity employing its intelligence? It is not necessary for my purposes here to adjudicate this dispute. It suffices to rely on what is generally accepted in scientific discourse, namely, it is the brain which makes distinctively human intelligence possible. When it comes to complex forms of cognition, a more complex brain accounts at least partially for the leap in human intelligence beyond more simple forms of life.

The leap is always tethered, of course. Even religious insight into transcendent reality never completely leaves its vegetative home. Rabbi Abraham Heschel reminds us, "much of the wisdom inherent in our consciousness is the root, rather than the fruit, of reason" (Heschel, 1951, p. 17).

Mental Activity, Including Rendering Sound Judgments

Up to this point I have tried to show that mammals in general, and humans in particular, share up to four traits of intelligence in continuity with brainless or stupid life forms: interiority, intentionality, communication, and adaptation. When our ancestors developed brains, intelligence became more complex. With the arrival of mental activity sponsored by brains, mammals and especially people developed the capacity for abstract intellectual reflection, advanced planning, and free will.

Perhaps at some legendary threshold in our evolutionary past our forbearers experienced abstract thinking. With abstract thinking they could anticipate alternative futures. With alternative futures in their minds, they needed to select which action to take and which to avoid. Abstract mentation, advanced planning, and free will most likely arrived together in a single package. With these three components to intelligence, our ancestors were required to make judgments. Sound judgments meant safer and more prosperous futures. The capacity for sound judgment was adaptive, no doubt.

Was this development in mental activity dictated by the brain? After all, the brain is composed of cells with intentionality. Yet, it is difficult to imagine that the computational complexity of the brain could, by itself, account for sound judgment. "The mind is more than a computational machine," avers Sante Fe Institute biologist, Stuart Kauffman. "Embodied in us, the human mind is a *meaning and doing organic system*" (Kauffman, 2008, p. 177, Kauffman's italics).

We observe in ourselves and in others that a reciprocity obtains in the relationship between brain and mind. The mind acts on the brain as much as the brain acts on the mind. We have already acknowledged that today's neuroscientists point out pre-reflective effects the brain has on our behavior. What we add here is the reflexive effects the mind has on the brain's mapping and intentionality. One might refer to the mind's influence on the brain as *top-down agency*. "*Top-down agency* refers to the ability to modulate behavior in relationship to conscious thought and intention," contends neurophilosopher Warren Brown (Brown, 1998, p. 117). In other words, our symbolic understanding and our abstract reasoning provide top-down influences on our reasoning process and, in addition, they direct our agency in the world. The self is an intentional agent who takes action and causes changes in the surrounding environment.

Physical cosmologist George Ellis helps to explain this by distinguishing between bottom-up causation as understood by the physicist from top-down causation which we experience when the mind makes decisions that influence brain activity. The mind selects and directs; the mind selects from among alternatives a single direction to be followed based upon the meaningfulness of an anticipated goal. Directed mental activity includes top-down causation as a complement to bottom-up causation. "Interlevel causation, both bottom-up and top-down, is key to brain function. Evolution has selected for it to occur. The underlying physics is channeled and constrained to enable this to happen" (Ellis, 2016, p. 346). Human intelligence is much more than merely brain function, in other words.

Among other things, top-down cognition enables us to imagine alternative futures, render a judgment, and then enlist the will to take a specific action. When asked, we can communicate the reasons for deciding what is best to do. "A *future orientation* is meant to denote the ability to run a conscious mental simulation or scenario of future possibilities for the actions of oneself and others, and to evaluate these scenarios in such a way as to regulate behavior and make decisions now with regard to desirable future events" (Brown, 1998, p. 117). Self-determination produced by top-down agency is not reducible to antecedent neuro-causation; top-down causation is not exhaustively reducible to bottom-up causation. Rather, intelligent human agency is the product of a personal self supervening on the brain by rendering a judgment followed by intentional action.

Genuine knowledge includes judgment. Sound judgment qualifies what counts as knowledge. Jesuit theologian Bernard Lonergan details the distinctions at work in human mentation. Knowledge may be located within consciousness; but it is not in itself consciousness. Knowledge is the result of a process wherein a self assesses experience, engages in active understanding, and renders judgments about what can be known and not known. "Consciousness is just experience, but knowledge is a compound of experience, understanding, and judging" (Lonergan, 1972, p. 106). Sound judgment belongs to intelligence when it is working to distinguish truth from falsity, appearance from reality, risk from safety, hope from fear.

It is via judgment that intelligent deliberation jumps the hurdle from falsity to truth. After our abstract reflection has accessed our perceptions and then filtered them through symbolic language and logical reasoning, we pause just before we take the leap to intellectual judgment. It is the leap of judgment that moves us from mere experience to knowledge. Knowledge is personally satisfying only if it is true knowledge. Knowledge by definition is knowledge of what exists, of what really is. Only via judgment does the human intellect know whether what it knows is in fact knowledge.

Here is the point of this discussion: when a person acts on the basis of intellectual judgment, the physical world becomes changed. Human intelligence becomes one cause among many in the physical history of the cosmos.

Intelligence without Consciousness, Value, or Free Will?

Note what is missing in this list of defining traits of intelligence: consciousness, value, and free will. The first observation I would like to make is this: we could not be human without consciousness, but we could be intelligent without consciousness. Human consciousness is fascinating, even if not necessary for intelligence. To the extent that the brain makes intelligence possible, consciousness functions to organize our brain's mental work space. If we think of our brain as a society of eighty-six billion neurons, some principle of organization is necessary for focused activity. "The function of consciousness," says Dehaene, "may be to simplify perception by drafting a summary of the current environment before voicing it out loud, in a coherent manner, to all other areas involved in memory, decision, and action" (Dehaene, 2014, p. 100). Consciousness arrives on the evolutionary stage along with mental activity, to be sure; but, in itself, consciousness does not define intelligence.

Secondly, what about value? "Human beings are able not only to detect but to be motivated by value," avers philosopher of science Thomas Nagel (Nagel, 2012, p. 112). When our ancestors first considered alternative futures, they needed a criterion – a value – for selecting one over the alternatives for action. Human beings weigh alternatives according to values, to be sure; yet, it appears to me that primitive intelligence arose without this capacity. Valuing arrives when abstract intellectual processing arrives, but in itself valuing does not define intelligence per se. At the human level, of course, we cannot grasp intelligence apart from the values which guide our decisions and actions.

This leads, thirdly, to the following observation: bottom-up physical processes are necessary, though not sufficient, to explain human free will. Key here is that the relevant physics includes indeterminate openness at the quantum level, that is, free will relies on randomness in the underlying physical domain. Quantum physicists tend to replace the Newtonian closed causal nexus with quantum randomness at the sub-atomic level. This basic physical randomness makes it possible for intelligent creatures to anticipate underdetermined alternative futures, select the desirable future from the undesirable according to a value system, and make decisions leading to the actions that will affect the external world. "The existence of both bottom-up and top-down causation in relation to the mind and brain removes the force of the bottom-up arguments against free will...and opens the way for more humane visions of the nature of humanity," avers Ellis (Ellis, 2016, p. 381).

Fourthly, I am not building free will into the definition of intelligence, at least not the contracausal version of free will. "The name *contracausal* reflects a philosophical theory that *really* free choices are not caused by anything" (Churchland, 2013, p. 179, Churchland's italics). Over against the contracausal account, I contend that free will belongs to the multi-level interaction between bottom-up and top-down causes described above by Ellis. In free agency, top-down causation is the focal factor. This makes the person exercising free will one cause among others. Free will and human agency belong inescapably to the causal nexus.

I argue elsewhere that free will is best described as *self-determination*, thereby avoiding appeal to acausality (Peters, 2003). In fact, this is the position I prefer: free will is a form of self-determination, where the free human person deliberates, decides, and acts to change his or her environment. But, is free will necessary to a definition of intelligence per se? No. Like valuing, free will belongs to a complex level of mental activity such as rendering a judgment. I have nothing against positing free will within human intelligence, to be sure. Nevertheless, I believe we can recognize intelligence at more primitive levels even when human free will is not at stake.

Three Possible Holes in My Argument's Boat

I can anticipate three holes in my argument's boat. The first hole takes us back to the evolutionary onset of life on Earth. It is not clear that simple life as we witness it today on Earth represents the earliest form of life. Stupid life may not represent the first transition from abiotic chemistry to living creatures themselves. LUCA, the last universal common ancestor, was already an advanced form of life, "having nucleic acids and proteins, as well as complex metabolic processes. In short, life as we know it represents a single example of a fairly advanced stage of life," say Carol Cleland and Shelley Copley (Cleland & Copley, 2006, p. 165).[8] The lack of evidence of the exact moment of transition from non-life to life in the fossil record is due to some missing files. "It is clear that these oldest traces of life record an evolutionary stage that is far beyond the origin of life or even very primitive cells. It may be difficult to find evidence of such earlier, intermediate life forms due to the lack of older, well-preserved rocks.... Thus, the most important stage in the history of life on Earth is missing. One can hope that, if life ever existed on Mars, rocks containing traces of life's earliest steps there will be discovered" (Cottin et al., 2015). One can only surmise that in these missing files we just might find early life that does not meet our criteria for intelligence. Should these missing files be found and the earliest life forms be shown to be unintelligent, then my hypothesis – where there's life there's intelligence – would fail. This may also falsify my maxim: it takes intelligence to beget intelligence. Discovering the missing link between prebiota and cellular biota could possibly poke a hole in the bottom of my hypothesis boat big enough to sink it.

The second hole might be drilled when determining the status we ascribe to the virus. Is a virus alive or not? With a rogue strand of DNA, for example, a virus lacks the trait of interiority, while it does exhibit intention. If one places the virus in the category of the living, then it would appear to be a form of life without the traits of intelligence I have adumbrated. Experts in virology shy away from declaring that a virus is a life-form, or even declaring it to be non-life. "Alive or dead is a stupid question," contends A. E. Boycott in his 1928 Presidential address to the Royal Society, "because it does not exhaust the possibilities. Our general notion of the universe leads us to expect that we shall meet with things that are not so alive as a sunflower, and not so dead as a brick" (cited by Johnson, 1982, p. 9). The virus may drill a small hole in the bottom of my hypothesis boat, but I don't believe it will sink.

The third hole has to do with what we might expect when searching for intelligent life on exoplanets. If the astrobiologist limits the search for biosignatures, the search might miss perceiving non-biological intelligence. Many of today's astrobiologists work with the assumption that evolution is progressive; this means that exoplanets with evolutionary progress lasting longer than that on Earth will likely have developed more advanced technology. "Other present technological civilizations in our galaxy probably developed much earlier than ours. Therefore, they should be much more advanced than our technology," says University of Arizona planetary scientist Robert Strom (Strom, 2015, p. 11). This means for some space speculators that a highly advanced technological civilization will be post-biological. That is, extraterrestrial intelligence may be located in highly advanced computers, not in flesh and blood.

Astrophysicist Martin Rees, for example, alerts space researchers to look for "intelligent machines" or "robotic fabricators" spread out in space (Rees, 2003, p. 172). Physicist turned astrobiologist Paul Davies similarly speculates that more advanced civilizations on exoplanets may have already evolved beyond biological bodies and inserted their intelligence into machines. If this is the case, then biosignatures would not lead us to intelligence.

My conclusion is a startling one. I think it very likely–in fact inevitable–that biological intelligence is only a transitory phenomenon, a fleeting phase in the evolution of intelligence in the universe. If we ever encounter extraterrestrial intelligence, I believe it is overwhelmingly likely to be post-biological in nature, a conclusion that has obvious and far reaching ramifications for SETI. *(Davies, 2010, p. 160)*

If Rees and Davies are correct, then my contention – *where there's life there's intelligence* – would fall short of providing an exhaustive association of life with intelligence. There might exist some intelligence apart from life. If extraterrestrial non-biological intelligence turns out to be post-biological, then the hole it drills in my argument would be shallow.

In sum, three drills could poke holes in this article's argument. Non-intelligent biology either on Earth or off-Earth would poke two holes, while the discovery of non-biological intelligence would poke the other one. As of this writing, however, insufficient empirical evidence exists for anyone to get out the buckets to start bailing.

Conclusion

The hypothesis of this article is this: *where there's life there is intelligence*. In support of this hypothesis, I have delineated seven traits belonging to an inclusive definition of intelligence: (1) interiority; (2) intentionality; (3) communication; (4) adaptation; (5) problem-solving; (6) self-reflection; and (7) judgment. It is clear that most mammals and certainly human beings exhibit all seven traits. I have shown that brainless microbes and simple organisms exhibit the first four traits. By establishing a spectrum of traits, all life from the simplest to the most complex can be dubbed intelligent, even though they differ in levels of complexity. The value of this argument is this: the astrobiologist searching our solar system should look for various degrees of intelligence, not *un*intelligent life.

Elspeth Wilson and Carol Cleland remind us that if we look only through our anthropocentric glasses, we may miss seeing the moral status of extraterrestrial creatures. If we look only through the lenses colored with human rationality, we may miss seeing the warrant for respecting unusual forms of alien intelligence. "Ultimately, those who want to hold up intelligence or rationality as a benchmark for moral status must identify and acknowledge the anthropocentric limitations of this standard" (Wilson & Cleland, 2015, p. 214).

The moral status of extraterrestrial life has not been on the agenda of this essay. Intelligence has. Like Wilson and Cleland, I have sought to find a path around the "anthropocentric limitations" of linking intelligence so closely to our human experience with mentation that we might miss seeing intelligence in different life forms.

Notes

1. I offer a special thank you to colleagues Ross Johnson and Margaret Race for their critical contributions to my writing of this article. I also thank Richard Procunier for his support of the Center for Theology and the Natural Science, which sponsors such research.
2. Here is an example of this assumption at work in astrobiology: "Unlike non-intelligent life, intelligent beings can have intentions" (Michaud, 2015, p. 289). What would happen if we discovered that simple life forms exhibit intention? University of Arizona astrobiologist Chris Impey, to cite another example, distinguishes between non-intelligent life on Earth from intelligent life by adding "thought" to "behavior." Thought adds two traits beyond mere behavior: "the ability to reflect on past experience and the power of abstraction" (Impey, 2007, p. 285). Thought is more than behavior, according to Impey. This is true, obviously. Yet, I hope to show here that behavior when intentional exhibits traits we think of as intelligent. Microbes might be thoughtless, but still intelligent.
3. I'm looking for intelligence in the middle, so to speak, in organic processes. This is a more modest search than that of Stuart Kauffman, who looks for mind in all physical processes at the quantum level. "Mind – consciousness, *res cogitans* – is identical with quantum coherent immaterial possibilities" (Kauffman, 2008, p. 209).
4. Howard Gardner delineates eight types of human intelligence: (1) bodily-kinesthetic; (2) interpersonal; (3) intrapersonal; (4) linguistic; (5) logical-mathematical; (6) musical; (7) naturalistic; and (8) spatial (Gardner, 2000). For our purposes here, it is sufficient to note that not all human intelligence is thought to take the form of cognitive capability directly correlated to the human brain.
5. In *The Bell Curve*, for example, the coauthors proffered an alleged connection between genetics and IQ, adding that variations in intelligence determine class status. They contended that "beyond significant technical dispute" are assertions such as: cognitive ability differentiates some human beings from others; IQ tests measure cognitive ability with sufficient accuracy; high IQ scores match with what we mean by *intelligence* or being *smart*; properly administered IQ tests are not demonstrably biased against social, economic, ethnic, or racial groups; and cognitive ability is substantially heritable. These purportedly undisputable assertions support their thesis: "cognitive ability is the decisive dividing force" between social classes (Herrnstein & Murray, 1994, p. 25). Distinguishing human groups by IQ results in the authors' recommendation that government financial support for early education be increased for the most intelligent – Ashkenazi Jews and East Asians (Japanese and Chinese) – while withdrawn from African Americans and Hispanics (Herrnstein & Murray, 1994, p. 548). American white people are of average intelligence, so they will not be smart enough to see that their tax money goes to benefit those with higher IQs. Significant for us here is to note how the concept of intelligence is generally employed in gradations to measure higher and lower among human beings. What I believe is needed is a minimum definition of intelligence to discern who's in and who's out. And it appears to us that life forms from microbial stages on are in. But, according to *The Bell Curve*, single-celled organisms would not be smart enough to receive pre-school aid.
6. In his chapter in this volume, "Autopoietic Systems and the Theology of Creation: On the Nature of Life," Alexander Maßmann refers to this spiral as the circle of causality (Maßmann, 2017).
7. Artificial Intelligence (AI) is not intelligent, at least at present. Complex, yes. Intelligent, no. AI does not have a mental life as humans do, even though some non-sentient machines can learn. AI is not built "these days" to create second generation intelligence and then become difficult to control, at least according to Yoshua Bengio, who is developing *deep learning* for AI. "Machine learning means you have a painstaking, slow process of acquiring information through millions of examples. A machine improves itself, yes, but very, very slowly, and in very specialized ways" (Knight, 2016).
8. A strong version of my hypothesis – where there's life there's intelligence – would need confirmation in Origin of Life Studies. "We suggest that what OoL [Origin of Life] studies all ultimately address is the onset of the various organizational phenomena that we associate with the living world. This unites all areas of research, from laboratory experimentation to Earth and planetary exploration, theory, and computation"(Scharf *et al.*, 2015, p. 1033).

References

Adler, P. N. & Nathans, J. (2016). The cellular compass. *Scientific American*, **314**(3), 66–71.

Albus, J. S. (1991). Outline for a theory of intelligence. *IEEE Transactions on Systems, Man, and Cybernetics*, **21**(3), 473–509.

Brown, W. S. (1998). Cognitive contributions to the soul. In W. S. Brown, N. Murphy, and H. N. Malony, eds., *Whatever Happened to the Soul? Scientific and Theological Portraits of Human Nature*, Minneapolis, MN: Fortress Press, pp. 99–126.

Bruneau, E. G., Jacoby, N. & Saxe, R. (2015). Empathic control through coordinated interaction of amygdala, theory of mind and extended pain matrix brain regions. *NeuroImage*, Web: http://www.sciencedirect.com/science/article/pii/S1053811915003250 Accessed 5/20/2016.

Brüstle, O. (2013). Miniature human brains. *Nature*, **501**(7467), 319–20.

Chang, Y.-W., Rettberg, L. A., Truener-Lange, A. *et al.* (2016). Architecture of the type IVa pilius machine. *Science*, **351**(6278), 1165.

Christian, D. (2011). *Maps of Time: An Introduction to Big History*, Berkeley, CA: University of California Press.

Christian, D. (2016). From mapping to meaning. In R. A. Culpepper and J. G. van der Watt, eds., *Creation Stories in Dialogue: The Bible, Science, and Folk Traditions*, Leiden: Brill, pp. 35–47.

Churchland, P. A. (2013). *Touching a Nerve: The Self as Brain*, New York, NY: W.W. Norton.

Cleland, C. E., & Copley, S. D. (2006). The possibility of alternative microbial life on Earth. *International Journal of Astrobiology*, **4**(3&4), 165–73.

Cottin, H., Kotler, J. M., Bartik, K. *et al.* (2015). Astrobiology and the possibility of life on Earth and elsewhere. *Space Science Review*, Web: http://link.springer.com/article/10.1007/s11214-015-0196-1#/page-1 Accessed: 3/8/2016.

Davies, P. (2010). *The Eerie Silence: Renewing Our Search for Alien Intelligence*, Boston, MA: Houghton Mifflin Harcourt.

Deacon, T. W. (2012). *Incomplete Nature: How Mind Emerged from Matter*, New York, NY: W.W. Norton.

Dehaene, S. (2014). *Consciousness and the Brain: Deciphering How the Brain Codes Our Thoughts*, New York, NY: Viking.

Ellis, G. (2016). *How Can Physics Underlie the Mind? Top-Down Causation in the Human Context*, Heidelberg: Springer.

Freeman, W. J. (2000). *How Brains Make Up Their Minds*, New York, NY: Columbia University Press.

Gardner, H. (2000). *Intelligence Reframed*, New York, NY: Basic Books.

Gluth, S. & Fontanesi, L. (2016). Wiring the altruistic brain. *Science*, **351**(6277), 1028–9.

Goodenough, U. & Deacon, T. W. (2006). The sacred emergence of nature. In P. Clayton and Z. Simpson, eds., *The Oxford Handbook of Religion and Science*, Oxford: Oxford University Press, pp. 853–71.

Hein, G., Morishima, Y., Leiberg, S. *et al.* (2016). The brain's functional network architecture reveals human motives. *Science*, **351**(6277), 1074–8.

Herrnstein, R. J. & Murray, C. (1994). *The Bell Curve: Intelligence and Class Structure in American Life*, New York, NY: Free Press.

Heschel, A. J. (1951). *Man is Not Alone: A Philosophy of Religion*, New York, NY: Farrar, Staus and Giroux.

Höffe, O. (2010). *Can Virtue Make us Happy?*, trans. D. R. McGauhey and A. Bunch, Evanston, IL: Northwestern University Press.
Impey, C. (2007). *The Living Cosmos: Our Search for Life in the Universe*, New York, NY: Random House.
Impey, C. (2010). *How It Ends: From You to the Universe*, New York, NY: W.W. Norton.
Johnson, R. T. (1982). *Viral Infections of the Nervous System*, New York, NY: Raven Press.
Kauffman, S. A. (2008). *Reinventing the Sacred: A New View of Science, Reason, and Religion*, New York, NY: Basic books.
Knight, W. (2016). Will machines eliminate us? *MIT Technology Review*, **119**(2), 30.
Kurzweil, R. (2015). Army ants' 'living' bridges suggest collective intelligence. *Kurzweil Accelerating Intelligence Newsletter* (November 25), Web: http://www.kurzweilai.net/army-ants-living-bridges-suggest-collective-intelligence?utm_source=KurzweilAI+Weekly+Newsletter&utm_campaign=a24cfc1674-UA-946742–1&utm_medium=email&utm_term=0_147a5a48c1-a24cfc1674–281997577 Accessed: 3/5/2016.
Lancaster, M. A., Renner, M., Martin, C.-A. *et al.* (2013). Cerebral organoids model human brain development and microcephaly. *Nature*, **501**(7467), 373–9.
Laird, D. W., Lampe, P. D. & Johnson, R. G. (2015). Cellular small talk. *Scientific American*, **313**(5), 70–7.
Legg, S. & Hutter, M. (2006). *A Collection of Definitions of Intelligence*, Web: http://www.vetta.org/documents/A-Collection-of-Definitions-of-Intelligence.pdf Accessed: 3/2/2016.
Lonergan, B. J. F. (1972). *Method in Theology*, New York, NY: Herder and Herder.
Lupisella, M. (2015). Life, intelligence, and the pursuit of value in cosmic evolution. In S. J. Dick, ed., *The Impact of Discovering Life Beyond Earth*, Cambridge: Cambridge University Press, pp. 159–74.
Malafouris, L. (2014). *How Things Shape the Mind: A Theory of Material Engagement*, Cambridge MA: Harvard University Press.
Marino, L. (2015). The landscape of intelligence. In S. J. Dick, ed., *The Impact of Discovering Life Beyond Earth*, Cambridge: Cambridge University Press, pp. 95–112.
Maßmann, A. (2017). Autopoietic systems and the theology of creation: on the nature of life. In A. Losch, ed., *What is Life? On Earth and Beyond*, Cambridge: Cambridge University Press, pp. 213–35.
Metzinger, T. (2009). *The Science of the Mind and the Myth of the Self*, New York, NY: Basic Books.
Michaud, M. A. (2015). Searching for extraterrestrial intelligence: preparing for an expected paradigm break. In S. J. Dick, ed., *The Impact of Discovering Life Beyond Earth*, Cambridge: Cambridge University Press, pp. 286–98.
Mix, L. J. (2009). *Life in Space: Astrobiology for Everyone*, Cambridge, MA: Harvard University Press.
Molhoek, B. (2016). *Reinhold Niebuhr's Theological Anthropology in Light of Evolutionary Biology: Science Shaping Anthropology Shaping Ethics*, unpublished Ph.D. dissertation, Graduate Theological Union.
Moser, M.-B. & Moser, E. I. (2016). Where am I? Where am I going? *Scientific American*, **314**(1), 26–33.
Nagel, T. (2012). *Mind and Cosmos: Why the Matrialist Neo-Darwinian Conception of Nature is Almost Certainly False*, Oxford: Oxford University Press.
Peters, T. (2003). *Playing God? Genetic Determinism and Human Freedom*, 2nd edn, London: Routledge.
Piaget, J. (1963). *The Psychology of Intelligence*, London: Routledge.
Rees, M. J. (2003). *Our Final Hour: A Scientist's Warning*, New York, NY: Basic Books.

Ricard, M., Lutz, A. & Davidson, R. J. (2014). Mind of the meditator. *Scientific American*, **311**(5), 39–45.

Scharf, C., Virgo, N., Cleaves II, H. J. *et al.* (2015). A strategy for origins of life research. *Astrobiology*, **15**(12), 1031–42.

Searle, J. (1983). *Intentionality: An Essay in the Philosophy of Mind*, Cambridge: Cambridge University Press.

Strom, R. G. (2015). We are not alone: extraterrestrial technological life in our galaxy. *Journal of Astrobiology and Outreach*, **3**(5), Web: http://www.esciencecentral.org/journals/we-are-not-alone-extraterrestrial-technological-life-in-our-galaxy-2332-2519-1000144.php?aid=63445 Accessed 3/22/2016.

Ward, P. & Kirschvink, J. (2015). *A New History of Life*, New York, NY: Bloomsbury Press.

Wilson, E. M. & Cleland, C. E. (2015). Guideposts for exploring our ethical and political responsibilities towards extraterrestrial life. In S. J. Dick. ed., *The Impact of Discovering Life Beyond Earth*, Cambridge: Cambridge University Press, pp. 207–21.

13

Life in the Universe, Incarnation and Salvation: A Conversation between Christianity and the Scientific Possibilities of Extra Terrestrial Life

JUAN PABLO MARRUFO DEL TORO, SJ

For ages, humans have asked probing questions about the nature of our existence: (1) "Are we alone in the universe?", (2) "Where do we come from?", (3) "Who made us?", and (4) "What is God's role in the universe?". We have tried to answer these questions using science, religion and philosophy.

We enjoy living in a world open to life, which frames the way we ask those questions. What would happen if we encountered life on other planets? Would this finding change or inform our theological perspectives about God and God's relationship to us?

In this article, I address the scientific issues concerning the origin of life and the possibility of finding intelligent life in the universe. After that, I discuss the possible implications of such a discovery for theology. I will specially address the issues of incarnation and salvation.

Life in the Universe

The consensus of cosmologists is that the universe began sometime between 11 and 14 billion years ago (Cayrel, Hill, Beers *et al.*, 2001; Krauss & Chaboyer, 2003). More recent data on cosmic microwave background radiation provided by Europe's Planck Mission establish it at 13.8 Ga (billion years ago).[1]

After about 10^{-32} seconds after the origin of the universe, atoms began to appear in great enough amounts so that gravity collapsed, forming particles, gas clouds, stars and galaxies. The rest of the elements formed as the stars evolved and produced energy by nuclear fusion of the light elements into heavier ones or by exploding into supernovas at the end of a star's life. These newly formed elements include the biogenic elements – necessary for life – such as carbon, nitrogen, oxygen, magnesium, sulfur and phosphorus.

Let us note that all life on Earth is carbon-based. Lynn Rothschild says we should assume that life on other planets will also be carbon-based because carbon is the fourth most common element in the universe and is capable of forming many compounds ranging from methane to DNA, which is the perfect storage medium for genetic code; further, atomic carbon and simple organic compounds have already been detected in space (Rothschild, 2004, p. 135; see Sagan, 1980, p. 46).

What is Life?

About 4 billion years ago a remarkably complicated process began on Earth as the biogenic elements present in our planet interacted with each other, giving rise to what we call "life." R. L. Sinsheimer defines life as having the following characteristics:

- a binding membrane that contains it which is flexible and self-made
- coordinated groups of catalysts (most likely proteins)
- it has DNA, a hereditary system to store information
- the ability to transform an external source of energy into suitable forms for driving reactions (breathing)
- a method of cell division that provides the new cells with hereditary instructions
- systems that regulate varied functions to allow adaptation and flexibility
- means of defense (Sinsheimer, 1996, pp. 35–9).

Sinsheimer's characteristics are widely accepted. If this is what is necessary to be considered "alive," what conditions are needed for such a life to exist?

Erwin Schrödinger, the physicist responsible for the equation that describes the statistical logic of quantum mechanics, also wrestled with the question. In his book *What is Life?* (Schrödinger, 1994), he asked two general questions about the physical nature of life. Schrödinger questioned the energetics of life and its chemistry in relationship with entropy. Terrence Deacon in his book *Incomplete Nature* describes the fundamental incongruity in the following way:

[Schrödinger's] first question concerned the energetics of life. Living organisms, individually and in the grand sweep of evolution, appear to make an end run around one of the most fundamental tendencies of the universe, the second law of thermodynamics. [This feature] demands an explanation of how such a counterintuitive trick is performed by the chemistry of organisms, when chemistry by itself seems to naturally follow the thermodynamic law to the letter. The second question was indirectly linked to the first. How could the chemistry of life embody the information necessary to instruct the development of organisms and maintain them on a path that defends against the incessant increase of entropy?
(Deacon, 2013, pp. 281–2)

Deacon explains Schrödinger's general answer to the first question: "living chemistry must somehow continuously stave off the pressures of thermodynamic decay, and to do this it needs to feed off of sources of free energy available in the outside world, which he enigmatically called 'negentropy' or sources of order. Schrödinger's general answer to the second question was that there must be a molecular storage medium available to organisms, which embody a chemically arbitrary and interchangeable but functionally distinguishable molecular pattern that can be preserved and copied and passed to succeeding generations" (Deacon, 2013, p. 281). Schrödinger speculated that at the heart of every organism, and indeed every living cell, there must be something akin to a crystal, because it would have to have repeatable units, and yet it would have to be non-crystalline in another way, since these units need to be slightly varying, unlike mineral crystals, which are highly regular. Schrödinger reasoned that in order to be able to store information: "each unit cell of this crystalline like molecule would need to exhibit one of a number of structurally different but energetically equivalent states, not unlike the way that the typographical characters of this sentence are interchangeable" (Deacon, 2013, pp. 281–2).

Deacon argues that this structure trend ought to be very similar to the description we presently have of DNA molecules. Schrödinger's crystal structure deeply influenced Watson and Crick's DNA model, which they proposed just a few years later. Although many biologists claim that Watson and Crick's model of DNA structure was the key to the secret of life because DNA is an intrinsic characteristic of life, this only addressed the second question proposed by Schrödinger, the means of storing of information and instructions. Deacon argues that the DNA model proposed by Watson and Crick did not address Schrödinger's first question, the issue with thermodynamics. Regarding this issue, Deacon writes: "For the most part, the vast majority of evolutionary and molecular biologists have followed the trail of molecular information, while a much smaller cadre of physicists and physical chemists have followed the trail of far-from-equilibrium thermodynamics. This is ultimately an untenable segregation, because information turns out to be dependent on thermodynamic relationships" (Deacon, 2013, p. 282). Deacon, as he tries to define what is life, places emphasis on the thermodynamic relationships, rather than only the biological composition and chemistry, as the majority of scientists do. Therefore, the interaction of life with its environment is just as important as the physical composition of it. We cannot separate one from the other. This is important because it tells us that interactions and relationships are also fundamental components of life.

Deacon performs a theoretical experiment to figure out what the minimal components for life are. His experiment is very interesting because rather than trying to gather elements to create life in a laboratory, he proposes to find out what is minimally necessary to define life by "starting with what we have, and then subtract features until the removal results in failure to function" (Deacon, 2013, p. 283). Deacon tries to find examples of minimal living cells in this top-down approach as he moves from complex organisms to more simple ones; he tries to produce "minimal cells, by stripping a simple bacterium of all but its most critical components. Though considerably simpler than naturally occurring organisms, these minimal cells still contain hundreds of genes and gene products" (Deacon, 2013, p. 283).[2] The most recent success in creating such a minimal organism was achieved in March 2016 by the J. Craig Venter Institute, which engineered a synthetic microbe, named Syn 3.0, by taking genes out of other microbes. Syn 3.0 contains only 473 genes (Service, 2016).

The other approach to try to understand the minimal physical requirements for life is the bottom-up approach, which, in the words of Deacon, "attempts to generate key system attributes of life from a minimized set of precursor molecular components. An increasing number of laboratory efforts are underway to combine molecular components salvaged from various organisms and placed into engineered cellular compartments called protocells." (See recent reviews, Rasmussen, Chen, Deamer *et al.* (2004) and Szathmáry (2005).) These are attempts to produce artificial cells capable of maintaining the physical processes required for self-replication or reproduction of the entire cell. So far, even the most complex protocells created are not able to sustain autonomous reproduction. It is possible that, as protocell research develops, and more complex protocells are fabricated, at some point they will be able to sustain autonomous reproduction. However, we are still very far from that point and we do not know which obstacles will be present down the road of investigating the physical properties of life. Interestingly enough, Deacon says "even the simplest protocells currently imagined are sufficiently complex to raise doubts that such systems could coalesce spontaneously from pre-organic substrates" (Deacon, 2013, p. 284).

So, how does life begin then, if not from inorganic materials? We are still far from finding out how to move from non-life to life. But even if scientists can ever do that, this may not give us all the answers we need. Deacon further argues: "Creating a stripped-down artificial bacterium or protocell that can reproduce itself will certainly provide important insights about life in general. It is quite likely that such cells could provide useful vehicles for replicable Nano machines functioning at a cellular scale. But protocells are unlikely to be useful models for the missing link between physics and biology" (Deacon, 2013, p. 285). Life is much more than physics, more than particles arranged the right way. We still have much to learn about what life is.

Necessary Conditions for Life

Since, at present, Earth contains the only known evidence of life in the universe, our own planet is our sole source of data for examining life; we can study the conditions that led to the origin of life on Earth as well as the present conditions that sustain various forms of life.

We do know of certain required conditions: the presence of liquid water, an energy source, enough biogenic elements, and time. Furthermore, according to Conway Morris (Morris, 2004), life also requires a planet with a shield against cosmic radiation, gravity must not be too strong, the rotation axis should be slightly tipped in relation to its orbit around the Sun, there must be large planets in the vicinity to divert asteroids, among other very unique characteristics. Here on Earth, we have all those characteristics and elements; however, we have not been able to reproduce the process of originating life. In his book *At Home in the Universe*, Stuart Kauffman claims that "life is a natural property of complex chemical systems, that when the number of different kinds of molecules in a chemical soup passes a certain threshold, a self-sustaining network of reactions – an autocatalytic metabolism – will suddenly appear" (Kauffman, 1996, p. 47). Kauffman developed this theory using mathematical models and computer-generated results. His theory is appealing to this topic as we explore the characteristics and origin of life. According to Kauffman, both the right amount of certain chemicals and the right conditions are needed. However, we have not been able to properly mix the original chemical components (biogenic) in the right proportions with the right conditions to produce a living thing. We may still be very unaware regarding some of the essential mechanisms of life.

According to Rothschild, "molecular sequence data hint that the last common ancestor for life may well have been a thermophile, that is, an organism that preferred to live at high temperatures" (Rothschild, 2004, p. 136). This may have been the only organism capable of surviving the harsh conditions on Earth four billion years ago.

Life on Mars could have originated in the same way; in fact, we know of Martian material that has been ejected from the surface of Mars and made its way to Earth, such as the Nakhla meteorite. The potential therefore exists that life forms could be transferred between the two planets. NASA is currently exploring such a theory.[3] If, on the other hand, we find life on Mars and conclude it is *not* related to Earth, then we would assume Mars had its own genesis, which could increase the possibility of a universe much more open to life. Besides Mars, there is also the possibility of microbial life on the moons of Jupiter and Saturn. If life is found on these moons, such life most likely could not have emerged by an exchange of materials with Earth, suggesting it originated independently.

The abundance and versatility of carbon in the universe is a very good indicator that carbon-based life is likely to be found in the universe if the right conditions are present. This has considerable implications for our question. To locate life in the universe, it is necessary to recognize the distribution of life on our planet. We need to ask ourselves: what are the limits of life on Earth? The discovery of life forms on Earth that inhabit places thought to be barren does not suggest that such a life form is present wherever we find similar environments. However, it can be an indication of what is possible.

Life has been discovered in environmental extremes, places previously thought to be uninhabitable. The organisms that live there are called *extremophiles*, which can thrive in the lowest pH, highest temperatures, and other environmental extremes such as high pressure, population density and parasite loading. Rothschild says that "many of these conditions could be obstacles for the long term development of carbon-based life forms" (Rothschild, 2004, p. 142). For example, proteins and nucleic acids lose their function at high temperatures. Some gases such as oxygen and carbon dioxide lose their solubility as the temperature increases too, making it very difficult to breathe. Yet there are some studies of organisms, such as the extremophile *Pyrolobus fumarii*, which live in environments of temperatures of 113 °C – well above the point of boiling water (Blochl, Rachel, Burggraf *et al.*, 1997, pp. 14–21). At these temperatures the lipid membranes that enclose the cell becomes fluid, losing the integrity of the cellular structure; yet, some organisms are able to survive under such conditions. We also have some examples of the opposite extremes such as trees, microbes, penguins and insects, all living in such cold environments in which the cellular membranes would be expected to become rigid, slowing down the biochemical reactions. A. Clarke has found that even cell lines can be readily preserved in liquid nitrogen (Clarke, 2003).

If life on Earth can be found under these extreme conditions, there is cause for hope that it can be found in places like Europa or Mars in which similar thermal conditions exist. The existence of extremophiles on Earth motivates us to continue our search for life amidst the harsh conditions of the universe.

Evolution of Life

In the second half of the twentieth century, there was a debate between Simon Conway Morris of the University of Cambridge and Stephen Jay Gould of Harvard University. Gould stated that if we could rewind the "tape" of the history of life on Earth and replay it, what we would see would be totally different than what we see now. The way life would organize itself again (if it ever does) would be completely unrecognizable (Gould, 1989, p. 323). On the other hand, Conway Morris said that evolutionary principles exist which constantly shape the way life evolves (Morris, 1998, p. 272). For Morris, evolution is not as random as Gould would assert. There are key elements that emerge from the evolutionary process, such as cellular organization from simple to more complex and specialized – a progression from which metabolism and intelligence originate. Morris argues that life in other planets would be very similar to the life we know on Earth: it would have eyes, high encephalization (intelligence), agriculture, viviparity, etc. Perhaps a different number of teeth and fingers.

Our planet is a perfect place for life to flourish and evolve. But is our universe like this? Scientist Martinez Hewlett says that the universe is biophilic (Hewlett, 2017, 6:7). For Hewlett, the universe loves life and has demonstrated this by the flourishing and resilience of life on our planet. Hewlett argues that the universe is biophilic because Earth is.

Intelligence

A special characteristic of some forms of complex life is intelligence. We are hopeful about finding proof of microbial life in the universe, but what about intelligent life? Does this have a special place in theology? Theologically speaking, we humans think we have a special place in creation precisely because we are capable of relating to God.

Rothschild says that intelligence can originate from activity. For example, a carnivorous predator is often associated with intelligence because complex behavior in reacting to changing situations requires flexible and rapid responses (Rothschild, 2004, p. 141). Consider mammals; Mark Springer suggests that there are four lines of placental mammals and that what we call intelligence can be found in three of them: in the Laurasiatheria line (which includes dogs and whales), the Afrotheria (which includes elephants), and in the Euarchonta and Glires (which includes humans). The Xenarthra line, which includes sloths, armadillos, anteaters, rabbits and mice, is not associated with the phenomenon we call intelligence and also does not contain carnivorous animals (Madsen, Scally, Douady *et al.*, 2001).

So, if we follow this argument, we can see how the ability to respond to a changing environment in order to survive – such as by hunting for food instead of just eating it from the ground or trees – can be a common factor in the evolution of intelligence in some beings. Further, Terrence Deacon in his book *Incomplete Nature* argues: "Intelligence is about making adaptively relevant responses to complex environmental contingencies, whether conscious or unconscious" (Deacon, 2013, p. 492). Deacon agrees with intelligence being a result of a changing environment.

On the other hand, Ted Peters has a much broader definition of intelligence. In another chapter of this volume, Peters argues that intelligence exists across the spectrum of life forms "from the simplest to the most complex" (Peters, 2017b). Peters argues that for life to be considered intelligent it has to have seven characteristics: "(1) interiority; (2) intentionality; (3) communication; (4) adaptation; (5) problem-solving; (6) self-reflection; and (7) judgment." According to Peters, "Most mammals and certainly human beings exhibit all seven traits. Yet, brainless microbes and simple life from the simplest to the most complex can be dubbed intelligent, even though they differ in levels of complexity." Peters basically argues that all life is intelligent because all life has, at least, the first four of the traits he defines for life to be intelligent. Peters is trying to expand our notion of intelligence to include all known forms of life. If we agree with Peters, then any form of life we encounter in the universe has some degree of intelligence.

In his book *The Symbolic Species*, Deacon argues that the human brain evolved as language evolved too. For Deacon, there is a huge abyss between our unique ability to communicate symbolically and the ability to communicate in other species (Deacon, 1998, p. 12). Deacon argues that this ability evolved with the brain and it is the source of our intelligence. Can we expect a similar situation in other species or other planets? Deacon says that the growth of the human brain is similar to that of other mammals (Deacon, 1998, p. 173). Further in his book, Deacon shows us that some chimpanzees can acquire some symbolic ability, as they can identify symbols with words and with actions. If we follow Deacon's argument, it is likely that our ability is not only a human quality or that it only exists here on this planet. As species in other places grow and evolve, it is possible that they can also communicate symbolically, giving rise to intelligence.

Merlin Donald, in his book *A Mind so Rare*, argues that the first hybrid minds on Earth (between animal and human) evolved when the first hominids moved from the mimetic to the mythic capacities. Donald argues that there are "major stages, or transitions, in [his] version of our cognitive *emergence*. Each of these transitions changed the nature of human consciousness in a major way" (Donald, 2001, p. 260). Donald speaks of a scenario of human evolution in which there exists tension between culture and conscious capacity, in which culture constantly pushes the conscious capacity to the edge, so that it expands constantly. The emergence of culture is what allowed the mind to keep adjusting constantly to the new realities of distributed cognition. As a result of such tension, we have the emergence of a symbolizing mentality (Donald, 2001, p. 260). This symbolizing mentality is, for Donald, the presupposition for the origin of what we call intelligence. So, if intelligence emerged from a co-evolution between consciousness and culture, can we expect this to happen elsewhere in the universe?

Intelligence seems to be, according to Donald, the result of evolution. Therefore, while we cannot expect to encounter human beings on other planets – as we are the product of evolution given the circumstances of the planet in which we live and the possibilities nature offers here) – we *can* encounter intelligent beings.

Searching for ETI (Extra Terrestrial Intelligence)

The scientist Francis Drake, in 1961, developed an equation to calculate the number of civilizations with whom contact is possible in our galaxy (Drake, 2003). Some of its factors are not known, which makes it very controversial, but it can help illuminate the possibility of encountering ETI.

The Drake equation is:

$$N = R_* \cdot f_p \cdot n_e \cdot f_l \cdot f_i \cdot f_c \cdot L$$

where:

N = the number of civilizations in our galaxy with which radio-communication may be possible
R_* = the average rate of star formation in the galaxy
f_p = the fraction of those stars that have planets
n_e = the average number of planets that can potentially have life
f_l = the fraction of planets in which life can develop life at some point
f_i = the fraction of planets in which intelligent life is likely to develop
f_c = the fraction of civilizations that could have technology that releases detectable signs of their existence into space
L = the length of time for which such civilizations release detectable signals into space.

As science progresses, we can better estimate some of the parameters of that equation. Since NASA's *Kepler* mission began looking for planets, we have produced a better estimate of the first three parameters. The last four, however, are basically unknown to us, rendering the values of N from practically zero to hundreds of thousands, thus giving us anywhere from a very pessimistic to a reasonably optimistic chance of contacting another civilization in the near future.

Owing to our lack of knowledge concerning some of the parameters, the Drake equation is useful for stimulating our curiosity and desire to explore the answers and for giving a better value to the unknown figures. It cannot, however, give us a precise answer to our question.

We know that our planet has a great variety of ecosystems, weather patterns, temperatures, different levels of exposure to sunlight, water and nutrients. In all of these different conditions, there is a great variety of life in different forms: bacteria, vegetation, protists, fungi, animals, some beings that breathe carbon dioxide and others that breathe oxygen. Some organisms thrive in extreme environments as mentioned above.

Some scientists, however, such as Alfred Wallace, argue against the existence of ETI; in fact, Wallace poses the theory of the rare Earth because of the principle of contingency. Wallace says that the evolution of life from abiotic chemistry is itself a random chance event, not a principle built into natural processes. For Wallace, what happened on Earth is not likely to be repeated on other worlds.[4] On the other hand, for Hewlett, what happens on Earth can happen elsewhere and, eventually, we will discover it (Hewlett, 2017, 6:7).

Both are valid arguments. Wallace's argument, although based in science, says that Earth is unique. We humans like to feel special and privileged among the other species on Earth and the universe. Nonetheless, Hewlett's argument gives more space for exploration, for growth and for our development, both scientific and spiritual. Given the most recent data we have from NASA's *Kepler* mission and the amount of exoplanets found in the past decade, our galaxy is now estimated to have between 100 and 400 billion exoplanets, at least 17 billions of these being the size of Earth (Cassan *et al.*, 2012). Though this number would be reduced if we included how many of those planets are in the habitable zone in their solar system, the figure is still significant. As of November 2014, about one in five Sun-like stars observed by NASA's spacecraft *Kepler* has an Earth-size planet in the habitable zone (Lewis, 2013). Going back to Hewlett, he asks, "Is it reasonable to expect [...] that we are alone in the universe? No, it is not reasonable" (Hewlett, 2017, 6:11). Let us not overemphasize the uniqueness of Earth but, rather, recognize it as one more planet among many inhabited worlds. This would require to have a detailed list and a deeper understandings of the criteria necessary for the evolution of life, and then for intelligent life, which we are still trying to understand. But we can be open to the possibilities that Earth is not the only place in the universe that has developed life. We already had a scientific revolution against geocentrism; we do not need another one.

We are still trying to understand what life is; however, I have argued that life is present in the universe, although we have not found it yet, and that the possibilities of extra terrestrial intelligent life are very real. Once life has originated, it evolves from simple to more complex, and that evolution tends toward intelligence. As it evolves, life goes through mutations by natural selection. It has to survive in spite of environmental changes, famine and predators, making it more and more complex and at some point, giving rise to language, intelligence, culture, art, civilization, technology and relationships.

I argue that there is life in the universe, and that there may even be intelligent life in other times and places. While the possibility of contacting ETI is very remote given the vastness of the universe and the obstacles of space and time, we can point to the fact that intelligent life does exist – as we ourselves know. Intelligent life exists on Earth and, thus, intelligent life is likely to originate and evolve more than once in the universe. Carbon and the rest of the biogenic elements are abundant. Even if we cannot make contact with ETI, our question about the implications of the Incarnation for life in the universe remains. Our search for intelligent life, and not even necessarily the discovery of it, informs our question about the implications of the Incarnation as salvation for the universe.

Our recent understandings for life and the search for life in the universe, especially intelligent life, have a deep impact in our theology, particularly the dogma of the Incarnation, in which God deeply relates to the creation. In the next section I explain some of those implications.

Theological Perspectives

The main questions on dealing with our search for ETI and the Incarnation are: Is Jesus of Nazareth the savior of the universe? Or is it possible to conceive of multiple Incarnations of the Logos? Our answers to these questions depend on our understandings of the Incarnation (salvation or revelation) and whether ETI needs to be saved or not.

If the Incarnation is about salvation, about God changing the universe from Creation to New Creation, then one Incarnation of the Logos may be enough. Ted Peters supports this perspective in what we call the "ontological argument" for the Incarnation (Peters, 2017a, 12:8). Or, on the other hand "is the Incarnation also about proclaiming the Good News so that people can come to believe in Christ and be saved?". Robert Russell defends this perspective, calling it a "revelational argument." In this case, many Incarnations are needed, one for every species in the galaxy that is intelligent and has moral capacity.

Robert Russell says: "From an ontological perspective, a single Incarnation of the divine Logos in Jesus of Nazareth is sufficient for the redemption of the universe" (Russell, 2017, 14:4). However, Russell says that it is very likely that ETI would have their own version of the Incarnation for several reasons. Russell makes clear the distinctions between the revelational and the ontological arguments as the reasons for the Incarnation as he writes: "The ontological view emphasizes the significance of the Incarnation as the act in which God redeems the world from such universal facts as sin and death and God's act begins the transformation of the world into the New Creation. The revelational view emphasizes the effect the Incarnation has on us personally, our coming to know Christ and our choosing to follow him in discipleship and newness of life" (Russell, 2017, 14:4).

Once he has made that distinction, Russell argues that the revelational argument requires the ontological: "There can indeed be redemption (an ontological act) without its revelation, in line with those supporting the possibility of universal salvation including the saving of those born before Jesus or who never heard the Good News. However, there cannot be a genuine, personal revelation of the Good News without the historical act by which God redeems the world; it would be incoherent to argue that we can encounter the Good News without there being any actual basis for it in human history" (Russell, 2017, 14:4). So, according to

Russell, the ontological view does not require the revelational because we can be saved without knowledge of it, but the revelational requires the ontological.

It is also important to distinguish between two kinds of "revelatory argument." One is the revelation of the Incarnation, which is the source of the issue between Russell and Peters; Russell refers to this as "special revelation." The other kind is the revelation of God in general, which is assumed but is irrelevant to the single vs. multiple Incarnations debate.

In sum, if Incarnation is about salvation (ontological argument), we could think of the Incarnation of Jesus of Nazareth as the one Incarnation for the cosmos. On the other hand, if we understand the purpose of the Incarnation as a relationship (the revelational argument) then it is possible to conceive of multiple Incarnations for the universe. Now, I summarize and discuss the possibilities of Incarnation and salvation for ETI and offer my conclusions.

Is Jesus of Nazareth the Only Savior of the Cosmos?

If we answer affirmatively, this means that the Incarnation of the Logos in Jesus of Nazareth (in first century Judea on Earth) is the only Incarnation of the Logos in all of space and time. Therefore, ETI do not have an Incarnation. This answer defends the supremacy of Jesus Christ as savior of the universe. Some theologians argue that this is the case. For them, the main issue is that Christology may be at risk of losing the uniqueness of Jesus of Nazareth as savior of the world if we consider the possibility of multiple Incarnations. These theologians oppose the idea of the existence of other worlds because they understand the sacrificial death of Jesus and his resurrection to be a unique and singular event, which was necessary only once. Likewise, Paul Davies rejects the possibility of multiple Incarnations because such a position is dramatically opposed to Christ as savior of humanity and ruler of the universe, as held by traditional Christologies. Davies assumes that, if ETI exist, then Christian revelation is bound to fail because that would mean that God has to become incarnate repeatedly, accommodating to every species. He describes this as the "Planet-Hopping Christ." Davies says that Christian theology is extremely vulnerable and threatened by the possibility of ETI. Davies writes: "Theologians and ministers of religion take a relaxed view of the possibility of extraterrestrials. They do not regard the prospect of contact as threatening to their belief systems. However, they are being dishonest. All the major world religions are strongly geocentric, indeed homocentric. Christianity is particularly vulnerable because of the unique position of Jesus Christ as God incarnate. Christians believe that Christ died specifically to save humankind. He did not die to save little green men" (Davies, 2000, p. 51). For Davies, the existence of ETI will make Christian theology crumble because we would have to be in a position of defending the primacy of Jesus of Nazareth in the universe, making our religion even more homocentric, or admit the idea of multiple Incarnations.

A problem with Davies' argument is that he assumes that the death of Christ is needed for salvation, and that ETI would also need to be saved by the incarnate God who goes to their planet with the purpose of dying in order to save them from their sin. Davies fails to realize that if we consider the purpose of the Incarnation to be salvation, the Incarnation in Jesus of Nazareth is enough because, in this case, he did die to save all creatures as St. Paul tells us: "He indeed died for all, so that those who live might no longer live for themselves but for him who for their sake died and was raised" (2 Corinthians 5:15). Contrary to Davies' argument, the acknowledgment of Jesus of Nazareth as savior of all creatures, including ETI, does not have to make Christianity homocentric, but rather more universal.

Contrary to Davies, we can think of the purpose of the Incarnation as establishing relationships, not only about salvation by the death of Christ. This is the point of Russell's revelatory argument: Incarnation is about proclaiming the Good News, not about dying on the cross.

Thomas O'Meara also criticizes theologians who oppose the idea of extraterrestrial beings. About them O'Meara writes: "[they] emphasize an exalted distance for God, the sinfulness of any free being, and a miraculous uniqueness of Jesus" (O'Meara, 2012, p. 13). We have seen how our traditional Christian theology places an emphasis on Christ being the king of the universe, Lord of the stars and of all creatures. This is mostly a poetic rhetoric. Christ's special role on Earth needs not be compromised by the possibility of ETI. Even if ETI does exist, we still can believe that the Word of God becomes flesh in Jesus of Nazareth. Just because the emergence of life in the universe is possible, our convictions about Jesus as Incarnation of the Logos should not be altered.

Wolfhart Pannenberg also supports these ideas; he also thinks that the discovery of ETI will not have shattering implications for our Christology. Contrary to Davies, Pannenberg writes:

It is hard to see [...] why the discovery of non-terrestrial intelligent beings should be shattering to Christian teaching. If there were such discoveries, they would of course, pose the task of defining theologically the relation of such beings to the Logos incarnate in Jesus of Nazareth, and therefore to us. But the as yet problematic and vague possibility of their existence in no way affects the credibility of the Christian teaching that in Jesus of Nazareth the Logos who works throughout the universe became a man and thus gave to humanity and its history a key function in giving to all creation its unity and destiny.

(Pannenberg, 1994, p. 76)

For Pannenberg, the Incarnation of the Word in Jesus Christ has universal implications. The universe was created by the Trinity, therefore, the presence of the Logos incarnate once in the universe, extends through space and time.[5] Incarnation, thus, means more than the Word in human form; it means that the universe embodies the Word, although the human Incarnation is particularly revelatory for the human species.

One possible approach is to think of the existence of ETI as a possibility and still defend the one Incarnation perspective. The main proponent of this possibility is Ted Peters who says: "God acts one way in the historical world of human beings and in a parallel way for the extraterrestrial worlds of non-humans" (Peters, 2017a, 12:8). For Peters, God does not have to be incarnate in order to communicate because God can use any other means that God considers appropriate. God has manifested God's self in different ways in humanity, such as "angelic visitations, burning bushes, dreams, the written Word, prayer and prophets" (Peters, 2017a, 12:9). God can communicate with ETI by any of these ways; yet despite the variety of forms in which God manifests and communicates, Peters tries to defend the supremacy of Jesus as the only Word of God incarnate in the cosmos. It seems that while the Incarnation was necessary for humans, other planets may have their own way of interacting with the divine. However, this argument is certainly not enough. God relates to the creatures in the deepest way God can. God would not limit God's revelation to dreams, visions and angelic visitations if the Logos can become flesh. These other revelations are general revelation; the issue here is the special revelation in the Incarnation. We may not be the most intelligent species in the universe and we have been witnesses of the Incarnation of the Logos. God can certainly relate at the same level to ETI species.

Peters concludes by arguing that the idea of one Incarnation and the existence of ETI cannot be in conflict: "What is misleading here is the assumption that the Christian religion is fragile, that it is so fixed upon its orientation to human beings centered on earth that an experience with extraterrestrial beings would shatter it. An alleged centrism renders Christianity vulnerable" (Peters, 2003, pp. 121f.). I argue that Christianity is not vulnerable to the possibility of the existence of ETI. Our religion, as stated above, can be universal and has to be open to new scientific possibilities, in this case, the possibility of extraterrestrial life. Peters advocates: "To the contrary, I find that when the issue of beings on other worlds has been raised it has been greeted positively . . . I advocate exotheology – that is, speculation on the theological significance of extraterrestrial life" (Peters, 2003). For Peters, theology can certainly grow by our search for ETI in a very positive way. Theology, and our understanding of how God relates to the universe can certainly be informed by science. Our faith can only grow when challenged.

Ted Peters further argues that the multiple Incarnation argument makes no sense because he understands the purpose of the Incarnation to be salvation, which is the ontological argument; he writes: "If the soteriological work of Jesus Christ in terrestrial history is efficacious cosmically, then this must include all creatures within God's creation, terrestrial and extraterrestrial alike" (Peters, 2017a, 12:9). This presumes that the salvation attained by Jesus Christ as the Logos incarnate on Earth, extends to all creatures. After all, the ontological argument does not need to include the revelational argument as Russell points out. There can be salvation without the revelation of it.

In sum, if we believe that the purpose of the Incarnation is salvation from sin, then we can conclude that the Incarnation of the Word in Jesus of Nazareth may be enough for the entire cosmos because the redemption brought about by Jesus of Nazareth can be understood as universal redemption. If ETI need salvation, then we would have to assume that what Jesus did on Earth is valid for all places and times because salvation does not need revelation. This understanding, to make sense, must presume no original sin on other planets, which is difficult to hold and impossible to verify. However, even if we defend the "one Incarnation" argument, our theology still can be re-envisioned by the possibilities of the existence of ETI. Although salvation does not need revelation, we must also consider that ETI on other worlds need an incarnate Logos to whom it can relate.

Is It Possible to Conceive of Multiple Incarnations?

If there is sin on all worlds, we need to explore the possibility of multiple Incarnations if the principal purpose of the Incarnation is to save creatures from the sin that separates them from God. Christ would then have to be fully incarnate in ETI, in order for ETI to also attain salvation. This is, certainly, assuming that ETI needs to be saved.

Further, if we understand the purpose of the Incarnation to be revelation, we can also think of multiple Incarnations because God would certainly reveal God's self to ETI.

About the possibility of multiple Incarnations, Aquinas writes: "The right way to manifest the unseen things of God is through things that are seen, and this is the purpose of the whole world" (Aquinas, III, I, sed contra). For Aquinas, if the right manifestation were to be seen, it would be fitting that the Word may become incarnate in other species, so that they may also witness God's love in that particular and unique way. Aquinas continues: "It is appropriate for the highest good to communicate itself to the creature in the highest possible way ... Clearly, it was right for God to become incarnate" (Aquinas, III, I, 1). Aquinas, in this argument, talks about why it is fitting for God to become incarnate in a human being, the highest possible way on Earth, but this does not deny the possibility of other high creatures in the universe, in which God could also become incarnate. Multiple Incarnations do not necessarily compete with each other.

For Aquinas, God emphasizes the higher forms of life, the life that can relate to God, such as angels and humans. For him, it is likely that the universe contains several of these forms of life. For Aquinas, the most important thing is a shared life with God. Aquinas writes: "Is it not suitable that God provides more for creatures being led by divine love to a natural good than for those creatures to whom that love offers a supernatural good" (Aquinas, I–II, 110, 2). Each free creature, therefore, receives grace from God to establish a relationship; this relationship is the source of hope for the resurrection. Through the Incarnation, God enters into the history God has created. Aquinas concludes, "the Incarnation was suitable to God because of the infinitely high level of his goodness intent on human salvation" (Aquinas, III, 1, 1, 2). For Aquinas, it is the divine love that draws the action of God in the universe. Aquinas continues: "The power of a divine person is infinite and cannot be limited to anything created" (Aquinas, III, 7, 3). This position argues for multiple Incarnations, but that they need not be for the purpose of sacrificial salvation.

Teilhard de Chardin also entertains the possibility of multiple Incarnations, as he writes: "It is infinitely probable that the conscious layer of the cosmos is not confined to a single point (our mankind) but continues beyond the earth into other stars and other times. In all probability mankind is neither 'unica' nor 'singularis', but is 'one among a thousand'" (Chardin, 1969, p. 44). Chardin acknowledges theologically the possibility of having other consciousness in the universe, other sentient and intelligent beings in which the conscious layer of the universe has also been bestowed. Further, Chardin ponders the question: "How, then, is it that, against all probability, this particular mankind was chosen as the center of the Redemption? And how, from that starting point, can Redemption be extended from star to star?" (Chardin, 1969, p. 44). Chardin, although he acknowledges that the question remains unanswered, says that:

The idea of an earth chosen arbitrarily from countless others as the focus of Redemption is one that I cannot accept; and on the other hand the hypothesis of a special revelation, in some millions of centuries to come, teaching the inhabitants of the system of Andromeda that the Word was incarnate on earth, is just ridiculous. All that I can entertain is the possibility of a multi-aspect Redemption which would be realized, as one and the same Redemption, on all the stars – rather as the sacrifice of the mass is multiplied, still the same sacrifice, at all times and in all places.
(Chardin, 1969, p. 44)

Here we see that Chardin seems to entertain the idea of the universality of Christ, and further, he poses the question of the possibility of multiple Incarnations as part of a single revelation–redemption: "Yet all the worlds do not coincide in time! There were worlds before our own, and there will be other worlds after it . . . unless we introduced relativity into time we should have to admit, surely, that Christ has still to be incarnate in some as yet unformed star? And what, then, becomes of 'Christus iam non moritur?' ['Christ being raised from the dead will never die again' (Romans 6:9)]. And what becomes, too, of the unique role of the Virgin Mary?" (Chardin, 1969, p. 44). Chardin wrestles with the same issue I address in this article, to which he says: "There are times when one almost despairs of being able to disentangle Catholic dogmas from the geocentricism in the framework of which they were born. And yet one thing in the Catholic creed is more certain than anything: that there is Christ 'in quo Omnia constant' ['In whom all things hold together' (Colossians 1:17)]. All secondary beliefs will have to give way, if necessary, to this fundamental article. Christ is all or nothing" (Chardin, 1969, p. 44). Chardin acknowledges that the search for ETI already helps us and impels us to rephrase the way we think about theology, in this case, the relationship between God and the universe.

We could think multiple Incarnations are possible, especially if we understand the purpose of the Incarnation as relational, as stated above and restated by Panikkar: "Christ did not come to 'teach' doctrines as much as he did to communicate life (John 10:10) and definitely to communicate himself, his own life, the life of the Father" (Panikkar, 1970, p. 116). If Christ comes to communicate life, he can do it anywhere and anytime there is life capable of communication, what we have defined as ETI.

O'Meara addresses the possibility of multiple relationships between God and the cosmos as he writes: "A variety of civilizations with billions of persons in the universe suggests a variety in number and in kind of intense relationships with the Trinity" (O'Meara, 2012, p. 73). The universe, as stated above, is really vast. We can only conclude that the love of God is just as vast, a love that seeks intimacy and relationship with all creatures. One Incarnation in one planet may be too limiting for God to show God's love for all. This is a challenging question.

We have been able to constantly bring to our imagination Jesus of Nazareth as the Incarnation for the universe. Through our devotions and prayers, we think of God, the Logos and the spirit as constantly relating to the universe, and we picture that as Jesus of Nazareth ruling over the cosmos. About this O'Meara writes: "The liturgical and devotional view of Jesus and Mary as presiding over the universe or over angels is not based on their human nature but upon their contact with the Word" (O'Meara, 2012, p. 50n12). However, we should also acknowledge the possibility of the ability of other beings to relate to God. This could prompt us to think that other creatures may have hypostatic union with God. O'Meara continues: "If other intelligent creatures have hypostatic or particularly intimate contacts with God, then a Jewish messiah and his mother would not by virtue of the Incarnation on our planet be unique or supreme in the cosmos" (O'Meara, 2012, p. 50n12). We would have a hard time in believing in the Logos if he did not look human. Other intelligent creatures would most likely have the same reaction.

If there are other intelligent beings in the cosmos, God's love, manifested through the Incarnation, would also be present to them. About this, O'Meara writes:

"Incarnation in a human being speaks to our race. While the possibility of extraterrestrials in the galaxies leads to possible Incarnations and alternate salvation histories, Incarnations would correspond to the forms of intelligent creatures with their own religious quests" (O'Meara, 2012, p. 48). Jesus of Nazareth was a historical person, who was fully human and would not take his humanity to other planets. Each Incarnation would have its own culture, race, history and identity. The possibility of multiple Incarnations does not, in any way, deny the uniqueness of Jesus Christ, who is the Incarnation of the Word for humanity.

God wants to relate to all, to manifest God's love for all. Is an Incarnation in all the species necessary to do that? It is also possible that God can manifest to ETI some other way, not necessarily an Incarnation. O'Meara writes:

If, however, there are other intelligent creatures but no Incarnations among them, then the union of the Logos and a terrestrial human would be a strong affirmation of the dignity of corporeal, intelligent life wherever it is found. Each Incarnation would spotlight and enhance one planet and, through that, a galaxy. Complementarity would invite civilizations with one or many Incarnations to teach each other. *(O'Meara, 2012, p. 50)*

The Incarnation of the Logos may be, according to this, unique to humans, but not limiting God's love for the universe, which can be manifested by other means. In fact, O'Meara reminds us that rather than limiting, multiple Incarnations will enrich our understanding of God in a "complementary" way.

Russell also argues to support the revelational argument stated above by claiming that an Incarnation in ETI is important given the immense distance to travel between stars and planets: "How would we communicate the terrestrial revelation of Jesus to the countless species of ETI strewn across the immeasurable distances of space. Needless to say, without such a communication they could not come to have faith and participate in the life of faith?" (Russell, 2017, 14:5). Also, the Incarnation would have to be physically, historically and culturally embodied in the species in which the Incarnation takes place in order to teach them how to love the divine on their own terms (Russell, 2017, 14:5–6). Russell is siding with the "revelational argument" which means that God reveals God's self to the many species in the cosmos. Russell specifically refers to the Incarnation as "special revelation," and not just about God revealing God's self in other key events, like burning bushes, angelic visitations, dreams, etc. According to Russell, "special revelation" means that God must reveal God's self through multiple Incarnations, and not just through multiple disclosures or theophanies. Russell proposes that many specific-species Incarnations are needed to communicate the Incarnation to all ETI such that they can each participate by faith.

Russell argues further that the "best alternative is multiple cosmic Incarnations" (Russell, 2017, 14:8). By this Russell assumes that God will become "incarnate in every species in this universe which achieves the level of rationality and moral capacity found in Homo sapiens" (Russell, 2017, 14:8). If we believe the Incarnation is about a manifestation of the love of God in the universe, then we can more easily imagine the Word made Flesh in the cosmos, not only in Jesus.

If the purpose of the Incarnation is relationship, then there is no need of sin or of redemption to have an Incarnation. In this case, we can conceive the possibility that God would relate to ETI as deep as God relates to us, in which case, the Word would also become incarnate in ETI. But we can also consider the possibility of multiple Incarnations in ETI if the purpose of the Incarnation is salvation; in this case, we have to assume that ETI have also sinned and need to be saved. If we ever find that Christian revelation has been shared on other worlds or with other creatures, this will change the way we understand ourselves and the way we understand God's relationship to the universe. It will change it in a good way because it will expand our knowledge and understanding, rather than reducing it.

According to Rahner, to think that other intelligent living beings in the cosmos have not received grace "does not do justice to the real and total relationship of God-spirit-grace" (quoted in O'Meara, 2007, p. 125). If ETI exist, their wisdom, freedom, and knowledge also make them the image and likeness of God. There may be other Incarnations, but their existence does not deny the importance of Jesus of Nazareth on Earth, his revelation and his message of love and of hope. O'Meara's words help draw all this together: "The universe is not a celestial house built by God for human beings and their religions. The feeling of vertigo in the cosmos, however, can further religious maturity: an enormous universe leads to an understanding of a greater God, a grasp of what 'infinity' might mean" (O'Meara, 2012, p. 92). God is about relationships, and the Incarnation is also about God's relationship to the universe. The more intelligent beings there are in the cosmos, the more we can find God in them and in us. Our theology, in the words of O'Meara, needs to strive for expansion, not for reduction (O'Meara, 2012, p. 16).

Besides the question of one Incarnation or many, we can ask whether there is a more nuanced way to think about the Incarnation. In the following section, I offer another possibility to think about the Incarnation while including the possibilities of ETI.

Deep Incarnation and the Cosmic Christ

As stated above, God becomes incarnate because God loves the universe, because God desires to relate to the universe. Panikkar argues that Christ cannot be contained only to Jesus, not even only to humanity, as further he explains: "Incarnation as an historical event cannot be considered a universal human fact" (Panikkar, 1970, pp. xvi–xvii). For Panikkar, Christ is the door that opens the relationships between God and the universe; Christ opens us to the relationship of the Trinitarian mystery.

Rather than choosing between one Incarnation or many, Niels Gregersen coined the phrase "Deep Incarnation" by which he means that the presence of the Logos reaches not only humanity, but every tissue of biological existence with its growth and decay.[6] For Gregersen, Deep Incarnation means that the Word became flesh (sarx), not just human (John 1:14) (Gregersen, 2001). Sean McDonagh expands this by saying that Jesus carried within himself "the signature of the supernovas and the geology and life history of the Earth" (McDonagh, 1986, pp. 118–19). This happens because the Logos, when it becomes flesh, becomes atoms, elements, and matter, originated in the stars. The Word, by becoming flesh on Earth, is part of the matter and history of this planet, too. The Incarnation, therefore, is a cosmic event at the most basic and deepest level. God becomes matter, atoms; the Word becomes part of creation. God is free to relate with the creation at deep levels, such as physical and relational. God does so as a self-expression of love, either in the single Incarnation of Jesus of Nazareth or in the possibility of multiple Incarnations throughout the universe.

Deep Incarnation states that the Word becomes flesh, in other words, matter. For Chardin, matter is a necessary feature of our relationship with the divine, as he writes:

Matter, above all, is not just the weight that drags us down, the mire that sucks us in, the bramble that bars our way. In itself, and before we find ourselves where we are, and before we choose, it is simply the slope on which we can go up just as well as go down, the milieu that can uphold or give way, the wind that can overthrow or lift up. Of its nature, and as a result of original sin, it is true that it represents a perpetual impulse towards failure. But by nature too, and as a result of the Incarnation, it contains the spur or the allurement to be our accomplice towards greater-being, and this counter-balances and even dominates the "fomes peccati" ("the tinder for sin"). *(Chardin, 2004, p. 67)*

The Incarnation, for Chardin, divinizes all matter, everything that is created, and not only humanity.

Deep Incarnation starts from the premise that all the creatures are connected. The elements we are made of are essentially the same, and we all are subject to evolution and the laws of physics. About this, theologian and biblical scholar Richard Bauckham writes:

In order to be related to all other species, humans do not need somehow and uniquely to sum up all other created natures in their own nature. Humans are a part of the interdependent web of life (and even of inanimate nature), and human history, though sometimes misunderstood as a process of growing independence from the rest of the natural world, has actually increased the scope and complexity of human interconnectedness with the whole of the natural world on Earth. Such interconnectedness also extends backward through the history of nature, as the human use of fossil fuels illustrates. *(Bauckham, 2015, p. 45)*

As was discussed above, life is made of matter, DNA, relationships, etc. Humans and all living things have those elements in common. The Word, by becoming human, becomes life, matter, and atoms. If we believe that "what is not assumed is not saved" (Gregory of Nazianzus, epistle 101), then all matter is saved because the Word has become matter in Jesus.

A Deep Incarnation means that the Logos experienced being matter, subject to pain, decay, and death. Deep Incarnation, therefore, entails a Deep Crucifixion and a Deep Resurrection.

Johnson poses the question: "Is the suffering solidarity of the crucified God limited to human beings? Or does it extend to the whole community of life of which human beings are a part? The logic of Deep Incarnation gives a strong warrant for extending divine solidarity from the cross into the groan of suffering and the silence of death of all creation" (Johnson, 2014, p. 205). So, if we follow the logic of the Deep Incarnation, this implies that Christ's suffering also extends to all of creation. This makes sense because pain, suffering and death are universal. Suffering is part of evolution (Russell, 2008, ch. 7). All creatures suffer and die. Jesus took part in this universal suffering. Eschatology says that suffering and death are not the end; rather, there is the promise of eternal life. About this, O'Meara says: "The gift of redemption and eternal life and the process of Incarnation are also extraterrestrial" (O'Meara, 2012, p. 58). Therefore, the Incarnation can be extended to all creatures of the cosmos, no matter if it is about revelation or about salvation.

As I discussed above, the Incarnation is about relationships and revelation. The Logos is capable of relating to all creatures, not only to humans. Richard Bauckham writes: "Just as Jesus' relatedness to other humans became, through his resurrection and exaltation, a universal relatedness to all other humans, so his wider relatedness to the rest of creation became a universal relatedness to the whole of creation" (Bauckham, 2015, p. 50). This can be possible because the risen Christ participates in the capacity of God to be fully present in the entire cosmos, but, through the Incarnation, Christ relates to the cosmos as a human being, while keeping his capacity to relate to other creatures. John Behr writes: "This coming of the Word to creation is not simply a matter of his presence within creation. It is, rather, a transformation of the creation's natural 'flux and dissolution,' its tendency to return to the nonbeing from which it was created, into an order and arrangements that bears the imprint of the Word and demonstrates the providence of God" (Behr, 2015, p. 91).

By virtue of our belonging to the universe, to matter, our mission is to approach closer and closer to our deepest relationship, to the truth of who we are, what the Incarnation of Christ makes us: capable of relation to God as the Word relates to God. This is the divinization of humanity, of the creatures of the earth and of the cosmos. About this, Richard Bauckham continues: "God is not incarnate in all other reality, but he is incarnate for all other reality... The Incarnation makes the incarnate One integrally part, not only of the human race, but also of the whole of this material reality – not only animals (with whom Jesus shares genetic continuity) but also plants and inanimate nature" (Bauckham, 2015, pp. 32–5). The Incarnation is, therefore, not separate from the rest of the cosmos. It is embedded into it giving it love, hope, dignity. Jesus is universal; his teachings of love, compassion, mercy and forgiveness apply to all creatures. But how does ETI engage in Jesus' teachings? Again, if the purpose of the Incarnation is relationship and revelation, then God reveals God's self to the entire universe through the Incarnation. The idea of Deep Incarnation does not at all discard the possibly of multiple Incarnations.

Russell also endorses the idea of Deep Incarnation as a possible bridge between the One-Incarnation argument from Peters and the multiple Incarnation proposal. Russell writes: "My claim is that the Word of God that is made the flesh common to every terrestrial creature is also the flesh that is common to all extraterrestrial life in the following sense: what makes it common is the universality of the physical processes that characterize flesh" (Russell, 2017, 14:9). Russell attempts to use Gregersen's notion of Deep Incarnation to reconcile the differences between the multiple Incarnation debate and the One-Incarnation by arguing that the meaning of flesh is not just earthly flesh, but that it pertains to the universe. Russell generalizes Gregersen's position of Deep Incarnation to all flesh in the universe. Russell concludes: "Thankfully, through Gregersen's vision, we can hold together ontological and revelational perspectives on Christology as two sides of the same Christological coin. In this way we arrive at a mutually satisfactory "win/win scenario, one which we might call 'the many-in-one Incarnation of the Logos in the universal flesh of life through the grace of its loving creator'" (Russell, 2017, 14:10). Bob Russell, by extending the flesh that the Logos becomes to all flesh, steers away from the problems of the multiple Incarnation debate while also acknowledging the supremacy of the Incarnation in Jesus of Nazareth. I agree with Russell and I defend the Deep Incarnation perspective; I also argue that the Deep Incarnation does not automatically dismiss the possibility of other Incarnations of the Logos in ETI species.

Deep Incarnation has its problems, too. The main issue with Deep Incarnation is to defend the uniqueness of Jesus Christ, while accepting his presence in the universe. As we do this, we tend to fall into anthropocentrism, which, in the words of Bauckham, is "one that supposes, not only that humans have a distinctive role within the creation... but also that it is through humans that the rest of creation must fulfill its purpose in relation to God." (Bauckham, 2015, p. 49). Deep Incarnation goes to the heart of Thomas Aquinas on salvation. Aquinas would say that all of non-human nature is saved through human salvation. However, another possibility is that other natures are saved through their own inherent relationship to the Logos. Deep Incarnation opens up the possibility of all creatures being saved through the one Incarnation in the Logos who becomes flesh, without discarding the possibility of multiple revelations of the Logos to other creatures, including Incarnation in ETI.

Even when we do not think that creation exists for the sake of humanity, which is a theme that has been taught in some theological traditions, we can get accustomed to thinking that humans are the only species capable of relating to God, placing ourselves in a privileged position within creation. Evolution and our recent understandings of life have helped to discredit these claims. So, besides Christological implications, scientific discovery helps us better to understand the role and position of humanity in the cosmos.

We creatures are made of the same elements; we have the presence of all the elements of life. We humans share organic and genetic material with all the living beings of the universe. When the Word becomes incarnate, it assumes all living matter, not just humanity. Gregersen reveals that Deep Incarnation, understood as the Incarnation of God in Christ that "reaches into the very tissue of biological existence and system of nature" (Gregersen, 2001, p. 205). By the word becoming flesh in Jesus, it not only becomes human flesh, it becomes organic matter, all systems of nature are encompassed in that because humans are made of the same elements of all other creatures. This idea encompasses some dilemmas, such as the importance of the historical Jesus, and the issue of the crucifixion and resurrection. According to Johnson, if there is a Deep Incarnation, there is also a Deep Crucifixion and a Deep Resurrection. Johnson writes: "Is this solidarity limited to human beings alone? All creatures come to an end; those with nervous systems know pain... Deep Incarnation in the event of the cross extends the respect of the Word/Wisdom made flesh all the way down into the groan of suffering and the silence of death"(Johnson, 2015, p. 146). If all creatures suffer, then the resurrection also gives hope of new life to all. A relationship with God, like all complete relationships, has to share in the hopes and in the suffering. Deep Incarnation gives this possibility to all creatures.

I have argued that, rather than debating between one Incarnation or many, the term "Deep Incarnation" can helps us understand the idea that the Logos relates constantly to the universe, and that through the Incarnation in Jesus of Nazareth, the Logos becomes not only human, but matter, atoms, a creature. This is the understanding of flesh (Sarx) that Gregersen talks about, which can be a better interpretation of the Incarnation if understood in terms of relationship and revelation. This understanding, of course, does not discard the possibility of multiple Incarnations in ETI, which the Logos can do as a way to continue the relationship with the universe.

Conclusions

We have talked about the possibility of life in the universe as well as the notions of creation, incarnation and salvation for humans and the cosmos given the probability of encountering ETI. All these understandings will have implications for our future encounter with ETI and our spiritual relationship with it; we need to address these implications now that our search for ETI has begun.

The universe seems to be open to life. The works of nature and evolution can help life to flourish; we are proof that there is intelligent life in the universe. Science and religion must be open to the possibility of life in other places. Asking questions and not assuming we already have the answers is how we have been able to explore our world and the universe. We need to keep asking the question about the possibility that life – either microbial, complex or intelligent – exists either in this galaxy or in others.

If confirmed, the existence of ETI can illuminate our understanding of God. God, according to the Christian tradition, creates the universe out of love; God loves creation and intends creation to return that love to God. This would mean that the incarnation would have taken place without the need for redemption. This action could conceivably happen to ETI.

As we have seen, our search for ETI prompts us to question our notions of life, incarnation and redemption.

We have not found ETI, but we are looking. The challenge we now face is to unite ourselves as a species, to include and accept each other, and to look at this search with hope, humility and openness. If we are willing to acknowledge the existence of ETI and search for it, we need to believe that God would not create a whole universe just for one sentient intelligent species that sinned and needed redemption. Redemption is not only healing from sin, but the gift of eternal life with God after death. Further, relationships with God go beyond a need for redemption. God creates and redeems because God loves. The decisions that Jesus made on Earth were impelled by his love for God and for his fellow humans. Let us assume that God loves ETI as much as God loves us, and wants to share that love with them. ETI may not have a notion of sin, but we all need a relationship with God. If we acknowledge this, then it will not be difficult to relate to ETI at a spiritual level. We may one day join hands with ETI and pray with the words that God taught us.

We will finally be able to relate spiritually with ETI if we can acknowledge that they are a species who are loved by God, a species in which incarnation and redemption are present. Hopefully, we will be willing to call them brothers and sisters in the Lord.

Notes

1 See http://www.esa.int/Our_Activities/Space_Science/Planck/Celebrating_the_legacy_of_ESA_s_Planck_mission (accessed April 14, 2016).
2 Deacon relies on Gil, Silva, Pereto et al. (2004, p. 518). See also Koonin (2000) quoted in Deacon (2013, p. 561).
3 See http://mars.nasa.gov/programmissions/overview/ (accessed April 12, 2016).
4 Quoted in Hewlett (2017).
5 We have other contemporary theologians who also defend the one Incarnation argument, such as Edgar Burns who writes: "the significance of Jesus Christ extends beyond our global limits. He is the foundation stone and apex of the universe and not merely the Savior of Adam's progeny" (see Burns, 1960, p. 286).
6 We could say that some early theologians dealt with this issue too, for example, Origen (184/185 – 253/254 AD) about whom Lyons writes: "Origen holds that angels and men have essentially the same nature and are consequently transformable into one another, his meaning is that the body of the soul united to the Logos acquires an angelic condition among the angels, just as among men it acquires a human condition" (Lyons, 1982, p. 139). Also, Dumitru Staniloae (1903–1993 AD) writes about the human condition as part of the cosmos: "The Human being is a hypostasis of the whole cosmic nature" (Staniloae, 1998, p. 294).

References

Aquinas, Thomas. *Summa Theologiae*.
Bauckham, R. (2015). The Incarnation and the Cosmic Christ. In N. H. Gregersen, ed., *Incarnation: On the Scope and Depth of Christology*, Minneapolis, MN: Fortress Press, pp. 25–58.
Behr, J. (2015). Saint Athanasius on 'Incarnation'. In N. H. Gregersen, ed., *Incarnation: On the Scope and Depth of Christology*, Minneapolis, MN: Fortress Press, pp. 79–98.
Blochl, E., Rachel, R., Burggraf, S. et al. (1997). *Pyrolobus fumarii*, gen. and sp. nov. represents a novel group of Achaea, extending the upper temperature limit for life to 113 °C. *Extremophiles*, **1**(1), 14–21.
Burns, J. (1960). Cosmolatry. *The Catholic World*, **191**(1), 283–287.
Cassan, A. et al. (2012). One or more bound planets per Milky Way star from microlensing observations. *Nature*, **481**, 167–9.
Cayrel, R., Hill, V., Beers, T. C. et al. (2001). Measurement of stellar age from uranium decay. *Nature*, **409**, 691–2.
Chardin, P. T. d. (1969). *Christianity and Evolution*, New York, NY: Helen and Kurt Wolff.
Chardin, P. T. d. (2004). *The Divine Milieu*, Portland, OR: Sussex Academic Press.
Clarke, A. (2003). Evolution and low temperatures. In L. Rothschild and A. Lister, eds., *Evolution on Planet Earth: The Impact of the Physical Environment*, London: Academic Press, pp. 187–208.
Davies, P. (2000). Transformations in spirituality and religion. In A. Tough, ed., *When SETI Succeeds: The Impact of High-Information Contact*,. Bellevue, WA: The Foundation For the Future, pp. 51–53.
Deacon, T. (1998). *The Symbolic Species*, New York, NY: W.W. Norton & Company.
Deacon, T. (2013). *Incomplete Nature: How Mind Emerged from Matter*, Kindle edn, New York, NY: W.W. Norton & Company.

Donald, M. (2001). *A Mind So Rare*, New York, NY: W.W. Norton & Company.
Drake, F. (2003). The Drake Equation revisited. *Astrobiology Magazine*, Web: http://www.astrobio.net/index.php?option=com_retrospection&task=detail&id=610&fid=14&pid=5 Accessed: January 2, 2016.
Gil, R., Silva, F. J., Peretó, J. & Moya, A. (2004). Determination of the core of a minimal bacterial gene set. *Microbiology and Molecular Biology Reviews*, **68**(3), 518–37.
Gould, S. J. (1989). *Wonderful Life: The Burgess Shale and the Nature of History*, New York, NY: W.W. Norton & Company.
Gregersen, N. (2001). The Cross of Christ in an Evolutionary World. *Dialog: A Journal of Theology*, **40**(3), 192–207.
Gregory of Nazianzus. *Epistle 101, To Cledonius the Priest Against Apollinarius.*
Hewlett, M. (2017). Yes, we'll meet them: A scientific argument for ETI. In T. Peters, M. Hewlett, J. Moritz and R. J. Russell, eds., *Astrotheology: Theology Meets Extraterrestrial Life*.
Johnson, E. (2014). *Ask the Beasts: Darwin and the God of Love*, Kindle edn., London: Bloomsbury Continuum.
Johnson, E. (2015). Jesus and the Cosmos: Soundings in Deep Christology. In N. H. Gregersen, ed., *Incarnation: On the Scope and Depth of Christology*, Minneapolis, MN: Fortress Press.
Kauffman, S. (1996). *At Home in the Universe: The Search for the Laws of Self-Organization and Complexity*, Oxford: Oxford University Press.
Koonin, E. V. (2000). How many genes can make a cell: The minimal-gene-set concept. *Annual Review of Genomics and Human Genetics*, **1**(1), 99–116.
Krauss, L. M. & Chaboyer, B. (2003). Age estimates of globular clusters in the Milky Way: Constraints on Cosmology. *Science* **299**(5603), 65–9.
Lewis, T. (2013). *Habitable Earth-Size Planets Common Across the Universe, Study Suggests*, Space.com Web: http://www.space.com/23456-habitable-earthlike-alien-planets-common-kepler.html Accessed November 30, 2014.
Lyons, J. A. (1982). *The Cosmic Christ in Origen and Teilhard de Chardin: A Comparative Study*, Oxford: Oxford University Press.
Madsen, O., Scally, M., Douady, C. J. *et al.* (2001) Parallel adaptive radiations in two major clades of placental mammals. *Nature*, **409**(6829), 610–14.
McDonagh, S. (1986). *To Care for the Earth*, Santa Fe, NM: Bear and Co.
Morris, S. C. (1998). *The Crucible of Creation: The Burgess Shale and the Nature of History*, New York, NY: W.W. Norton & Company.
Morris, S. C. (2004). *Life's Solution*, Cambridge: Cambridge University Press.
O'Meara, T. F. (2007). *God in the World: A Guide to Karl Rahner's Theology*, Collegeville, MN: Liturgical Press.
O'Meara, T. F. (2012). *Vast Universe: Extraterrestrials and Christian Revelation*, Collegeville, MN: Liturgical Press.
Panikkar, R. (1970). *Christophany the Fullness of Man*, Maryknoll, NY: Orbis Books.
Pannenberg, W. (1994). *Systematic Theology Vol. 2*, Grand Rapids, MI: Eerdmans.
Peters, T. (2009). Astrotheology and the ETI myth. *Theology and Science*, **7**(1), 3–29.
Peters, T. (2003). Exotheology: Speculations on extraterrestrial life. In T. Peters, ed., *Science, Theology, and Ethics*, Burlington, VT: Ashgate, pp. 121–28.
Peters, T. (2017a). One incarnation or many? In T. Peters, M. Hewlett, J. Moritz and R. J. Russell, eds., *Astrotheology: Theology Meets Extraterrestrial Life*.
Peters, T. (2017b). Where there's life there's intelligence. In A. Losch, ed., *What is Life? On Earth and Beyond*. Cambridge: Cambridge University Press, pp. 236–59.

Rasmussen, S., Chen, L., Deamer, D. *et al.* (2004). Transitions from nonliving to living matter. *Science* **303**(5660), 963–5.
Rothschild, L. J. (2004). Life in the Universe: an astrobiological perspective. In C. Impey and C. Petry, eds., *Science and Theology: Ruminations on the Cosmos*, Vatican City: Vatican Observatory Publications, pp. 129–148
Russell, R. J. (2008). *Cosmology: From Alpha to Omega*, Kindle edn., Minneapolis, MN: Fortress Press.
Russell, R. J. (2017). Many incarnations or one? In T. Peters, M. Hewlett, J. Moritz and R. J. Russell, eds., *Astrotheology: Theology Meets Extraterrestrial Life*.
Sagan, C. (1980). *Cosmic Connection: an Extraterrestrial Perspective*, Cambridge: Anchor.
Service, R. F. (2016). Synthetic microbe lives with fewer than 500 genes. *Science*, Web: http://www.sciencemag.org/news/2016/03/synthetic-microbe-lives-less-500-genes. Accessed: April 3, 2016.
Schrödinger, E. (1994). *What is life? Mind and Matter*, Cambridge: Cambridge University Press.
Sinsheimer, R. L. (1996). What is life? A closer look. *Engineering & Science*, **3**, 35–7.
Staniloae, D. (1998). *Orthodox Dogmatic Theology*, Brookline, MA: Holy Cross Orthodox Press.
Szathmáry, E. (2005). Life: In search of the simplest cell. *Nature*, **433**(7025), 469–70.

14

Talking Lions, Intelligent Aliens, and Knowing God – Some Epistemological Reflections on a Speculative Issue

TAEDE A. SMEDES

Introduction

What *really* drives the search for exoplanets and for extraterrestrial life? Why invest such enormous amounts of money in a research project that until quite recently relied on nothing more than speculation by the most creative minds on Earth? If one would like to discuss the role of faith in science, SETI, and other scientific endeavors that aim to find signs of life elsewhere in the universe are still plausible candidates for doing so. It seems to be much more than mere curiosity that drives this kind of research. Of course, curiosity plays a role, but isn't there more at stake here? Isn't there a *deeper* drive? And if so, what is it?

It is often said that the discovery of life on other planets will have a dramatic impact on our culture. No doubt, it would have an enormous impact in the natural sciences. Many biologists believe that life on Earth is a mere fluke, an accident, a freak of nature. And, as a consequence, many biologists express a certain skepticism towards claims about extraterrestrial life. The famous biologist Ernst Mayr reportedly refused to discuss even the possibility of extraterrestrial life in classes or public debates because of his belief that life on Earth was a mere accident (Gingerich, 2006, p. 34). The discovery that life exists elsewhere would have Earth-shattering consequences for such a biological outlook. And what if life is a "cosmic imperative," as the Belgian biochemist and Nobel Prize winner Christian De Duve (1995) used to call it, and is found throughout the universe? Obviously, the discovery that the universe is teeming with life would be a magnificent and glorious but also revolutionary scientific discovery.

But how would, for instance, philosophy and theology be affected by such findings? Would philosophers or theologians get all excited about the discovery of microbial life on Mars? Of course, if microbial life were to be found on Mars, that could mean that life on Earth originated in space and somehow, at a point in the distant past, was brought to Earth, perhaps by comets. But even if life originates in space, so what? It is my personal estimate that most philosophers and theologians would not get too excited by the discovery of microbial or bacterial life elsewhere in the universe.

I do believe, on the other hand, that the search for life elsewhere in the universe eventually is driven by the conceivable possibility, however remote, and perhaps even the *hope* that there is *intelligent life* elsewhere. By "intelligent life" I mean life that shares at least two characteristics with us humans: being intelligent (though it is notoriously difficult to describe exactly what is meant by intelligence; I leave it undefined for now) and being self-conscious (although some may argue that being intelligent already *assumes or implies* being self-conscious). The point is: if *life* has emerged elsewhere in the universe, that could mean that at least on some planets *intelligence* could have evolved as well. Many scientists argue that the evolution of intelligent life may be extremely rare, but still we know it is a possibility, since it happened at least once in cosmic history, here on Earth.

I believe then that there are ultimately two questions that drive the scientific quest for extraterrestrial life: where do we, humans, come from? And secondly, are humans unique in the cosmos, especially with regard to intelligence and consciousness? Finding microbial life is certainly interesting and may help to answer the first question, that of the origin of humanity. But what we really want to know, and that is the reason we pay scientists a lot of money to find out, is whether or not human beings are the only beings in the universe that live in advanced civilizations, that have developed science and technology with which they study the universe. Are there other beings like us that ponder their existence and think about the meaning of life? In other words, we want to know whether there are beings that we consider intelligent. We want to know whether there may be others out there, other intelligent beings with which we could potentially communicate, beings who, like ourselves, create meaning. Intelligent beings we can somehow relate to. Other beings like us.

But notice my formulation: "We want to know whether there are beings that *we* consider intelligent." In effect, what we want to know is whether there are beings out there that at least somewhat resemble ourselves. We always start from the situation that *we* are familiar with, and that means, with humans. The interest in extraterrestrial life thus is always from the outset framed in an anthropocentric manner, by which I mean that otherness is reduced to sameness.[1] The human frame of reference that we are so familiar with is already presupposed. Even SETI, the Search for Extraterrestrial Intelligence, presupposes an anthropocentric frame of reference, since it takes what we know about human evolution, the evolution of human societies and civilizations and the evolution of human technology as the framework for the search for technological signs of extraterrestrial intelligence.

On the one hand, this human framework in a sense limits scientific research, since it presupposes that alien technology more or less works in the same way as human technology, which is rather speculative. On the other hand, as scientists connected to SETI have repeatedly formulated it: one has to start somewhere, and not searching at all would result in not finding anything at all (unless they find us first). In a sense, the search of SETI can be compared to fishing with a net with meshes of a certain size. This means it is only useful to catch fish that have a specific size, but it would be unwise or premature to conclude that if you did not catch any fish, the pond apparently contains no fish at all. (It could be teeming with fish that are too small to be caught by the net.) In other words, just because our technology is unable to detect intelligent civilizations out there, that does not mean that they are not there. I believe this reason is a valid, if ad hoc, argument for conducting the kind of research that SETI is doing. It may be based on many presumptions that can be questioned, but one who never places a bet can be certain that he never wins.

Theology

But what if intelligent extraterrestrial aliens were to be found? One of the questions that is still often asked is: if extraterrestrial intelligence were found to exist, what would this imply for Christian theology, for example? Would their existence entail a threat to Christian theology?

Although the issue of extraterrestrial intelligence has been on the theological and philosophical agenda for centuries, the issue received more urgency since the first exoplanet was discovered in 1992. Atheists often claim that the discovery of alien life would be detrimental to Christian faith. The "Peters ETI Religious Crisis Survey," which was conducted in 2007, adopted this claim as its working hypothesis.[2] The survey showed rather convincingly that whereas non-believers thought that Christian theology would suffer a tremendous blow with the discovery of extraterrestrial life, Christian believers actually widely agreed to the contrary. They were convinced that Christian theology has the resources to cope with the presence of extraterrestrial life.

In recent years, quite a few books have been published that aim to explore in what way Christian doctrine would be impacted by the discovery of extraterrestrial intelligence (see e.g. Dick, 2000; O'Meara, 2012; Weintraub, 2014; Wilkinson, 2014). Most of these works agree that the discovery of intelligent life elsewhere in the universe would have a major impact on theology, in the sense that such a discovery would ask for a reinterpretation of at least some of its core doctrines (most notably that of original sin and Christology), but that it would probably not lead to a reconceptualization of Christian theology as a whole.[3]

But here another reduction of otherness to sameness emerges. By stating that Christian theology somehow allows for the presence of intelligent life elsewhere in the universe, or that the existence extraterrestrial intelligent beings can be taken up within the theological–hermeneutical frame of reference of Christian theology, it is presupposed that the otherness of extraterrestrial intelligence can be reduced to sameness: to that with which we are familiar, in this case Christian theology. So here also the anthropocentric attitude toward reality is presupposed.

But what is more, since one implicit claim of the Christian faith is that Christianity is objectively and universally true, this means that by taking up extraterrestrial intelligent beings within the framework of Christian theology, it is implicitly claimed that those beings are part of God's plans with the universe as envisioned in Christian theology – in effect, these beings are Christianized, whether they are aware of it or not and irrespective of whether they themselves would agree to such an assessment or not.

The issue that is on the table is the following: is the implicit reduction of otherness to sameness in theology warranted? In this chapter, I will argue it is not. In the next stage of this rather skeptical chapter, I will argue that if intelligent extraterrestrial beings exist, then with regard to the knowledge we now have, it is plausible to assume that mutual understanding will be close to impossible. This holds even if we are somehow able to recognize extraterrestrial beings as life-forms and as intelligent. Moreover, recent scientific findings suggest that human expressions of religion (and especially human views of God) are so closely bound to human biology, that even if we would be able to detect and understand intelligent extraterrestrial beings, it is questionable that intelligent extraterrestrial beings will have religion, or at least a religion that as such is recognizable to us. So let us turn to the details of the arguments for these, admittedly, rather large claims and their implications.

Ontology

In many science fiction books, movies, and TV series, aliens are portrayed as having humanoid form, meaning they somehow resemble humans. But even in cases where the aliens are not humanoid, such as in the brilliant but gory *Alien* films (starring Sigourney Weaver), the bloodthirsty alien still has a recognizable head, two hands or claws and two feet, and it is highly intelligent.[4] In our culture the *Star Trek* series has probably been the most influential. *Star Trek* takes place in a universe in which most aliens – Vulcans, Klingons, Bajorans, and many others – are highly humanoid. Some have even described the *Star Trek* aliens as "one small step" aliens, meaning:

just one change from middle-class Californians: Vulcans are middle-class Californians who talk about logic a lot; Bajorans are middle-class Californians who talk about spirituality a lot; Klingons are middle-class Californians who talk about honour and war a lot.
(Cohen & Stewart, 2004, p. 53)

Do we have reason to believe that intelligent aliens will look like us? Most scientists say no. The point is that "[t]he human form evolved *because it works on Earth*"

(Cohen & Stewart, 2004, p. 148). In different environments, evolution would make sure that life-forms that emerged there would adapt their form to the environment they inhabit. The conditions pertaining to Earth allowed the human form to emerge as the one that works for us, but it is implausible to assume that life-forms on distant planets would develop the same humanoid form. In fact, many biologists share the view of the late paleontologist Stephen Jay Gould, that the evolutionary history of life on Earth is riddled with contingencies, so that if one would rewind the tape of life, humans would probably not be part of the scenery any longer.

Admittedly, there are dissonant voices too. Paleontologist Simon Conway Morris defends a phenomenon he calls "evolutionary convergence" (Conway Morris, 2003). Conway Morris argues that if the tape of life would be rewound, at least something human-like would probably emerge again. Conway Morris's view is controversial, but he backs up his case that evolutionary convergence has happened many times in evolution on Earth with a substantial amount of empirical data, so that biologists are not able to dismiss it that easily. Yet most biologists are hesitant to assent to Conway Morris' claim that humans or human-like beings are the inevitable outcome of the evolutionary process.[5]

My point with the foregoing is that there are biological indications that the physical human form is supervenient on the conditions pertaining to Earth. If so, then the way alien life-forms look is probably supervenient on the conditions pertaining to their world. And if such life-forms are so vastly different from what we know, we may not even be able to recognize them as life-forms in the first place.[6]

Epistemology

A similar skeptical argument can be made for human epistemology. Most of us are aware that our minds are closely linked to the workings of our physical brains. And yet, we often tend to forget that because of that link, our epistemology is closely linked to our biological–evolutionary heritage. And thus, as evolutionary epistemology and evolutionary psychology have argued, our concepts of "understanding" and "meaning" implicitly presuppose a human or humanoid ontology. Obviously this is a claim I need to explain in more detail.

Suppose I am having a conversation with a friend. For that conversation to be meaningful, I have to be able to understand what she is saying and I have to be able to respond appropriately to her, so that in turn I enable her (or so I presume) to understand me. Now humans live in quite different cultures (even between the Dutch and Belgians there is cultural gap even though Dutch is spoken in both countries), but in most cases humans from different cultures (eventually) are able to understand each other. One fundamental condition for this understanding is that humans share the same ontology, that is: they are human beings. With all the variety that exists between them they share at least the same evolutionary–biological history that makes them *Homo sapiens*.

However, I cannot say that I *understand* a dog or a cat – except in a metaphorical way – because the cat and I do not share the same ontology. A cat's biological constituency is totally different from mine. There may be a certain degree of understanding, that may show itself as a way of meaningful behavior in relation to each other (I give the cat food whenever she behaves in a way that I interpret as "she feels hungry," and when she eats the food I take it that I was right). So there is a certain "behavioral alignment" or "synchronization of behavior" between me and my cat, but from a philosophical perspective it is futile to speak about understanding between us in any truly epistemological sense. The point is that the difference in epistemology is rooted in a difference in ontology, in this case a difference in biology and neurology.

So, not only our ontology, but also our epistemology is supervenient on the conditions pertaining to Earth, which means that the ways we perceive the world and conceptualize it, and the manners in which the human brain has developed and now functions, are closely bound to the environment we inhabit. As the philosopher Gerhard Vollmer argues, the environment we inhabit is "that section of the real world to which man is adapted in perception, experience, and action" which, Vollmer continues, can be called

> the mesocosm. It is a world of medium dimensions: medium distances and time periods, small velocities and forces, low complexity. Our intuition (...) is adapted to this world of medium dimensions. Here our intuition is useful, here our spontaneous judgments are reliable, here we feel at home. *(Vollmer, 2005, pp. 260–1)*

Note that although "mesocosm" sounds like an ontological term, depicting a segment of reality, Vollmer uses it as an epistemological term. Just as humans inhabit an ecological niche, so we also inhabit a "cognitive niche" which Vollmer calls the mesocosm. Every creature has its own cognitive niche; the mesocosm is the cognitive niche that is inhabited by humans. Vollmer thus acknowledges that it is an anthropocentric term (cf. Vollmer, 2003, pp. 77–85).

The consequence is that if we want to understand a concept, phenomenon, structure, theory, etc., that concept, phenomenon, structure, theory, or whatever, has to undergo a transformation such that it fits our cognitive niche, that is: the mesocosm. We have to be able to reduce the difference to sameness. The mesocosm lays down the rules and conditions that must be fulfilled if we want to claim that we *understand* something. In Kantian terms: the mesocosm is *transcendental*: it is the set of conditions that makes knowledge and understanding *possible*, but, equally important, it also denotes the *limitations* of our knowledge. Everything that we know and can claim to know, is part of the mesocosm, and everything that is outside our mesocosm (a) can either become knowable due to some transformation that creates a fit in our mesocosm, or (b) remain unknowable if such a transformation is not possible.

Now perhaps something like this ontological difference (the ecological niche we inhabit) that entails an epistemological difference (the cognitive niche we inhabit, or the mesocosm) was what Wittgenstein had in mind when he wrote the famous yet elusive words: "If a lion could speak, we could not understand him" (Wittgenstein, 1958, p. 223). I take it that Wittgenstein's remark was not about lions, but about understanding, and that the issue that he wanted to raise was whether the lifeworld of the lion – its ecological and cognitive niche – overlaps with ours to such an extent that a sufficient degree of shared concepts exists so that we could understand the lion (Sandis, 2012, pp. 146–147).[7]

The point of course is that there is a difference between understanding the speech of a lion and, for instance, the speech of a Chinese man. Whereas the Chinese man shares our ecological and cognitive niche, we trust that there are enough commonalities and shared concepts, so that there is some way to transform the meaning of his language into another language.[8] But what about the lion? As there is no overlap between the lion's ecological and cognitive niche and ours, it is not certain that our trust that there are enough shared concepts to allow for understanding, is somehow warranted. Wittgenstein's quote thus seems to express skepticism concerning our ability to understand what the lion was saying *even if the lion could talk*.

Yet, admittedly, between the lion and a human being there is at least a shared geographical niche: humans and lions inhabit the same planet and at least some humans share the same part of the planet with lions. Assuming our previous arguments, we could plausibly infer that the difference between an extraterrestrial intelligent life-form somewhere on a distant planet, and Earth-bound humans, would be even greater than between a lion and a human. It is not implausible to speculate that the difference between an extraterrestrial intelligent life-form and a human is such that they are *incommensurable*: in that case, the ecological and cognitive niches are so vastly different as to ensure that there are no commonalities and shared concepts whatsoever.

Admittedly, it may be a step too far to say that *therefore* it is impossible to come to a mutual understanding or at least meaningful communication. We simply do not know. But it is fairly safe to conclude that such an understanding would not come naturally, and that it would require a for now unforeseeable effort on both parties. The point is that for such a mutual understanding to occur, it should be possible for the cognitive niche of the extraterrestrial to be somehow transformed so as to overlap with the cognitive niche of humans. Whether that is possible – also considering the fact that the cognitive niche supervenes on the ecological niche, which in turn supervenes on the condition of the respective home planet – is for the moment pure speculation, logical possibilities well-contained in the realm of science fiction.

The Cognitive Science of Religion

Let us take the above argument one step further and into the realm of religious epistemology and cognition. The idea that human epistemology is closely tied to human biology has in recent years spurred interest in the question to what extent human cognizing about religion also is closely tied to our biology. This is the area of the *cognitive science of religion* (hereafter abbreviated as "CSR"), an interdisciplinary field of study that emerged in the 1990s as a branch of the cognitive sciences. The CSR draws upon experimental results from the cognitive sciences and insights from evolutionary epistemology and evolutionary psychology to explain how culturally recurrent features of human minds and their natural and social environments shape, constrain, and generate religious thought and action. The CSR thus attempts to give a naturalistic account of religion.[9]

One of the basic claims of the CSR is that religion is a byproduct of cognitive processes. Belief in gods or God "arises through the natural, ordinary operation of human minds in natural ordinary environments" (Barrett, 2004, p. 124). The CSR thus argues that religious belief arises in humans as a byproduct of the normal workings of human minds in ordinary environments, much like the way the ability to speak a language or facial recognition arises, or even perhaps like the way people learn to appreciate music and works of art. If a child grows up in an environment where its innate abilities to speak a language are triggered, the child will under normal circumstances eventually begin to speak. Facial recognition is triggered whenever a child meets human faces (and, as it turns out, the human ability to recognize faces becomes active almost immediately after the child is born). Something similar seems to be the case in religious cognition: in certain situations humans apparently are triggered to think about supernatural entities such as gods, angels, demons, ghosts or the spirits of ancestors.

As scientists in the field of CSR emphasize, this is a natural trait and not (as many atheists would have it) an aberration or malfunction of the human cognitive process. Religious cognition is said to be closely linked to the normal workings of the human mind in ordinary environments. Religion uses the same cognitive mechanisms that are used in social contexts, for example. Moreover, the cognitive mechanisms involved in religious cognition seem more or less universal, meaning that the same cognitive mechanisms seem at play irrespective of cultural differences. Thus research suggests that even in remote areas where inhabitants have hardly had any contact with other human beings, the way these inhabitants think about their gods or spirits seems to be the more or less similar other, well-known religions, indicating that the same cognitive mechanisms are at work. Apparently, the enormous cultural differences that may be present among different humans apparently do not completely overrule the fact that all humans share the same ecological and cognitive niche: the mesocosm.

Theology Again

What this suggests is that religious cognition and belief is another human trait that is constrained by biology. This means that our human ways of thinking and talking about the Transcendent seem to be constrained by the way the human cognitive apparatus functions, and that has a long evolutionary history. Most specifically, research in the CSR indicates that thinking about God in personal and human terms is one of the key features of the god concepts of at least the Jewish, Christian, and Muslim religions. The god concepts of these religions are anthropomorphic, and clearly based upon human person-concepts, so that gods are considered open to all kinds of social interaction, such as petitioning, bargaining, etc. The Old Testament speaks about God as having human features, such as emotions (anger, remorse, kindness) and the ability to act physically in the world.

My use of the term "anthropomorphism" here is not meant in a pejorative way, but I do admit that from a theological and philosophical perspective the anthropomorphic ways of thinking about God have often triggered moments of criticism, similar to the Presocratic philosopher Xenophanes' criticism of the naïve, anthropomorphic god concepts which he ridiculed:

the Ethiopians say that their gods are snub-nosed and black, the Thracians that theirs have light blue eyes and red hair. But if cattle and horses or lions had hands, or were able to draw with their hands and do the works that men can do, horses would draw the forms of the gods like horses, and cattle like cattle, and they would make their bodies such as they each had themselves. *(Kirk, Raven & Schofield, 1983, p. 169)*

Xenophanes is often considered as being the first philosophical atheist, though this is wrong. The criticism of anthropomorphic views of the gods was no reason for Xenophanes to abandon god concepts altogether. Instead he argued for a theology in which the idea of "one god, greatest among gods and men, in no way similar to mortals either in body or in thought" was central (Kirk, Raven & Schofield, 1983, p. 169). It seems, therefore, that Xenophanes was drawn by his conclusions towards a form of apophatic (i.e. negative) theology, according to which it makes more sense to say what God is not, rather than to say what God is (i.e. positive or cataphatic theology). Many theologians in the Christian, Jewish, and Muslim traditions have emphasized the apophatic moment in theology, although in much "popular" religion this apophatic moment is often lost.

To return to our discussion concerning intelligent, alien life-forms: in science fiction films and series, the gods that are worshipped by aliens are eerily reminiscent of human gods. The Klingon gods from *Star Trek* are violent warrior gods (they are often reminiscent of the gods as portrayed in the *Edda*), whereas the gods of the more spiritual Bajorans are benign and often come close to the Christian god concept. There is no reason whatsoever to assume that these ideas have any plausibility to them. What happens is that the way god concepts function in human societies is projected onto alien societies. Such projections assume that the cognitive systems of those aliens functions similar to those of humans. In other words, the otherness of the aliens is assumed to be reducible to sameness.

By now it should have become clear that I am skeptical about this assumption. The idea that physical form and function and intelligence everywhere in the universe have evolved in more or less similar ways, regardless of the biological context, has no scientific footing. One can of course imagine an as yet unknown natural law that ensures that intelligence throughout the universe develops in similar ways (perhaps similar to Conway Morris's idea of evolutionary convergence), but we have no indication that such a law actually exists. However, we do have strong indications from our own natural history that biology enables but also constrains epistemology (or formulated differently: that epistemology supervenes on biology).

If it is the case that epistemology supervenes on biology on earth, it is at least possible that this also holds for intelligent alien life elsewhere in the universe. Now, if that much is granted, then just as in the case of Wittgenstein's aphorism about the speaking lion, it may be that there are not enough shared concepts between the human religious discourse and an alien religious discourse to allow for mutual understanding.

All this actually raises an issue that so far I have not found discussed in the theological and philosophical literature concerning the possibility of extraterrestrial life. If god concepts supervene on the cognitive functioning of human beings, and if cognitive functioning in turn supervenes on biological properties (without being reducible to them), then this raises the possibility that having god concepts or religion in general is an aspect that is limited to humans only. In other words, it cannot be ruled out that intelligent extraterrestrial beings, since they have emerged in a completely different ecological niche, as a result have such a different cognitive system, that they have *no religion and no god concepts* at all. And I do not mean that they have religions and god concepts that we are unable to recognize as such, but I mean: it is possible that they lack religion and god concepts altogether. We cannot rule out the possibility that religion and god concepts are unique features of human beings.

Perhaps one could summarize what I am hinting at by paraphrasing Wittgenstein: "If a human would pray, an alien could not understand him." I am aware that this may be slightly baffling. Even though scholars of religion have no generally recognized definition of "religion," we seem to intuitively recognize religious practices and concepts whenever we encounter them. It sounds rather counterintuitive to think that extraterrestrial beings, if they are worthy of the adjective "intelligent" have no concept of religion and lack god concepts. "Surely such beings, if they are intelligent, *must* have something that functions like a religion," one could reply. But what is the logical status of this "must"? We often simply assume, due to the universality of religion among humans, that if aliens are to be considered intelligent, they will also have (something like) religion. But that assumption may just be plain wrong if it turns out that cognition supervenes on biological properties.

Remember that this reply assumes a reduction of otherness to sameness, in this case a reduction of the otherness of alien intelligence to what humans can recognize as such. It assumes some kind of necessity that intelligent extraterrestrial beings have more or less similar concepts for ideas and practices that we are familiar with. It is obvious that this assumption is what breathes life into science fiction. Narrative characters, be they human or alien, have to be recognizable to us to be interesting. They have to have some familiarity to them so that we can identify ourselves with them, to feel empathy for them, and that means that they somehow have to be similar to humans in some important respects. But remember that this is science fiction. There is no actual cosmological law of nature, as far as we know, that ensures that intelligence develops in more or less similar ways irrespective of physical and biological context.

But this seems to raise a problem for theology. For, if religion turns out to be a purely human concept, does not this fact imply that religion is a human invention, a construct that is the product of the contingent evolution of the human cognitive system? I fully agree that if religion is a product of the human mind, even a byproduct of normal cognition in normal environments, then religion is indeed a human invention.

However, does it follow necessarily that *because* religion is a product of the human mind, *therefore* religion is an illusion or a delusion? True, our humanity *configures* the way we think and talk about God, and this means it limits our imagination. But this configuration also makes it possible to talk about (and for believers to experience a personal relationship with) the transcendent God. The human ear is configured to hear only a specific range of sound, but that does not entail that sound itself is an illusion.

Evolutionary epistemology and the cognitive science of religion may slowly be in the process of uncovering the biological rootedness of religious belief. But from the fact that religious belief is biologically rooted and constrained no definite conclusion can be drawn about either the illusory or the truthful character of religious belief. Admittedly, it confirms that we cannot think and talk about God outside or without our human frame of reference, and as such it confirms the "projection thesis" that gained prominence (or rather notoriety) in modernity via the works of Ludwig Feuerbach and Sigmund Freud – but the atheistic conclusions these thinkers draw from the projection thesis are not entailed by it.

Perhaps we can agree that religion and religious concepts are human configurations. Perhaps they are inventions or even delusions. Who can say? But is it not equally possible that they are the uniquely human responses to authentic experiences of the Transcendent?

Conclusion

Whatever conclusion one is drawn to, the reflections of this paper point to a theological notion that is fundamental to the Judeo-Christian tradition, namely that if we absolutize our human language and our human images of God, we tend to "domesticate" God's transcendence (cf. Placher, 1996). If we take our religious concepts too seriously, we reduce God's otherness to sameness, we reduce it to what feeble human minds can comprehend and grasp. If our minds can grasp only what is accessible from our mesocosm, then it would be rather shortsighted and foolish to conclude that the mesocosm entails all that exists. Theologically, speaking about God's transcendence functions to express and stimulate a sense of God's otherness or *alterity*. It is meant to stimulate an awareness that God may be totally other than we are able to imagine. But we lose that sense of God's *alterity* when we take our human ways of thinking and talking about God too seriously.

As such, the reflections in this paper in a sense have been an exercise in what could be called *apophatic astrobiology*. The skepticism that I have expressed in this paper is meant to express or stimulate a sensitivity for "epistemically modest" modes of thinking. Moreover, I hope to have shown that both apophatic astrobiology and apophatic theology ultimately turn out to be also exercises in philosophical and theological anthropology: these discourses mirror our humanity, they are about the limits to our human understanding. Limits that we may not be able to overcome.

In the end, if we accept in faith that our religions are humanly configured responses to experiences of the Transcendent, that they are human tools, means to an end – then this opens up the possibility that the Transcendent has revealed itself to intelligent extraterrestrial aliens too, be it perhaps in ways that we are literally unequipped to imagine and understand. In the end, it may equally hold true that if an alien could pray, we could not understand him.

Notes

1 Note that I do *not* mean by "anthropocentrism" that the human frame of reference or even human existence is the *most valuable*. I use "anthropocentrism" here in an epistemological and hermeneutical way, indicating the frame of reference that is always presupposed in every human activity, including science and religion.
2 See http://www.counterbalance.org/etsurv/index-frame.html.
3 Darwin's evolutionary theory already problematized the notion of original sin, since one of its implications is that since there has never existed a first human being, a historical Adam and Eve have never existed. As can be seen from contemporary theology, this implication hardly led to a conceptual revolution within theology, even though most theologians agree that the implications of Darwin's theory for the notion of original sin (and thus the meaning of the Incarnation) are quite substantial.
4 Though the alien had many animal-like features as well – the fear it instills in most viewers probably comes from the ambivalence it embodies: extremely intelligent but also "beastly brutal." Interestingly, *Prometheus*, the prequel to the *Alien* trilogy, links the aliens to a "fall" of mankind and even hints at a link with humans.
5 Morris (2003, p. 328) writes: "Contrary to received wisdom and the prevailing ethos of despair, the contingencies of biological history will make no long-term difference to the outcome. Yet the existence of life itself on the Earth appears to be surrounded with improbabilities. To reiterate: life may be a universal principle, but we can still be alone." In other words, if life emerges, it will probably follow similar paths, but the contingencies surrounding the emergence of life may be so that it is extremely rare and possibly only to be found on Earth.
6 Unless it is true, as some "ancient aliens" thinkers hold, that humans do not originate from Earth but are brought here by aliens from another world. I will not take such ideas into serious consideration.
7 Sandis's entire paper is relevant to the issues I raise in this chapter.
8 I do admit that the original language may have a surplus of meaning that is lost in the transformation (i.e. translation).
9 Some good introductions to the CSR are: Boyer, 2001; Barrett, 2004; and McCauley, 2011.

References

Barrett, J.L. (2004). *Why Would Anyone Believe in God?*, Lanham, MD: AltaMira Press.
Boyer, P. (2001). *Religion Explained: The Evolutionary Origins of Religious Thought*, New York, NY: Basic Books.
Cohen, J. & Stewart, I. (2004). *What Does a Martian Look Like? The Science of Extraterrestrial Life*, London: Ebury Press.
Conway Morris, S. (2003). *Life's Solution: Inevitable Humans in a Lonely Universe*, Cambridge: Cambridge University Press.
De Duve, C. (1995). *Vital Dust: Life As a Cosmic Imperative*, New York, NY: BasicBooks.
Dick, S., ed. (2000). *Many Worlds: The New Universe, Extraterrestrial Life & The Theological Implications*, Philadelphia, PA/London: Templeton Foundation Press.
Gingerich, O. (2006). *God's Universe*, Cambridge, MA: Belknap Press.
Kirk, G.S., Raven, J.E. & Schofield, M. (1983). *The Presocratic Philosophers*, 2nd edn, Cambridge: Cambridge University Press.
McCauley, R.N. (2011). *Why Religion is Natural and Science is Not*, Oxford: Oxford University Press.

O'Meara, T.F. (2012). *Vast Universe: Extraterrestrials and Christian Revelation*, Collegeville, MN: Liturgical Press.

Placher, W. (1996). *The Domestication of Transcendence: How Modern Thinking about God Went Wrong*, Louisville, KY: Westminster John Knox Press.

Sandis, C. (2012). Understanding the lion for real. In A. Marques and N. Venturinha, eds., *Knowledge, Language and Mind: Wittgenstein's Thought in Progress*, Berlin: de Gruyter.

Vollmer, G. (2003). *Was können wir wissen? Band 1: Die Natur der Erkenntnis. Beiträge zur Evolutionären Erkenntnistheorie*, Stuttgart: S. Hirzel Verlag.

Vollmer, G. (2005). How is it that we can know this world? New arguments in evolutionary epistemology. In V. Hösle and C. Illies, eds., *Darwinism & Philosophy*, Notre Dame, IN: University of Notre Dame Press, pp. 259–74.

Weintraub, D.A. (2014). *Religions and Extraterrestrial Life: How Will We Deal With It?*, Cham: Springer.

Wilkinson, D. (2014). *Science, Religion, and the Search for Extraterrestrial Intelligence*, Oxford: Oxford University Press.

Wittgenstein, L. (1958). *Philosophical Investigations*, Oxford: Blackwell.

Conclusion

15

What is Life? On Earth and Beyond: Conclusion

ANDREAS LOSCH

So, what is life? Raising the question at all means going beyond science to a certain degree, which deems the question basically answered: it's just physics and chemistry. It is my impression, that this, however, is more an understandable delimitation against vitalism than a token of deeper understanding. At least, the historical dimension is of the essence when talking about life, so it's "physics and chemistry with a history". And wouldn't it suit the scientific spirit not to rest with the current conclusions, but to go on and ask: what *exactly* makes the difference between living and non-living phenomena? If it is all about the same matter, the organization or relation of the matter is the one that changes. Is life an *in-formational phenomenon*, then?

Let's recapitulate what we learned about life.

What was at the beginning? How did life originate? We can start with Marie-Christine Maurel's take on philosopher Karl Popper that the origin and the evolution of life coincide, there was no start from nothing.[1] Maurel discusses whether there was a gradual evolution or whether the rhythm of evolution is irregular, as in Stephen J. Gould's theory of punctuated equilibrium, interpreting the Cambrian explosion. This may well depend on the changing environmental circumstances over time. Opportunism, active in symbiosis as well as genetic take-over, appears to be a fundamental property of the living. This opportunism allows new properties to arise that are inherent to a situation, and for Maurel these are similar to the propensities of Karl Popper.

Does life make a difference? The difference between living and non-living had once been seen as fundamental, the term "organic chemistry" witnessing that belief. From the perspective of science, however, continuity is more important. Antonio Lazcano tells the story of how the seemingly deep chasm separating the living from the non-living started to close, demonstrating the continuity between chemistry and biology. Ernst Haeckel's idea of a "unity of all nature and the unity of her laws of development" is an expression of this, and enlarged Darwin's evolutionary framework to a cosmic scale. Obviously, astrobiology in general and particularly this book (*On Earth and Beyond*) owe him something. Life is not characterized by any other properties than the opportunism mentioned, and special about it is just the combination of its properties as the outcome of a historical process. "Non-teleological, non-progressive explanations are required to refine our understanding of the properties of the subcellular constituents of life and their interactions" (pp. 90–1 in this book).

One question of this book was: how do science and belief (of all sorts) interact? In his time, Haeckel, although very popular with the public, was quite disputed amongst scientists; the resistance within empirical science against his "metaphysics" was enormous (Schröder, 2008, p. 14). Nevertheless, he may be seen as an example of a fruitful interaction between science and, actually, *non*-belief, extending beyond Laplace's methodological claim of "no need of the God hypothesis" to explain planetary movements.[2] It is a lesson of history that scientists need to leave aside their "Weltanschauung" in scientific methodology. Nevertheless, interaction on a motivational level cannot be avoided. Here, belief and non-belief can play a fruitful role. While even zealous atheism will be closer to science than creationism bluntly ignoring scientific results, an anti-religious stance could also potentially result in overlooking something important. Also, it is dangerous to allow scientific-sounding assumptions to take the place of religious beliefs.

This is why Christian Weidemann sets himself apart from Christian DeDuve, who perceives the chance hypothesis of the origin of life as too miraculous or creationist. Weidemann argues that it should not be dismissed too quickly. The origin of life might be a fluke, which would also answer the Fermi paradox "Where is everybody?" (all the aliens). Weidemann's bottom line is that proponents of the conviction that life is a common phenomenon in the observable universe should at least acknowledge this conviction for what it is: an act of faith. The term "faith" is not meant here in a narrow religious sense, but, nevertheless, one could ask whether the term is appropriate, because the hypothesis that there is life at locations in the universe other than Earth could potentially be empirically verified one day. This is also why Immanuel Kant changed his mind about the matter. Formerly his prime example for a matter of belief, he later regards life on other worlds as a matter of opinion, potentially accessible to science (Losch, 2016, p. 265). Because of the vastness of the universe, it remains, however, a question of whether it could ever be falsified.

Will the attempt to define life lead to anything? Is the question "What is life?" meaningful at all? Maybe it is more important for the philosopher than for the scientist? We face an embarrassment of riches; many divergent accounts of "life" have been proposed. Our ordinary notion of life is simply too broad to allow for an analysis in terms of necessary and sufficient conditions, and the concept of life may fulfill roles that go beyond what can be investigated by the sciences. Claus Beisbart's approach, however, narrows down the concept of life to a scientifically perceived so-called "s-life" that strikes a fair compromise between the four desiderata of *similarity* (to what is to be explained), *exactness*, *fruitfulness* and *simplicity*, which (according to Carnap) weigh in in an explication. How to balance these factors needs to be answered in close collaboration between philosophers and scientists, Beisbart is convinced. He does not offer a concrete proposal for the definition of life, more guidelines on how to approach in dialogue with science. There are scientists like Stephen Hawking who see the role of philosophy in our days diminished to analysis of language (1989, p. 185). I guess Hawking misunderstands the importance of this task of language analysis. What is desirable indeed, however, is a close conversation between philosophy and science.

Can philosophy indeed contribute to science? Michel Morange, both biologist and philosopher, underlines the importance of this conversation. From science we learn that organisms do not have special nature but are a category constructed by humans before any form of science existed.

I would only caution that this categorical distinction, even if human-made, might be justified in its proper domain, as we perceive the world quite differently as our living environment ("Lebenswelt") than as part of science's necessarily reductionist approach. Science could never tell us about orientation, how to act and what to do; this is beyond its capacity. Nevertheless, if we would act without being scientifically informed, we would leave out a great deal of information about reality we have available for our orientation. If science tells us that the differences between the living and the non-living are only gradual, we still have to decide which grade of difference makes a qualitative or even categorical difference *for us*.

In any case, Morange is convinced that there is something that deserves more thorough scientific study: organisms have a combination of characteristics that are never found together in inanimate objects. This could be a concrete contribution of philosophical reflection to stimulate scientific research. In any case, the border between living and non-living seems to be more porous than expected.

Can theology contribute as constructively to science as philosophy? Certainly this whole book, to some extent, is an attempt at evaluating this question, but of course it is specifically addressed in the editor's own chapter. The Introduction and, to some degree, the Conclusion are dedicated to preventing scientists and philosophers from developing false perspectives of theology and to correcting mistakes and inconsistencies in their presentations of theological and religious issues. I take it for granted that there can also be misperceptions of scientific issues within theology; the better scientific information of theology is actually the main road of the asymmetrical science and religion discourse. Yet how can theology contribute something to science itself? I had to admit that the ideas of Welker that I took up listed only contributions of theology to the interface between science and theology, so more to the established field of "science and religion" than to science itself. I added to this the idea that theology can remind science that the completeness of knowledge has not yet been achieved, yet this general demand for a humble approach is maybe not necessary for a decent scientist. It is maybe more a reaction to the popularization of science and the bold claims of various popular writers, such as those I hinted at by Stephen Hawking (ending with "... then we would know the mind of God", 1989, p. 185). The boldest claim I made, however, is that theology in one of its dimensions can maybe serve in an auxiliary function for science, pointing out the existing gaps in the scientific account of the universe, and sometimes even propose a preliminary "filler". This can, of course, be misunderstood. I do not mean the artificial construction of gaps like the attempts of creationism or the "intelligent design" movement to find inexplicabilities in science, interpreting them as spaces of God's action. What I think is that there are, of course, a lot of gaps left in the scientific account of the world, and we humans have a certain tendency to "triumphantly ignore" their presence, especially when science's advance just closed another one. Here, theology in dialogue with science could maybe be a pointer to some of what is missing, knowing its "assistance" may one day not to be needed anymore. Nevertheless, I want to stress that this temporary theology-of-the-gaps is, of course, not theology's proper function, but is still the best it can contribute to science as an academic discipline.

Another way of relating theology to science is reinterpreting scientific findings in light of a wider theological or philosophical perspective. Alexander Maßmann undertakes such an approach, when he reminds us of the Cartesian origins of the scientific method, asking if we should not regrind our mechanistic lenses to see more clearly "what life has been like all along" (p. 213 in this book), and to rediscover mind in nature. With Evan Thompson and Hans Jonas he utilizes a certain strand of philosophy to claim the presence of teleology in the living world. For him, physical self-organization is the external dimension of cognition, and cognition is the internal aspect of self-organization. Likewise, the theologian can regard the human soul as inward dimension of the person inseparable from his or her physical systematic organization.

These are bold claims. Teleology is probably important for the description of human beings, but is it indeed a general feature of life? Biologists (such as Ernst Mayr) would acknowledge some sort of *teleonomy* in individuals, viz. systems operating on the basis of a program of coded information (Mayr, 1965, p. 42), but remembering vitalistic fallacies they would be careful to avoid teleological thought: "... there is no evidence that would support teleology in the sense of various vitalistic or finalistic theories" (Mayr, 1965, p. 43). Well, maybe a philosophical approach is less constrained by the scientific method. Philosophically, it might even be a question of whether there is something like design in nature, because of the idea of a creator God or when assuming superintelligences as creators of our universe (Walker & Ćirković, 2006). The history of science, however, is a story of the evacuation of teleological thought from research. The reader should therefore be aware that, with this chapter, she or he dwells in the domain of theology beyond the domain of science. Nevertheless, this approach aims to be a *scientifically informed* interpretation.

This applies for Ted Peters' chapter as well. His concept of intelligence is as innovative as it is broad (without consciousness, values or free will), and it allows him to view all forms of life as more or less intelligent. So here is a theologian who to some extent discards the uniqueness of human beings and allows for intelligence to be widespread in nature. For Peters, what happens at the human level is in continuity with all levels of life in our biological evolution, a general stance that scientists would embrace, I assume. Yet one could ask what Peters is achieving with his generalization of the concept of intelligence in particular. Certainly it makes sense to leave our anthropocentric perspective behind when we search for extraterrestrial life, but can his concept really "steer astrobiological eyes to see what might surprise us"? Would such an approach, for instance, interpret the *Viking* experiments differently? Both Maßmann's and Peters' approaches are, while highly original in their hypotheses, maybe very courageous and challenging ways of viewing a dialogue between science and religion.

As mentioned in the Introduction, this book tries to balance speculation and critical thought. Some of our authors clearly lean more towards speculation than others who favour a more sober approach. For Morange, there is no strong link between life and consciousness, and the idea that computers will soon start to think belongs to the realm of dreams (p. 108 in this book). We allow ourselves in this book some dreaming, then, which is certainly possible when meditating on the vast realm of possibilities in the starry sky above us. Indeed, everything else could be Geocentrism. I also believe that literary discourse is indeed a "treasure-trove of potentially useful scientific hypotheses", as Milan Ćirković puts it (p. 161). We only have to be aware that most of them will be proven trash when facing reality. Not all of them, though. There was recently a *Nature* Science Fiction Special, celebrating the birth of H. G. Wells 150 years ago and the advent of Star Trek 50 years ago (Issue 537). What should be remembered is not only that some technologies have become real, but also the great role such fiction plays in motivating (becoming) scientists. So, who knows? There could be superintelligent artificial intelligence on Earth itself one day. According to Susan Schneider, who is referring to Nick Bostrom here (Bostrom, 2014), quite sophisticated artificial intelligence is already under development.

Likewise, alien life out there could be postbiological. Could we control it or even understand its computations? Yet we do not have to go so far; even something as tiny as fossilized microbial life on Mars would be a sensation and would contribute to the "astrobiological revolution" – which at the moment, however, could appear more like a revolution in (exo)planetary research with a lot of fuzz about it. David Wilkinson, himself a supporter of dealing with the recent developments substantially, warns against too hastily equating "Earth-like planets to habitable Earth-like planets and then to inhabited Earth-like planets and then to inhabited by intelligent beings Earth-like planets" (Wilkinson, 2016, p. 421). It should be acknowledged that there is also the story of the discovery of extremophiles that can dwell in almost any environment on Earth, so there's more to astrobiology than just the discovery of exoplanets. Couldn't some sort of extreme microbial life exist in other places of our solar system as well?

What should we look for in space? We cannot avoid some anthropocentric bias, but we humans are also the ones who can speculate beyond the bounds of our experience. Life on Earth is the only example of life we know, and as the laws of the Universe are still supposed to function the same everywhere, it is certainly not a bad start to search for life as we know it. That's why we look for water, but actually there is plenty of water in our solar system already. Still, no other life has been detected.

At the end of our excursion into the origins, development and prospect of Life on Earth and beyond I would claim that the question "What is life?" is still open. It is partially answered by science as being physics and chemistry with a history, but this does not seem to be the whole story. The relation of its constituents makes the difference between life and non-life, but what sort of relation this actually is has to be subjected to further research. Is it the key to the "secret of life"? Theology would rather be fine with it, one dares to assume. It would certainly interpret an aspect of this relationality as relatedness to the transcendental (cf. Jackelén, 2015). Yet this is more on a very fundamental level, like the question "what is it that breathes fire into the equations?" (Hawking, 1989, p. 184) than an area science has to take care of.

Another important remaining question is whether we need teleological language when describing life, or is this only wishful thinking, a projection of our cultural perception on nature? Maybe one could suggest teleology to be a particular *interpretation* of the biological phenomena, but I doubt it can be regarded as part of science. Even when this book is a witness of a dialogue between science, philosophy and theology, this does not mean that there is mutual agreement on everything. The method of science is a quite skeptical enterprise, and hence there will be a Skeptical Afterword after this Conclusion.

The previously mentioned opportunism of the living also means that life depends on its environment. Maybe one should take a broader approach and view the whole biosphere as a living entity? Just as we ourselves are host to quite a number of microorganisms, one could maybe regard our whole planet as alive as well, hosting all its diverse life forms (Grinspoon, 2016). Will there be other planets carrying life, then? As Joshua Krissansen-Totton and David Catling beautifully stated at the end of their chapter: we may find that the universe is a desolate and sterile place. But perhaps, we will instead discover exoplanets teeming with life, so that one day we will tell our children "see, *that one*, that star has a planet just like Earth." (p. 52 in this book). So theology should be prepared, and Juan Pablo Marrufo del Toro (among many others) develops this ancient theme of astrotheology (Derham, 1715). But maybe, again, we face anthropocentric bias here and extraterrestrial intelligent life lacks God concepts altogether, as Taede Smedes points out. This has to be taken into account, for sure, although one has to be aware that the idea of areligious aliens (saving humankind from religion) drove SETI research from the start (Peters, 2009, pp. 4–5; Wilkinson, 2013, pp. 125–129).

What is life? There is nothing special about its matter, so the evolution of the special relation of its constituents might be the key to understand it. Unfortunately, we still do not know much about the beginnings of this evolution. If one day we would find traces of life beyond Earth we might learn a great deal more about it. And about ourselves.

Notes

1 Leibniz's classic question "Why is there something rather than nothing?" remains open, of course, and the theological idea of *creatio ex nihilo* naturally aims at this fundamental level of existence of anything. There is even a distinct Hebrew word ("*bara*") used in the Bible for this fundamental level of creation, reserved for God's action. Creation*ism* gets this wrong, maybe, but we should care about creationism only because of science (and theology) *education*, not for proper academic reasons.
2 It is a matter of scholarly dispute if Laplace did not only support methodological atheism in science but also was himself atheistically orientated, cf. Henrich, 2010, p. 168.

References

Bostrom, N. (2014). *Superintelligence. Paths, Dangers, Strategies*, 1st edn, Oxford: Oxford University Press.
Derham, W. (1715). *Astro-theology*, London: Printed for W. Innys.
Grinspoon, D. (2016). *Earth in Human Hands. The Rise of Terra Sapiens and Hope for our Planet*, New York, NY: Grand Central Publishing.
Hawking, S. (1989). *A Brief History of Time. From the Big Bang to Black Holes*, Bantam Paper edn, New York, NY: Bantam Books.

Henrich, J. (2010). *Die Fixierung des modernen Wissenschaftsideals durch Laplace*, Berlin: Akademie Verlag. Available online at http://www.oldenbourg-link.com/doi/book/10.1524/9783050051314.

Jackelén, A. (2015). Life: an ill-defined relationship. In D. Evers, M. Fuller, A. Jackelén and K.-W. Sæther, eds., *Issues in Science and Theology: What is Life?*, Cham: Springer International Publishing, pp. 69–85.

Losch, A. (2016). Kant's wager. Kant's strong belief in extra-terrestrial life, the history of this question and its challenge for theology today. *International Journal of Astrobiology* **15**(04), 261–70.

Mayr, E. (1965). Cause and effect in biology. In D. Lerner, ed., *Cause and Effect*, New York, NY: Free Press, pp. 33–50.

Peters, T. (2009). Astrotheology and the ETI myth. *Theology and Science* **7**(1), 3–29.

Schröder, T.M. (2008). *Naturwissenschaften und Protestantismus im Deutschen Kaiserreich*, Stuttgart: Franz Steiner Verlag.

Walker, M.A. & Ćirković, M. M. (2006). Astrophysical fine tuning, naturalism, and the contemporary design argument. *International Studies in the Philosophy of Science* **20**(3), 285–307.

Wilkinson, D. (2013). *Science, Religion, and the Search for Extraterrestrial Intelligence*, 1st edn, Oxford: Oxford University Press. Available online at http://lib.myilibrary.com/detail.asp?id=497741.

Wilkinson, D. (2016). Searching for another Earth. The recent history of the discovery of exoplanets. *Zygon* **51**(2), 414–30.

16

A Skeptical Afterword

ANTONIO LAZCANO

The possibility of extraterrestrial life is a legitimate scientific and philosophical issue that has been addressed, sometimes in considerable detail, since Pre-Socratic times by naturalists and philosophers. At times in which the nature of stars was unknown and evolutionary biology had not yet been developed, these early speculations rested on the idea of a uniform Universe and had little or no empirical basis. Despite claims on the contrary, there is little or no epistemological continuity of these ideas with current attempts to study life in the Universe, nor did such early conjectures lead to major religious disputes. It is true that the German philosopher and astrologer Philipp Melanchthon, like a few other Renaissance thinkers, was troubled by theological issues related to the salvation of hypothetical extraterrestrial humanities. However, he was basically an isolated case. Contrary to popular belief, for instance, Melanchthon's contemporary Giordano Bruno was sent to the stake not because he believed in the plurality of inhabited worlds, but rather because his heretical views questioned the divinity of Christ and included other radical proposals that threatened the basic tenets of the Catholic Church during the politically charged atmosphere of the Protestant Reformation.

Religions are part of the socio-cultural framework that shapes the environment in which scientific ideas develop and, as shown by their comparative histories, have influenced one another. It is difficult to envision, for instance, the development of the nineteenth-century evolutionary perspective without the Judeo-Christian lineal vision of time, which played a key role in the development of historical narratives. By the same token, Darwin's secular non-teleological arguments were conveniently interpreted by theists as a natural outcome of the law-abiding Universe created by the rational God they advocated. As stated at Darwin's funeral by one of his pallbearers, the Reverend Frederic William Farrar, "[t]his man, on whom for years bigotry and ignorance poured out their scorn, has been called a materialist. I do not see in all his writings one trace of materialism... I read in every line the healthy, noble, well-balanced wonder of a spirit profoundly reverent, kindled into deepest admiration for the works of God..." and added that "[a]nd because these false antagonisms have been infinitely dangerous to faith, over Darwin's grave let us once more assure the students of science that, for us, the spirit of mediaeval ecclesiasticism is dead" (Glick, 2010, p. 116).

Many believers must have been relieved with the fact that with one known exception, Darwin never addressed in public the issue of the origins of life, much less that of inhabited planets. The same was not true of some of his more faithful followers. During the second half of the nineteenth century astronomers, geologists and naturalists started to discuss the possible links between planetary environments and hypothetical extraterrestrial forms of life. Exasperated by Camille Flammarion's flamboyant writings, in 1899 Ernst Haeckel described him as "equally distinguished by exuberant imagination and brilliant style, and by a deplorable lack of critical judgment and biological knowledge...". Haeckel argued that if life existed elsewhere, then lower organisms were probably similar to the terrestrial ones, but not the higher life forms, which may not even be vertebrates, adding that "perhaps some higher animal stem, which is superior to the vertebrate in formation, higher beings have arisen [on other planets] who far transcend us earthly men in intelligence" (Crowe, 1986, p. 427).

During the 50 years that followed, very few engaged in discussions of the religious significance of extraterrestrial life. This is easy to understand. With few exceptions, until the 1960s it was believed that the emergence of the biosphere had been a unique process involving billions and billions of years, microbes were perceived as pathogens and not as ancestors of extant life, the formation of planetary systems was believed to be rare, and even unmanned space travel was considered unlikely. The age of space exploration prompted by the launching of *Sputnik*, together with the publicity surrounding the attempts by radioastronomers and theoretical physicists to detect signals from hypothetical extraterrestrial civilizations, prompted some well-known American clerics like Theodore M. Hesburgh and Krister Stendahl to address the theological implications of other inhabited worlds. However, these discussions remained on the outer orbits of academia and were largely ignored outside the USA, where for a long time religion has played a much more conspicuous role in public life than in other societies.

Today our approaches to the issue of life in the Universe have changed dramatically. Neither the formation of planets nor the emergence of the biosphere are seen as the result of inscrutable random events, but rather as natural outcomes of evolutionary events. However, the search for extraterrestrial life remains high on speculation and low on facts. Given the reports on liquid water in other Solar System bodies, the presence of a wide array of extraterrestrial organic molecules, and the high number of extrasolar planetary systems, it is easy to understand why the events that led to the emergence of life have been interpreted by some as an inevitable process – and yet, historical contingency, which introduces a major element of chance and uncertainty in our evolutionary narratives, cannot be discounted.

Although it is tempting to assume that the emergence of life is an unavoidable process that maybe continuously taking place throughout the Universe, it is still to be shown that it exists (or has existed) in places other than the Earth. Evolutionary processes are largely unpredictable, and the origin of life was not a continuous, unbroken chain of progressive transformations steadily proceeding to the first living systems. Many prebiotic cul-de-sacs and false starts probably took place. While it may be true that the transition to life from non-living systems did not require a rather narrow set of environmental constrains, we cannot discount the possibility that even a slight modification of the primitive environment could have prevented the appearance of life on our planet and the appearance of intelligent life. As the French philosopher Pascal once remarked, had Cleopatra's nose been different, the course of history may have changed (Lazcano, 2004).

However unpalatable this conclusion may be, life may be a rare and even unique phenomenon in the Universe. Perhaps not surprisingly, some of the most distinguished contemporary evolutionists like George Gaylord Simpson, Ernst Mayr, Lynn Margulis and Theodosius Dobzhansky, were also some of the harshest critics of the teleological schemes advocated by those convinced of the abundance of intelligent life in the Universe. Their healthy skepticism is a lesson we must keep in mind.

References

Crowe, M. J. (1986). *The Extraterrestrial Life Debate*, Cambridge: Cambridge University Press.

Glick, T. F. (2010). *What about Darwin?* Baltimore, MD: The John Hopkins University Press.

Lazcano, A. (2004). Astrobiology: towards an understanding of the origin of life in the Universe. In: R. P. Norris and F. H. Stootman, eds., *Bioastronomy: Life among the Stars: IAU Symposium*, San Francisco, CA: Astronomical Society of the Pacific, pp. 245–253.

Index

Anthropic Principle, 134, 140
anthropocentrism, 160, 171, 172, 281, 299
astrobiology, 5, 6, 8, 44, 51, 97, 98, 100, 106, 156, 157, 158, 159, 160, 161, 162, 163, 164, 165, 166, 167, 168, 169, 170, 171, 172, 174, 178, 179, 180, 183, 188, 191, 192, 194, 195, 256, 298, 304, 308
autopoiesis, 7, 213, 214, 216, 217, 220, 222, 223, 228, 229, 230, 232

bacteria, 20, 21, 61, 62, 63, 84, 88, 103, 104, 105, 108, 169, 241, 268
Bible, 2, 205, 224, 225, 309
biosignatures, 16, 38, 40, 44, 45, 47, 48, 51, 58, 71, 254
brain, 23, 75, 109, 181, 182, 184, 185, 186, 187, 188, 194, 237, 240, 243, 244, 245, 247, 248, 249, 250, 251, 252, 253, 256, 267, 292

Carnapian explication, 5, 112, 124, 125, 126, 128, 305
cell, 7, 16, 18, 19, 20, 21, 23, 28, 62, 63, 79, 81, 83, 90, 180, 215, 216, 219, 228, 229, 232, 240, 241, 242, 245, 246, 247, 261, 262, 263, 265
chance hypothesis, 5, 132, 134, 136, 137, 139, 141, 142, 144, 145, 146, 147, 148, 149, 152, 304
coevolution, 85, 245, 267
cognitive science, 109, 161, 179, 184, 186, 188, 294, 298
complexity, 30, 75, 88, 89, 99, 132, 143, 149, 236, 237, 246, 248, 250, 255, 267, 279, 292
consciousness, 8, 97, 108, 109, 191, 192, 193, 194, 196, 213, 217, 223, 224, 246, 249, 250, 251, 252, 256, 267, 275, 288, 307
contingency, 4, 90, 106, 164, 220, 268, 313
control problem, 6, 178, 183, 188, 189, 190, 195
creation, 2, 7, 8, 21, 27, 80, 100, 108, 183, 188, 193, 204, 205, 208, 213, 214, 215, 227, 228, 229, 230, 231, 232, 233, 266, 270, 272, 274, 279, 280, 281, 282, 283, 309
creationism, 2, 6, 22, 204, 205, 304, 306, 309

definition, 26, 37, 89, 113, 114, 115, 116, 118, 119, 120, 121, 128, 129, 168, 188, 236, 237, 238, 240, 244, 252, 255, 256, 297
analytic, 116, 117, 118, 119, 120
descriptive, 116, 118, 119, 120, 121, 124, 125
explicit, 125
dialogue, 2, 4, 6, 7, 97, 204, 206, 207, 209, 213, 217, 233, 305, 307, 308
direct imaging, 3, 32, 34, 36, 40, 41, 45, 51
diversity, 20, 21, 23, 30, 103, 142, 143, 164
DNA, 19, 25, 28, 87, 88, 100, 101, 102, 104, 201, 221, 229, 254, 261, 262, 280

Earth, 1, 2, 3, 4, 5, 6, 14, 15, 16, 22, 30, 31, 32, 33, 34, 36, 37, 38, 39, 40, 41, 42, 43, 44, 45, 46, 47, 49, 50, 51, 52, 58, 59, 60, 61, 66, 71, 79, 87, 105, 109, 118, 134, 137, 138, 139, 141, 142, 145, 149, 150, 151, 152, 158, 159, 160, 163, 164, 165, 168, 170, 178, 179, 180, 182, 183, 184, 185, 188, 190, 193, 194, 195, 213, 236, 237, 243, 246, 254, 255, 256, 261, 263, 264, 266, 267, 269, 271, 272, 274, 278, 279, 283, 287, 288, 290, 291, 292, 299, 304, 307, 308, 309, 313
conditions on, 15, 59, 152, 264
primitive, 24, 87
rare, 167, 168, 169, 172, 268
evolution, 2, 4, 5, 8, 14, 15, 16, 18, 19, 20, 21, 22, 25, 26, 27, 30, 79, 83, 84, 86, 87, 88, 90, 99, 105, 106, 121, 134, 139, 149, 151, 152, 163, 166, 170, 180, 181, 188, 193, 202, 203, 204, 205, 217, 219, 221, 222, 231, 232, 238, 240, 241, 243, 246, 254, 255, 261, 265, 266, 267, 269, 279, 280, 283, 291, 297, 307, 309
Darwinian, 1, 30, 44, 111, 117, 124, 163
human, 267, 288
of life, 3, 22, 90, 139, 141, 142, 149, 163, 205, 265, 268, 269, 288, 303
precellular, 75, 85, 86, 87, 88
evolutionary epistemology, 291, 294

Index

exoplanets, 3, 31, 32, 33, 35, 36, 37, 38, 40, 41, 42, 45, 46, 47, 48, 51, 52, 111, 254, 269, 308, 309
  search for, 8, 287
extraterrestrial, 8, 48, 49, 50, 147, 160, 165, 182, 194, 255, 272, 273, 274, 280, 290, 293, 297, 298

Fermi paradox, 50, 148, 149, 152, 304
fossil, 4, 13, 19, 27, 52, 57, 60, 61, 62, 66, 67, 68, 165, 170, 253, 279

geology, 57, 279
God, 2, 6, 8, 17, 152, 194, 203, 205, 206, 207, 208, 209, 214, 215, 222, 223, 225, 227, 228, 230, 231, 233, 260, 266, 270, 271, 272, 273, 274, 275, 276, 277, 278, 279, 280, 281, 282, 283, 290, 294, 295, 297, 298, 304, 306, 307, 309, 311

habitability, xi, 37, 39, 41, 44, 167, 168, 169
habitable zone, 37, 38, 39, 51, 135, 142, 166, 169, 269
Holy Spirit, 7, 225, 227, 230, 231, 232, 233

incarnation, 8, 260, 269, 270, 271, 272, 273, 274, 275, 276, 277, 278, 279, 280, 281, 282, 283, 284, 299
information, 15, 16, 18, 20, 25, 35, 41, 44, 58, 68, 100, 101, 102, 104, 137, 138, 187, 191, 192, 194, 219, 221, 233, 237, 242, 245, 249, 256, 261, 262, 305, 306
  genetic, 25, 100, 101, 102, 103, 222
  medium for, 181, 182
intelligence, 7, 48, 51, 50, 151, 162, 178, 179, 181, 184, 186, 187, 188, 189, 191, 192, 193, 194, 236, 237, 238, 239, 240, 241, 242, 244, 245, 246, 247, 249, 250, 251, 252, 254, 255, 256, 265, 266, 267, 269, 288, 296, 307, 312
  artificial, 6, 50, 108, 178, 182, 185, 241, 256, 307
  definition of, 236, 237, 238, 239, 241, 253, 266
  extraterrestrial, 8, 47, 48, 50, 163, 182, 190, 254, 255, 288, 289, 290, 297
  human, 108, 183, 189, 190, 237, 238, 239, 240, 243, 244, 247, 248, 249, 250, 251, 252, 253
  search for, 157, 172
  sign of, 108, 245

Jesus, 225, 270, 271, 272, 273, 274, 276, 277, 278, 279, 280, 281, 282, 283, 284

life, 3, 5, 7, 14, 15, 16, 17, 18, 19, 22, 27, 30, 31, 37, 38, 41, 43, 44, 45, 47, 48, 51, 52, 57, 58, 59, 60, 61, 71, 78, 79, 80, 82, 83, 84, 85, 86, 88, 89, 90, 97, 98, 99, 100, 101, 102, 103, 104, 106, 107, 108, 109, 111, 112, 113, 114, 115, 116, 117, 120, 121, 122, 123, 124, 126, 127, 128, 132, 133, 134, 136, 137, 138, 142, 144, 146, 147, 148, 159, 161, 163, 165, 166, 167, 170, 172, 181, 182, 184, 185, 188, 191, 192, 195, 201, 202, 203, 206, 207, 209, 213, 214, 215, 218, 222, 224, 225, 226, 228, 229, 230, 231, 232, 233, 236, 237, 238, 239, 241, 242, 243, 244, 245, 246, 247, 250, 260, 261, 262, 263, 264, 266, 276, 283, 291, 303
  artificial, 18, 19, 20, 102, 104, 108, 118, 185, 263
  beginning of, 23, 30, 48, 78, 84, 156, 263, 290
  concept of, 15, 16, 97, 99, 100, 104, 115, 119, 120, 121, 125, 127, 129, 201
  definition of, xi, 1, 4, 5, 16, 17, 75, 87, 90, 91, 102, 111, 112, 115, 116, 117, 118, 119, 120, 121, 122, 124, 125, 127, 128, 129, 208, 304, 305
  detection of, 31, 32, 41, 44, 47
  domains of, 20, 24
  emergence of, 5, 86, 133, 135, 139, 141, 143, 145, 146, 148, 272, 313
  eternal, 280, 283
  explication of, 124, 127
  extraterrestrial, 2, 3, 4, 5, 8, 9, 44, 58, 74, 143, 144, 148, 158, 159, 163, 165, 196, 236, 255, 259, 273, 281, 287, 288, 289, 293, 296, 307, 309, 310, 311, 312
  forms of, 21, 24, 25, 58, 59, 70, 108, 246, 250, 265, 267, 275, 307
  history of, 4, 15, 20, 61, 164, 254, 265
  human, 31, 98, 208, 225, 230
  indicator of, 1, 43, 45, 46, 51, 65, 66, 69, 70, 264, 287
  intelligent, 7, 50, 51, 52, 134, 141, 142, 143, 144, 145, 163, 180, 185, 190, 236, 246, 254, 255, 260, 266, 269, 270, 277, 283, 288, 289, 293, 296, 309, 313
  microbial, 1, 3, 6, 23, 59, 60, 71, 141, 166, 179, 236, 264, 287, 288, 308
  nature of, 1, 4, 81, 85, 86, 87, 88, 89, 90, 97, 103, 111, 261
  on Earth, 22, 41, 43, 57, 90, 106, 132, 133, 136, 137, 139, 141, 142, 143, 144, 146, 147, 149, 150, 156, 163, 165, 232, 238, 253, 260, 265, 308
  on exoplanets, 35, 41, 43, 44, 254
  prokaryotic, 59, 62, 66, 68
  search for, 41, 51, 60, 100, 157, 172, 265, 269, 270, 288, 308

macromolecules, 14, 19, 24, 83, 100, 101, 104, 108, 147, 148
Mars, 1, 3, 9, 37, 39, 58, 60, 71, 72, 156, 160, 163, 165, 174, 254, 264, 265, 287, 308
mechanism, 3, 18, 20, 28, 83, 88, 89, 148, 149, 163, 213, 217, 218, 232, 233, 242
mesocosm, 292, 293, 295, 298
metabolism, 1, 15, 16, 23, 25, 43, 88, 89, 101, 102, 103, 111, 117, 118, 120, 129, 181, 182, 215, 219, 227, 229, 264, 265
microbes, 1, 60, 61, 62, 64, 66, 67, 68, 69, 78, 236, 255, 263, 265, 267, 312
microorganisms, 1, 57, 59, 60, 61, 63, 70, 84, 102, 103, 156, 231, 309
mind, 7, 80, 108, 187, 207, 213, 214, 215, 217, 219, 222, 223, 232, 233, 243, 244, 249, 250, 251, 253, 256, 267, 294, 297, 306

mind (*cont.*)
  of God, 206, 306
  theory of, 8, 239, 248, 249
molecular biology, 4, 7, 98, 99, 100, 105, 124
morphology, 20, 23, 58, 61, 63, 65
multiverse, 134, 136, 139, 140, 143, 150

ontology, 291, 292
origin, 3, 5, 13, 14, 15, 18, 19, 21, 22, 23, 26, 62, 67, 68, 76, 77, 79, 80, 81, 82, 83, 87, 89, 103, 142, 149, 151, 164, 267, 288, 303
  of life, 3, 5, 13, 14, 15, 17, 18, 19, 20, 22, 24, 27, 41, 51, 52, 75, 78, 79, 81, 83, 84, 86, 87, 88, 89, 90, 97, 98, 99, 100, 101, 102, 106, 109, 111, 132, 133, 134, 141, 142, 144, 145, 146, 147, 148, 149, 150, 152, 253, 260, 264, 303, 304, 312, 313
  of mind, 109
  of the universe, 14, 260

philosophy, xi, 2, 4, 7, 9, 16, 75, 97, 108, 114, 115, 117, 128, 157, 164, 173, 201, 202, 204, 206, 213, 214, 215, 217, 224, 260, 287, 305, 306, 308
  of mind, 151, 189, 194, 195, 242
  of science, 213, 215, 252
prebiotic
  environment, 20, 24, 89
  synthesis, 4, 14, 27, 84, 87, 90
Principle of Mediocrity, 140, 141, 143, 151
protocell, 19, 20, 263

reductionism, 75, 87
religion, 2, 6, 8, 9, 170, 194, 202, 204, 207, 208, 209, 260, 271, 273, 278, 283, 290, 294, 295, 296, 297, 298, 299, 305, 307, 309, 312
reproduction, 16, 20, 21, 78, 111, 117, 124, 185, 215, 228, 243, 263
ribozyme, 20, 21, 89, 90
RNA, 20, 22, 24, 25, 88, 89, 101, 103
  world, 89, 101

salvation, 207, 260, 269, 270, 271, 272, 273, 274, 275, 277, 278, 280, 281, 283, 311

science, xi, 2, 4, 5, 6, 7, 8, 9, 13, 14, 15, 16, 18, 19, 20, 28, 30, 31, 75, 76, 77, 82, 86, 90, 97, 99, 107, 108, 114, 115, 121, 128, 139, 147, 157, 158, 159, 160, 162, 163, 170, 171, 172, 173, 201, 203, 204, 205, 206, 207, 208, 209, 210, 213, 214, 215, 217, 222, 228, 232, 236, 260, 268, 273, 287, 288, 290, 294, 299, 303, 304, 305, 306, 307, 308, 309
  history of, 117, 122, 158, 166
science-fiction, 160, 161, 162, 183, 296, 297
self-organization, 7, 18, 26, 89, 205, 213, 217, 218, 220, 228, 230, 306
SETI, 3, 8, 47, 48, 49, 50, 51, 52, 156, 157, 163, 165, 167, 171, 173, 179, 180, 182, 188, 190, 255, 287, 288, 289, 309
superintelligence, 2, 6, 178, 179, 180, 181, 182, 183, 184, 185, 186, 188, 189, 192, 193, 194, 195, 307
symbiosis, 27, 28, 242, 303
synthetic biology, 4, 19, 98, 104, 107, 178, 202

teleology, 7, 213, 217, 220, 222, 223, 232, 243, 306, 308
theology, xi, 2, 6, 7, 8, 9, 194, 201, 203, 204, 205, 206, 207, 208, 209, 210, 213, 217, 230, 231, 233, 260, 266, 270, 271, 272, 273, 274, 276, 278, 287, 289, 290, 295, 297, 298, 299, 305, 306, 308, 309

universe, 3, 7, 14, 30, 31, 44, 50, 52, 99, 100, 132, 134, 135, 136, 137, 138, 140, 143, 144, 150, 151, 152, 156, 161, 163, 164, 165, 171, 172, 179, 185, 190, 191, 192, 194, 206, 209, 214, 219, 223, 232, 247, 254, 255, 260, 261, 264, 265, 266, 267, 269, 270, 271, 272, 273, 274, 275, 276, 277, 278, 279, 280, 281, 282, 283, 284, 287, 288, 289, 290, 296, 306, 307, 309
  infinite, 141, 143
  life in the, 6, 8, 44, 179, 183, 190, 260, 263, 266, 269, 270, 283
  observable, 48, 135, 144, 149, 304

virus, 4, 16, 20, 22, 82, 103, 104, 105, 229, 254
vitalism, 3, 16, 17, 18, 98, 99, 202, 206, 210, 217, 303